Making Sense of the
Social World

4
edition

Related Titles in Research Methods and Statistics From SAGE

Other SAGE Titles of Interest

Making Sense of the
Social World
METHODS OF INVESTIGATION
4
edition

Daniel F. Chambliss
Hamilton College

Russell K. Schutt
University of Massachusetts Boston

Los Angeles | London | New Delhi
Singapore | Washington DC

Los Angeles | London | New Delhi
Singapore | Washington DC

FOR INFORMATION:

SAGE Publications, Inc.
2455 Teller Road
Thousand Oaks, California 91320
E-mail: order@sagepub.com

SAGE Publications Ltd.
1 Oliver's Yard
55 City Road
London EC1Y 1SP
United Kingdom

SAGE Publications India Pvt. Ltd.
B 1/I 1 Mohan Cooperative Industrial Area
Mathura Road, New Delhi 110 044
India

SAGE Publications Asia-Pacific Pte. Ltd.
33 Pekin Street #02-01
Far East Square
Singapore 048763

Acquisitions Editor: Jerry Westby
Associate Editor: Megan Krattli
Assistant Editor: Rachael Leblond
Editorial Assistant: Erim Sarbuland
Production Editor: Eric Garner
Copy Editor: Gretchen Treadwell
Typesetter: C&M Digitals (P) Ltd.
Proofreader: Susan Schon
Indexer: Sheila Bodell
Cover Designer: Bryan Fishman
Marketing Manager: Erica DeLuca
Permissions Editor: Karen Ehrmann

Printed in Canada

Library of Congress Cataloging-in-Publication Data

Chambliss, Daniel F.

Making sense of the social world: methods of investigation / Daniel F. Chambliss, Russell K. Schutt.—4th ed.

p. cm.

Includes bibliographical references and index.

ISBN 978-1-4522-1771-0 (pbk.)

1. Social problems—Research. 2. Social sciences—Research. I. Schutt, Russell K. II. Title.

HN29.C468 2013
361.1072—dc23   2011049045

This book is printed on acid-free paper.

12 13 14 15 16 10 9 8 7 6 5 4 3 2 1

Brief Contents

On the Study Site

Detailed Contents

6. Causation and Experimental Design 103

7. Survey Research 128

10. Qualitative Data Analysis   205

11. Evaluation Research   234

On the Study Site

About the Authors

Daniel F. Chambliss, PhD, is the Eugene M. Tobin Distinguished Professor of Sociology at Hamilton College in Clinton, New York, where he has taught since 1981. He received his PhD from Yale University in 1982; later that year, his thesis research received the American Sociological Association's Medical Sociology Dissertation Prize. In 1988, he published the book *Champions: The Making of Olympic Swimmers,* which received the Book of the Year Prize from the U.S. Olympic Committee. In 1989, he received the American Sociology Association's Theory Prize for work on organizational excellence based on his swimming research. Recipient of both Fulbright and Rockefeller Foundation fellowships, he published his second book, *Beyond Caring: Hospitals, Nurses, and the Social Organization of Ethics,* in 1996; for that work, he was awarded the ASA's Elliot Freidson Prize in Medical Sociology. His research and teaching interests include organizational analysis, higher education, social theory, and comparative research methods.

Russell K. Schutt, PhD, is Professor and Chair of Sociology at the University of Massachusetts, Boston, where he received the 2007 Chancellor's Award for Distinguished Service. Since 1990, he has also been Lecturer on Sociology in the Department of Psychiatry (Beth Israel-Deaconess Medical Center) at the Harvard Medical School. He completed his BA, MA, and PhD degrees at the University of Illinois at Chicago and was a Postdoctoral Fellow in the Sociology of Social Control Training Program at Yale University. In addition to seven editions of the text on which this brief edition is based, *Investigating the Social World: The Process and Practice of Research,* and four other coauthored versions—for the fields of social work, criminal justice, psychology, and education—he is the author of the new book, *Homelessness, Housing, and Mental Illness,* and of *Organization in a Changing Environment,* coeditor of *The Organizational Response to Social Problems,* and coauthor of *Responding to the Homeless: Policy and Practice.* He has authored and coauthored numerous journal articles, book chapters, and research reports on homelessness, mental health, organizations, law, and teaching research methods. He currently directs an evaluation of a Massachusetts Department of Public Health coordinated care program. His primary research focuses on social factors that shape the impact of housing, employment, and services for severely mentally ill persons and on the service preferences of homeless persons and service personnel. He has also studied influences on well-being, satisfaction, and cognitive functioning; processes of organizational change and the delivery of case management; decision making in juvenile justice and in union admissions; political participation; media representations of mental illness; and HIV/AIDS prevention.

Preface

If you have been eager to begin your first course in social science research methods, we are happy to affirm that you've come to the right place. We have written this book to give you just what you were hoping for—an introduction to research that is interesting, thoughtful, and thorough.

But what if you've been looking toward this course with dread, putting it off for longer than you should, wondering why all this "scientific" stuff is required of students who are really seeking something quite different in their major? Well, even if you had just some of these thoughts, we want you to know that we've had your concerns in mind, too. In *Making Sense of the Social World,* we introduce social research with a book that combines professional sophistication with unparalleled accessibility: Any college student will be able to read and understand it—even enjoy it—while experienced social science researchers, we hope, can learn from our integrated approach to the fundamentals. And whatever your predisposition to research methods, we think you'll soon realize that understanding them is critical to being an informed citizen in our complex, fast-paced social world.

🔲 Teaching and Learning Goals

Our book will introduce you to social science research methods that can be used to study diverse social processes and to improve our understanding of social issues. Each chapter illustrates important principles and techniques in research methods with interesting examples drawn from formal social science investigations and everyday experiences.

Even if you never conduct a formal social science investigation after you complete this course, you will find that improved understanding of research methods will sharpen your critical faculties. You will become a more informed consumer, and thus a better user, of the results of the many social science studies that shape social policy and popular beliefs. Throughout this book, you will learn what questions to ask when critiquing a research study and how to evaluate the answers. You can begin to sharpen your critical teeth on the illustrative studies throughout the book. Exercises at the end of each chapter will allow you to find, discuss, critique, and actually do similar research.

If you are already charting a course toward a social science career, or if you decide to do so after completing this course, we aim to give you enough "how to" instruction so that you can design your own research projects. We also offer "doing" exercises at the end of each chapter that will help you try out particular steps in the research process.

But our goal is not just to turn you into a more effective research critic or a good research technician. We do not believe that research methods can be learned by rote or applied mechanically. Thus, you will learn the benefits and liabilities of each major research approach as well as the rationale for using a combination of methods in some situations. You will also come to appreciate why the results of particular research studies must be interpreted within the context of prior research and through the lens of social theory.

Organization of the Book

The first three chapters introduce the why and how of research in general. Chapter 1 shows how research has helped us understand how social relations have changed in recent years and the impact of these changes. Chapter 2 illustrates the basic stages of research with studies of domestic violence, Olympic swimmers, and environmental disasters. Chapter 3 introduces the ethical considerations that should guide your decisions throughout the research process. The next three chapters discuss how to evaluate the way researchers design their measures (Chapter 4), draw their samples (Chapter 5), and justify their statements about causal connections (Chapter 6).

As we present the logic of testing causal connections in Chapter 6, we also present the basics of the experimental designs that provide the strongest tests for causality. In Chapter 7, we cover the most common method of data collection in sociology—surveys—and in Chapter 8, we present the basic statistical methods that are used to analyze the results of the quantitative data that often are collected in experiments and surveys. Here we examine the results of the 2010 General Social Survey to see how these statistics are used.

Chapters 9 and 10 shift the focus from strategies for collecting and analyzing quantitative data to strategies for collecting and analyzing qualitative data. In Chapter 9, we focus on the basic methods of collecting qualitative data: participant observation and ethnography, intensive interviews, and focus groups. We also introduce approaches such as ethnomethodology and netnography. In Chapter 10, we review the logic of qualitative data analysis and several specific approaches: narrative analysis, conversation analysis, and grounded theory. We also describe historical and comparative methods, as well as the mixed-methods style of research. Chapter 11 explains how you can combine different methods to evaluate social programs. Chapter 12 covers the review of prior research, the development of research proposals, and the writing and reporting of research results.

Distinctive Features of This Edition

In making changes for this edition, we feel we have advanced even further in pursuit of our goal of making research methods one of your most enjoyable and engaging courses. We have incorporated valuable suggestions from many faculty reviewers and students who have used the book over the several years since it was first released. As in the previous three editions, this book has also benefited from advances in its parent volume, Russell Schutt's *Investigating the Social World: The Process and Practice of Research* (now in its seventh edition).

New material on the role of social media in research. To reflect recent developments in social media, we have incorporated not only a variety of examples of research based on digital technologies, but also extended discussions of how such media modify research techniques both in survey and in qualitative methods.

New sections on historical/comparative research and mixed methods. At the suggestion of our reviewers, we have expanded the coverage of macro forms of research and demonstrated the advantages of mixing methods to balance the strengths and weaknesses of different approaches.

Updated methods and examples. Research methods is a continually evolving field, and we have included new kinds of research such as audit studies, an innovative form of field experiment that has major public policy implications. We have also updated our examples throughout the text, often using reviewers' suggestions of research that is simultaneously informative, instructive, and often provocative, to make for stimulating reading.

Major clarification of difficult topics. Again prompted by reviewers, we have rewritten and clarified sections on famously difficult topics: units and levels of analysis, basic statistical analyses, and quantitative analyses.

New end-of-chapter material. We have revised many discussion questions to ensure that they will facilitate in-class discussion, and we have added new questions on research ethics after each chapter.

Updated appendix on secondary data resources. Appendix C on sources of secondary data has been updated, and we have added a random numbers table, Appendix B, to allow practice in sampling.

Updated information on the impact of cell phones and the web on survey response. Chapter 7 on survey research provides the latest information on the impact of increasing use of cell phones and the web on survey practice.

As in the first three editions, our text also offers other distinctive features:

Brief examples of social research. In each chapter, these illustrate particular points and show how research techniques are used to answer important social questions. Whatever your particular substantive interests in social science, you'll find some interesting studies that will arouse your curiosity.

Integrated treatment of causality and experimental design. We have combined the discussions of causation and experimental design in order to focus on the issues that are most often encountered during research in sociology, criminal justice, education, social work, communications, and political science.

Realistic coverage of ethical concerns and ethical decision making. Like the parent volume, *Investigating the Social World,* this text presents ethical issues that arise in the course of using each method of data collection, as well as comprehensive coverage of research ethics in a new chapter.

Engaging end-of-chapter exercises. We organize the exercises under the headings of discussing, finding, critiquing, and doing, and end with questions about ethics. New exercises have been added, and some of the old ones have been omitted. The result is a set of learning opportunities that should greatly facilitate the learning process.

Software-based learning opportunities. The text's website (www.sagepub.com/chambliss4e) includes review exercises to help you master the concepts of social research, a set of articles that provide examples of different methods, and a portion of the 2010 General Social Survey (GSS) so you can try out quantitative data analysis (if your school provides access to the SPSS statistical package). Appendix C provides an introduction to SPSS.

Aids to effective study. Lists of main points and key terms provide quick summaries at the end of each chapter. In addition, key terms are highlighted in boldface type when first introduced and defined in the text. Definitions of key terms can also be found in the glossary/index at the end of the book. The text's website (www.sagepub .com/chambliss4e) offers more review questions. An instructor's manual includes more exercises that have been specially designed for collaborative group work in and outside of class. Appendix A, Finding Information, provides up-to-date information about using the Internet, and Appendix C lists secondary data sources.

Acknowledgments

First, we would like to thank Jerry Westby, publisher and senior editor at SAGE, our main managerial contact and source of encouragement as we developed our text. Other members of the SAGE team helped in multiple ways: Megan Krattli provided expert assistance with the review of the reviews, Eric Garner smoothly guided the manuscript through the production process, and Gretchen Treadwell did a fine job of copyediting.

The reviewers for this edition helped us to realize the potential for the revision. We are very grateful for the wisdom and critical acumen of the following:

James David Ballard, California State University, Northridge

Irenee R. Beattie, University of California, Merced

Rebecca Brooks, Ohio Northern University

Julie A. Kmec, Washington State University

Dianne Mosley, Texas Southern University

Sookhee Oh, University of Missouri-Kansas City

Margaret Platt Jendrek, Miami University (Oxford)

Xiaohe Xu, The University of Texas at San Antonio

And for previous editions:

Matthew M. Caverly, University of Florida

Jin Young Choi, Sam Houston State University

Cristina Bodinger-deUriarte, California State University-Los Angeles

Kellie J. Hagewen, University of Nebraska-Lincoln

Jerome L. Himmelstein, Amherst College

Karen McCue, University of New Mexico

Kate Peirce, Texas State University

Travis N. Ridout, Washington State University

Nick Sanyal, University of Idaho

Steve Swinford, Montana State University

Felicia P. Wiltz, Suffolk University

Reviewers of the first edition were the following:

Sandy D. Alvarez, Indiana State University

Julio Borquez, University of Michigan-Dearborn

Matthew W. Brosi, Oklahoma State University

Keith F. Durkin, Ohio Northern University

Juanita M. Firestone, University of Texas at San Antonio

Dena Hartley, University of Akron

Laura Hecht, California State University-Bakersfield

Ann Marie Kinnell, The University of Southern Mississippi

Manfred Kuechler, Hunter College

Vera Lopez, Arizona State University

Ed Nelson, California State University-Fresno

Colin Olson, University of New Mexico

Kristen Zgoba, Rutgers University

Reviewers of the first edition proposal were the following:

Diane C. Bates, Sam Houston State University

Mark Edwards, Oregon State University

David Folz, University of Tennessee-Knoxville

Ann Marie Kinnell, University of Southern Mississippi

Ronald Perry, Arizona State University

Chenyang Xiao, Washington State University

David Zefir, Plymouth State College

We are grateful to Sunshine Hillygus, director of Harvard University's Program on Survey Research, for sharing with us the latest findings about survey response rates. Elizabeth Schneider's contributions to Appendix A draw on her comparable work with Russ Schutt (her husband) for the sixth edition of *Investigating the Social World*. We thank Kate Russell for writing a new track for the interactive exercises with links to SAGE journal articles, Megan Reynolds and Kathryn Stoeckert for their work on previous sets of exercises, Amanda Colligan and Robyn Brown for their help on the eBook, and we thank VPG Integrated Media for the online programming.

We also have some personal thank-yous:

Dan Chambliss: I wish gratefully to acknowledge the assistance, in many areas, of Marcia Wilkinson, who as typist, transcriber, organizer, and administrative aide and daily conscience is simply irreplaceable. My students at Hamilton College have been a blessing throughout: Chris Takacs helped to design and create several of the new exhibits, solving intellectual problems through graphic displays; Shauna Sweet told me where the book was good and where it wasn't, clarified the regression effect, and showed me how people actually should read contingency tables; Katey Healy-Wurzburg, in one of many moments of intellectual brilliance, explained the underlying rhetorical problem of the ecological fallacy; and Erin Voyik, as a teaching assistant in my Methods class, laid out for me time and again what students do and don't actually understand, and enjoy, about social research methods. Many others, students and colleagues alike, have contributed without recognition; let's just say that all of intellectual life is communal, and we fully appreciate that fact. And finally, I hope that my wife, Susan Morgan, enjoyed, at least vicariously, the thrills I felt in working on this book as much as I enjoyed sharing them with her.

Russ Schutt: I am grateful to the many reviewers of the previous six editions of *Investigating the Social World*, as well as to the many staff and consultants at SAGE who helped to make that text a success. My thanks to Kate Russell, the outstanding graduate student who checked web exercises and updated statistical examples throughout the text and for the website. I also want to express my appreciation to my many research collaborators, with whom I have shared so many fascinating and educational experiences and from whom I have learned so much, and for the many fine students at the University of Massachusetts, Boston, who continue to renew my enthusiasm for quality teaching. Most importantly, I thank my wife, Elizabeth Schneider, for her ongoing support and love, and my daughter, Julia, for being such a marvelous young woman.

Dan and Russ: Finally, Dan wants to say that Russ Schutt is a wonderful coauthor, with whom he is honored to work: totally responsible and respectful, hardworking, serious in his scholarship but without a trace of arrogance. His generous personality has allowed this collaboration to sail along beautifully. Russ adds that Dan is the perfect model of the gentleman and scholar, whose research savvy and keen intelligence are matched to a warm and caring persona. We both like to think that our talents are almost entirely complementary. We are immensely grateful for the chance to work together.

Dan:
To my sweetheart Susan and to the gifts she brought to me—
Sarah, Daniel, Anne, and Rebecca
Russ:
To Beth and Julia

CHAPTER 1

Science, Society, and Social Research

Facebook and other online social networking services added a new dimension to the social world in the early years of the new millennium. If you saw the film, *The Social Network,* you know that Mark Zuckerberg started Facebook in 2004, as a service for college students like himself, and that he didn't stop there. By the end of 2010, Facebook had more than 550,000,000 members—one out of every 12 people in the world and almost half the United States population—and was adding about 700,000 new members each day (Grossman 2010). The social world will never be the same.

Do you know what impact online social networking has had? Has it helped you to keep in touch with your friends? To make new friends? Does it distract you from reading your textbooks and paying attention in class? Is it reducing your face-to-face interactions with other people? And, is your experience with Facebook similar to that of other people, including those in other countries? Overall, are computer-mediated forms of communication enriching your social ties or impoverishing them?

That's where social researchers begin, with questions about the social world and a desire to find an answer. But social research differs from ordinary thinking by focusing on people outside our immediate experience, and by the use of systematic research methods. Keith Hampton, Oren Livio, and Lauren Sessions Goulet (2010) asked whether people who make wireless Internet connections in public spaces reduce their engagement with others in those spaces. Rich Ling and Gitte Stald (2010) asked whether different technology-mediated forms of communication like cell phones and social networking differ in their effects on social ties. Kevin Lewis and others (2008) asked if some types of students were more likely to use Facebook than others. Although their research methods ranged from observing others (Hampton & Gupta 2008) to conducting a survey on the web (Ling & Stald 2010), and analyzing Facebook records (Lewis, Christakis, Gonzalez, Kaufman, & Wimmer 2008), each set of researchers carefully designed a research project about their question and presented their research findings in a published report.

Using research methods to investigate questions about the social world results in knowledge that can be more important, more trustworthy, and more useful than personal opinions or individual experiences. By the chapter's end, you should know what is "scientific" in social science and appreciate how the methods of science can help us understand the problems of society.

Video Link 1.1
Watch more on
social research.

▣ Learning About the Social World

Read each of the following questions and jot down your answers. Don't ruminate about the questions or worry about your responses: *This is not a test;* there are no "wrong" answers.

1. Does social networking software help you stay in touch with your friends?

2. What percentage of Americans are connected to the Internet?

3. How many close friends does an average American have?

4. Does wireless access (Wi-Fi) in public places like Starbucks decrease social interaction among customers?

5. Do both cell phones and e-mail tend to hinder the development of strong social ties?

6. How does Internet use vary across social groups?

You probably didn't have any trouble answering the first question, about your own experiences. But the second question and the others concern *the social world*—the experiences and orientations of people other than yourself. To answer them, we need to combine answers from many different people and perhaps from other sources. Studying research methods should help you learn what criteria to apply when evaluating these different answers and what methods to use when seeking to develop your own answers.

Let's compare your answers to Questions 2 through 6 with findings from research using social science methods.

1. The 2009 Current Population Survey, by the U.S. Bureau of the Census of approximately 54,000 households revealed that 63.5% of the U.S. population had a broadband connection to the Internet at home (a total of 68.7% were connected to the Internet at home). This percentage has increased rapidly since broadband first came into use in 2000.

2. In 2004, the average American had 2.08 discussion partners, or "confidants"—close friends we could say. This average had declined from 2.94 in 1985, so you might speculate that it is even lower by now (McPherson, Smith-Lovin, & Brashears 2006:358).

3. After observing Internet use in coffee shops with wireless access in two cities, Hampton and Gupta (2008) concluded that there were two types of Wi-Fi users: some who used their Internet connection to create a secondary work office and others who used their Internet connection as a tool for meeting others in the coffee shop.

Audio Link 1.1
Listen to how social media impacts research.

4. Based on surveys in Norway and Denmark, Ling and Stald (2010) concluded that mobile phones increase social ties among close friends and family members, while e-mail communication tends to decrease the intensity of our focus on close friends and family members.

5. Internet use differs dramatically between social groups. As indicated in Exhibit 1.1, in 2007 Internet use ranged from as low as 32% among those with less than a high school education to 88% among those with at least a bachelor's degree—although Internet use has increased for all education levels since 1997 (Strickling 2010). Internet use also increases with family income and is higher among non-Hispanic whites and Asian Americans than among Hispanic Americans and non-Hispanic black Americans (Cooper & Gallager 2004:Appendix, Table 1). People who are under 30 are most likely to use the Internet (93% of those age 12 to 29), as compared to those who are middle aged (81% among those age 30 to 49 and 70% of those age 50 to 64) or older (38% of those age 65 and older) (Pew Internet 2010). In their study at one college, Lewis et al. (2008) found that black students had more Facebook friends than whites, while female students posted more Facebook pictures than male students.

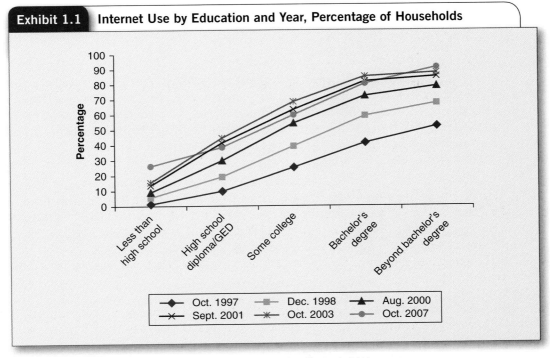

Exhibit 1.1   **Internet Use by Education and Year, Percentage of Households**

Legend:
- Oct. 1997
- Dec. 1998
- Aug. 2000
- Sept. 2001
- Oct. 2003
- Oct. 2007

Note: Reports available online at www.census.gov/population/www/socdemo//computer/html

Could you have predicted the results of this research? How do these answers compare with those you gave? Do you think your personal experience may have distorted your guesses? That would be completely normal; it's how we normally think.

People come to different conclusions about the social world for another reason: It's easy to make errors in logic, particularly when we are analyzing the social world, in which we ourselves are conscious participants. We can call some of these errors *everyday errors*, because they occur so frequently in the nonscientific, unreflective conversations that we hear on a daily basis.

Our favorite example of such errors in reasoning comes from a letter to Ann Landers, the newspaper advice columnist (sadly, now deceased). See if you can spot the problems here: The letter was written by a woman who had just moved, with her two pet cats, from an apartment in the city to a house in the country. In the city, she had not let the cats go outside, but she felt guilty about keeping them locked up. Upon arrival at the country house, she let the cats out—but they tiptoed cautiously to the door, looked outside, then went right back into the living room and lay down.

The woman concluded that people shouldn't feel guilty about keeping their cats indoors, since even when they have the chance, cats don't really want to play outside.

Did you spot this person's errors in reasoning?

- *Overgeneralization*—She observed only two cats, both of which were previously confined indoors. Maybe they aren't like most cats.

- *Selective or inaccurate observation*—She observed the cats at the outside door only once. But maybe if she let them out several times, they would become more comfortable with going out.

- *Illogical reasoning*—She assumed that other people feel guilty about keeping their cats indoors. But maybe they don't.

- *Resistance to change*—She was quick to conclude that she had no need to change her approach to the cats. But maybe she just didn't want to change her own routines and was eager to believe that she was managing her cats just fine already.

You don't have to be a scientist or use sophisticated research techniques to avoid these four errors in reasoning. If you recognize them and make a conscious effort to avoid them, you can improve your own reasoning. In the process, you also will be taking the advice of your parents (or minister, teacher, or other adviser) not to stereotype people, to avoid jumping to conclusions, and to look at the big picture. These are the same kinds of mistakes that the methods of social science are designed to help us avoid.

Let's look at each kind of error in turn.

Overgeneralization

Overgeneralization occurs when we unjustifiably conclude that what is true for some cases is true for all cases. We are always drawing conclusions about people and social processes from our own interactions with them, but sometimes we forget that our experiences are limited. The social (and natural) world is, after all, a complex place. Maybe someone made a wisecrack about the ugly shoes you're wearing today, but that doesn't mean that "everyone is talking about you." Or there may have been two drunk-driving accidents following fraternity parties this year, but by itself, this doesn't mean that all fraternity brothers are drunk drivers. Or maybe you had a boring teacher in your high school chemistry class, but that doesn't mean all chemistry teachers are boring. We can interact with only a small fraction of the

> **Overgeneralization:** Occurs when we unjustifiably conclude that what is true for some cases is true for all cases.

individuals who inhabit the social world, especially in a limited span of time; rarely are they completely typical people. One heavy Internet user found that his online friendships were "much deeper and have better quality" than his other friendships (Parks & Floyd 1996). Would his experiences generalize to yours? To those of others?

Selective or Inaccurate Observation

We also have to avoid **selective** or **inaccurate observation**—choosing to look only at things that are in line with our preferences or beliefs. When we dislike individuals or institutions, it is all too easy to notice their every failing. For example, if we are convinced that heavy Internet users are antisocial, we can find many confirming instances. But what about elderly people who serve as Internet pen pals for grade school children or therapists who deliver online counseling? If we acknowledge only the instances that confirm our predispositions, we are victims of our own selective observation. Exhibit 1.2 depicts the difference between selective observation and overgeneralization.

> **Selective (inaccurate) observation:** Choosing to look only at things that are in line with our preferences or beliefs.

Our observations can also simply be inaccurate. When you were in high school, maybe your mother complained that you were "always staying out late with your friends." Perhaps that was inaccurate; you only stayed out late occasionally. And when you complained that she "yelled" at you, even though her voice never actually increased in volume, that, too, was an inaccurate observation. In social science, we try to be more precise than that.

Such errors often occur in casual conversation and in everyday observation of the world around us. What we think we have seen is not necessarily what we really have seen (or heard, smelled, felt, or tasted). Even when our senses are functioning fully, our minds have to interpret what we have sensed (Humphrey 1992). The optical illusion in Exhibit 1.3, which can be viewed as either two faces or a vase, should help you realize that even simple visual perception requires interpretation.

Exhibit 1.2   **The Difference Between Overgeneralization and Selective Observation**

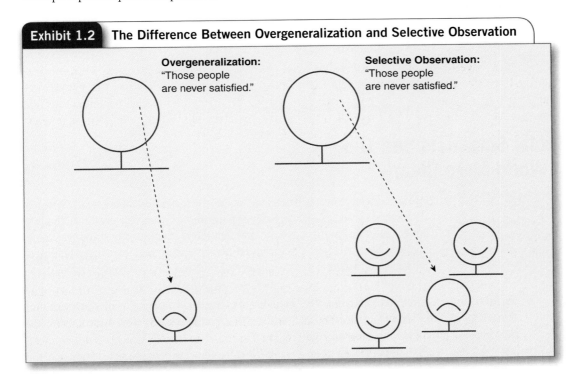

Exhibit 1.3 An Optical Illusion

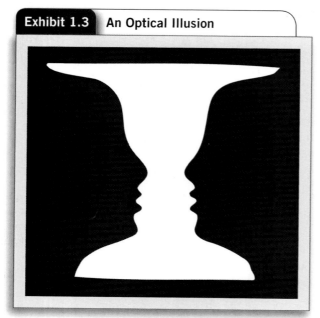

Illogical reasoning: The premature jumping to conclusions or arguing on the basis of invalid assumptions.

Resistance to change: The reluctance to change our ideas in light of new information.

Illogical Reasoning

When we prematurely jump to conclusions or argue on the basis of invalid assumptions, we are using **illogical reasoning**. For example, we might think that people who don't have many social ties just aren't friendly, even if we know they have just moved into a community and started a new job. Obviously, that's not logical. On the other hand, an unquestioned assumption that everyone seeks social ties or benefits from them overlooks some important considerations, such as the impact of childhood difficulties on social trust and the exclusionary character of many tightly knit social groups. Logic that seems impeccable to one person can seem twisted to another—but people having different assumptions, rather than just failing to "think straight," usually cause the problem.

Resistance to Change

Resistance to change, the reluctance to change our ideas in light of new information, is a common problem. After all, we know how tempting it is to make statements that conform to our own needs rather than to the observable facts (I can't live on that salary!). It can also be difficult to admit that we were wrong once we have staked out a position on an issue (I don't want to discuss this anymore.). Excessive devotion to tradition can stifle adaptation to changing circumstances (This is how we've always done it, that's why.). People often accept the recommendations of those in positions of authority without question (Only the president has all the facts.). In all of these ways, we often close our eyes to what's actually happening in the world.

Can Social Scientists See The Social World More Clearly?

Science: A set of logical, systematic, documented methods for investigating nature and natural processes; the knowledge produced by these investigations.

Can social science do any better? Can we see the social world more clearly if we use the methods of social science? **Science** relies on logical and systematic methods to answer questions, and it does so in a way that allows others to inspect and evaluate its methods. So social scientists develop, refine, apply, and report their understanding of the social world more systematically, or "scientifically," than the general public does.

Journal Link 1.1
Read about our changing perspectives of community.

- **Social science** research methods reduce the likelihood of overgeneralization by using systematic procedures for selecting individuals or groups to study so that the study subjects are representative of the individuals or groups to which we wish to generalize.

- To avoid illogical reasoning, social researchers use explicit criteria for identifying causes and for determining whether these criteria are met in a particular instance.

- Social science methods can reduce the risk of selective or inaccurate observation by requiring that we measure and sample phenomena systematically.

- Scientific methods lessen the tendency to answer questions about the social world from ego-based commitments, excessive devotion to tradition, or unquestioning respect for authority. Social scientists insist: Show us the evidence!

> **Social science:** The use of scientific methods to investigate individuals, societies, and social processes; the knowledge produced by these investigations.

Social Research in Practice

Although all social science research seeks to minimize errors in reasoning, different projects may have different goals. The four most important goals of social research are (1) description, (2) exploration, (3) explanation, and (4) evaluation. Let's look at examples of each.

Research|Social Impact Link 1.1
Read about social research in practice.

Description: How Often Do Americans "Neighbor"?

During the last quarter of the 20th century, the annual (biennial since 1996) General Social Survey (GSS) has investigated a wide range of characteristics, attitudes, and behaviors. Each year, more than 1,000 adults in the United States complete GSS phone interviews; many questions repeat from year to year so that trends can be identified. Robert Putnam often used GSS data in his famous Bowling Alone investigation of social ties in America.

Survey responses indicate that "neighboring" has been declining throughout this period. As indicated in Exhibit 1.4 (Putnam 2000:106), the percentage of GSS respondents who reported spending "a social evening with someone who lives in your neighborhood . . . about once a month or more often" was 60% for married people in 1975 and about 65% for singles. By 1998, the comparable percentages were 45% for married people and 50% for singles. This is **descriptive research** because the findings simply describe differences or variations in social phenomena.

Audio Link 1.2
Listen to other research by Putnam.

Exploration: How Do Athletic Teams Build Player Loyalty?

> **Descriptive research:** Research in which social phenomena are defined and described.

Organizations like combat units, surgical teams, and athletic teams must develop intense organizational loyalty among participants if they are to maximize their performance. How do they do it? This question motivated Patricia and Peter Adler (2000) to study college athletics. They wanted to explore this topic without preconceptions or fixed hypotheses. So Peter Adler joined his college basketball team as a "team sociologist," while Patti participated in some team activities as his wife and as a professor at the school. They recorded observations and comments at the end of each day for a period of 5 years. They also interviewed at length the coaches and all 38 basketball team members during that period.

Careful and systematic review of their notes led Adler and Adler (2000) to conclude that intense organizational loyalty emerged from five processes: (1) domination, (2) identification, (3) commitment, (4) integration, and (5) goal alignment. We won't review each of these processes here, but the following quote indicates how they found the process of integration into a cohesive group to work:

> By the time the three months were over [the summer before they started classes] I felt like I was there a year already. I felt so connected to the guys. You've played with them, it's been 130 degrees in the gym, you've elbowed each other, knocked each other around. Now you've felt

Research|Social Impact
Link 1.2
Read more about
exploratory research.

a relationship, it's a team, a brotherhood type of thing. Everybody's got to eat the same rotten food, go through the same thing, and all you have is each other. So you've got a shared bond, a camaraderie. It's a whole houseful of brothers. And that's home to everybody in the dorm, not your parents' house. (p. 43)

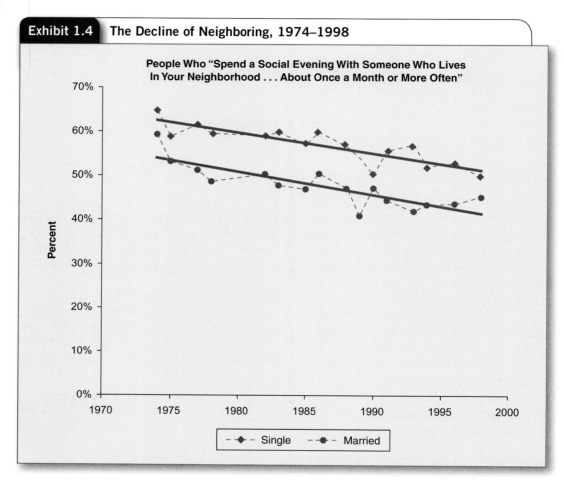

Exhibit 1.4 **The Decline of Neighboring, 1974–1998**

People Who "Spend a Social Evening With Someone Who Lives In Your Neighborhood . . . About Once a Month or More Often"

Legend: Single, Married

Participating in and observing the team over this long period enabled Adler and Adler (2000) to identify and to distinguish particular aspects of such loyalty-building processes, such as identifying three modes of integration into the group: (1) unification in opposition to others, (2) development of group solidarity, and (3) sponsorship by older players. They also identified negative consequences of failures in group loyalty, such as the emergence of an atmosphere of jealousy and mistrust, and the disruption of group cohesion, as when one team member focused only on maximizing his own scoring statistics.

In this project, Adler and Adler did more than simply describe what people did—they tried to explore the different elements of organizational loyalty and the processes by which loyalty was built. **Exploratory research** seeks to find out how people get along in the setting under question, what meanings they give to their actions, and what issues concern them. You might say the goal is to learn "what's going on here?"

Exploratory research: Seeks to find out how people get along in the setting under question, what meanings they give to their actions, and what issues concern them.

Research in the News

TWITTER USERS FLOCK TOGETHER

Like "birds of a feather that flock together," Twitter users tend to respond to others who express sentiments similar to their own. Johan Bolen from the University of Indiana reviewed 6-months-worth of tweets from 102,009 active Twitter users. Twitter users who expressed happy moods tended to retweet or reply to others who were happy, while those who indicted loneliness engaged more with other lonely Twitterers.

Source: Bilton, Nick. 2011. Twitter users flock together. *The New York Times,* March 21:B7.

Explanation: Does Social Context Influence Adolescent Outcomes?

Often, social scientists want to explain social phenomena, usually by identifying causes and effects. Bruce Rankin at Koc University in Turkey and James Quane at Harvard (Rankin & Quane 2002) analyzed data collected in a large survey of African American mothers and their adolescent children to test the effect of social context on adolescent outcomes. Their source of data was a study funded by the MacArthur Foundation, Youth Achievement and the Structure of Inner City Communities, in which face-to-face interviews were conducted with more than 636 youth living in 62 poor and mixed-income urban Chicago neighborhoods.

> **Explanatory research:** Seeks to identify causes and effects of social phenomena and to predict how one phenomenon will change or vary in response to variation in another phenomenon.

Explanatory research like this seeks to identify causes and effects of social phenomena and to predict how one phenomenon will change or vary in response to variation in another phenomenon. Rankin and Quane (2002) were most concerned with determining the relative importance of three different aspects of social context—neighborhoods, families, and peers—on adolescent outcomes (both positive and negative). To make this determination, they had to conduct their analysis in a way that allowed them to separate the effects of neighborhood characteristics, like residential stability and economic disadvantage, from parental involvement in child rearing and other family features, as well as from peer influence. They found that neighborhood characteristics affect youth outcomes primarily by influencing the extent of parental monitoring and the quality of peer groups.

Journal Link 1.2
Read about how social interventions can affect well-being.

Evaluation: Does More Social Capital Result in More Community Participation?

The "It's Our Neighbourhood's Turn" project (Onze Burrt aan Zet, or OBAZ) in the city of Enschede, the Netherlands, was one of a series of projects initiated by the Dutch Interior and Kingdom Relations ministry to increase the quality of life and safety of individuals in the most deprived neighborhoods in the Netherlands. In the fall of 2001, residents in three of the city's poorest neighborhoods were informed that their communities had received funds to use for community improvement and that residents had to be actively involved in formulating and implementing the improvement plans (Lelieveldt 2003:1). Political scientist Herman Lelieveldt (2004:537) at the University of Twente, the Netherlands, and others then surveyed community residents to learn about their social relations and their level of local political participation; a second survey was conducted 1 year after the project began.

Lelieveldt wanted to evaluate the impact of the OBAZ project—to see whether the "livability and safety of the neighborhood" could be improved by taking steps like those Putnam (2000:408) recommended to increase "social capital," meaning that citizens would spend more time connecting with their neighbors.

It turned out that residents who had higher levels of social capital participated more in community political processes. However, not every form of social capital made much of a difference. Neighborliness—the extent to which citizens are engaged in networks with their neighbors—was an important predictor of political participation, as was a feeling of obligation to participate. By contrast, a sense of trust in others (something that Putnam emphasizes) was not consistently important (Lelieveldt 2004:535, 547–548): Those who got more involved in the OBAZ political process tended to distrust their neighbors. When researchers focus their attention on social programs like the OBAZ project, they are conducting **evaluation research**—research that describes or identifies the impact of social policies and programs.

Certainly many research studies have more than one such goal—all studies include some description, for instance. But clarifying your primary goal can often help when deciding how to do your research.

> **Evaluation research:** Research that describes or identifies the impact of social policies and programs.

How Well Have We Done Our Research?

Social scientists want validity in their research findings—they want to find the truth. The goal of social science is not to reach conclusions that other people will like or that suit our personal preferences. We shouldn't start our research determined to "prove" that our college's writing program is successful, or that women are portrayed unfairly in advertisements, or that the last presidential election was rigged, or that homeless people are badly treated. We may learn that all of these are true, or aren't; but our goal as social scientists should be to learn the truth, even if it's sometimes disagreeable to us. The goal is to figure out how and why some part of the social world operates as it does and to reach valid conclusions. We reach the goal of **validity** when our statements or conclusions about empirical reality are correct.

In *Making Sense of the Social World: Methods of Investigation,* we will be concerned with three kinds of validity: (1) measurement validity, (2) generalizability, and (3) causal validity (also known as internal validity). We will learn that invalid measures, invalid generalizations, or invalid causal inferences result in invalid conclusions.

> **Validity:** The state that exists when statements or conclusions about empirical reality are correct.

Measurement Validity

Measurement validity is our first concern, because without having measured what we think we've measured, we don't even know what we're talking about. So when Putnam (2000:291) introduces a measure of "social capital" that has such components as number of club meetings attended and number of times worked on a community project, we have to stop and consider the validity of this measure. Measurement validity is the focus of Chapter 4.

Problems with measurement validity can occur for many reasons. In studies of Internet forums, for instance, researchers have found that some participants use fictitious identities, even pretending to be a different gender (men posing as women, for instance) (Donath 1999). Therefore, it's difficult to measure gender in these forums, and researchers could not rely on gender as disclosed in the forums when identifying differences in usage patterns between men and women. Similarly, if you ask people, "Are you an alcoholic?" they probably won't say yes, even if they

> **Measurement validity:** Exists when an indicator measures what we think it measures.

are; the question elicits less valid information than would be forthcoming by asking them how many drinks they consume, on average, each day. Some college men may be hesitant to admit to watching reruns of *South Park* on television 6 hours a day, so researchers use electronic monitoring devices on TV sets to measure what programs people watch and how often.

Encyclopedia Link 1.2
Read an overview
of the importance of
generalizability.

Generalizability

The **generalizability** of a study is the extent to which it can inform us about persons, places, or events that were not directly studied. For instance, if we ask our favorite students how much they enjoyed our Research Methods course, can we assume that other students (perhaps not as favored) would give the same answers? Maybe they would—but probably not. Generalizability is the focus of Chapter 5.

Generalizability is always an important consideration when you review social science research. Even the huge, international National Geographic Society (2000) survey of Internet users had some limitations in generalizability. Only certain people were included in the sample: people who were connected to the Internet, who had heard about the survey, and who actually chose to participate. This meant that many more respondents came from wealthier countries, which had higher rates of computer and Internet use, than from poorer countries. However, the inclusion of individuals from 178 countries and territories does allow some interesting comparisons among countries.

There are two kinds of generalizability: sample and cross-population.

Sample generalizability is a key concern in survey research. Political polls, such as the Gallup Poll or Zogby International, may study a sample of 1,400 likely voters, for example, and then generalize the findings to the entire American population of 80 million likely voters. No one would be interested in the results of political polls if they represented only the tiny sample that actually was surveyed rather than the entire population.

Cross-population generalizability occurs to the extent that the results of a study hold true for multiple populations; these populations may not all have been sampled, or they may be represented as subgroups within the sample studied. We can only wonder about the cross-population generalizability of Putnam's findings about social ties in the United States. Has the same decline occurred in Mexico, Argentina, Britain, or Thailand?

> **Generalizability:** Exists when a conclusion holds true for the population, group, setting, or event that we say it does, given the conditions that we specify; it is the extent to which a study can inform us about persons, places, or events that were not directly studied.

> **Sample generalizability:** Exists when a conclusion based on a sample, or subset, of a larger population holds true for that population.

> **Cross-population generalizability (external validity):** Exists when findings about one group, population, or setting hold true for other groups, populations, or settings (Exhibit 1.5).

Causal Validity

Causal validity, also known as **internal validity**, refers to the truthfulness of an assertion that A causes B. It is the focus of Chapter 6.

Most research seeks to determine what causes what, so social scientists frequently must be concerned with causal validity. For example, Gary Cohen and Barbara Kerr (1998) asked whether computer-mediated counseling could be as effective as face-to-face counseling for mental health problems—that is, whether one type of counseling leads to better results than the other. They could have compared people who had voluntarily experienced one of these types of treatment, but it's quite likely that individuals who sought out a live person for counseling would differ, in important ways, from those who sought computer-mediated counseling. Younger people tend to use computers more; so do more educated people. Or maybe less sociable people would be

> **Causal validity (internal validity):** Exists when a conclusion that A leads to, or results in, B is correct.

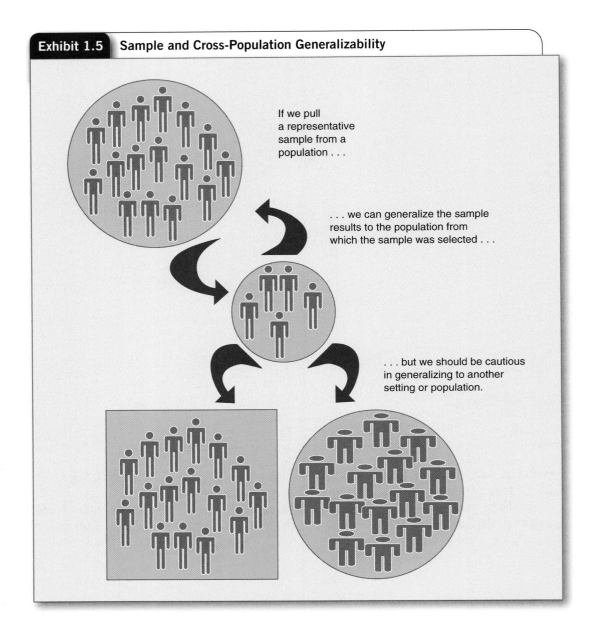

Exhibit 1.5   Sample and Cross-Population Generalizability

If we pull a representative sample from a population . . .

. . . we can generalize the sample results to the population from which the sample was selected . . .

. . . but we should be cautious in generalizing to another setting or population.

Journal Link 1.3
Read about data based on a representative sample.

more drawn to computer-mediated counseling. Normally it would be hard to tell if different results from the two therapies were caused by the therapies themselves or by different kinds of people going to each.

So Cohen and Kerr (1998) designed an experiment in which students seeking counseling were assigned randomly (by a procedure somewhat like flipping a coin) to either computer-mediated or face-to-face counseling. In effect, people going to one kind of counseling were just like people going to the other; as it happens, their anxiety scores afterwards were roughly the same. There seemed to be no difference. (Exhibit 1.6). By using the random assignment procedure, Cohen and Kerr strengthened the causal validity of this conclusion.

On the other hand, even in properly randomized experiments, causal findings can be mistaken because of some factor that was not recognized during planning for the study. If the computer-mediated counseling sessions were conducted in a modern building with all the latest amenities, while face-to-face counseling was delivered in a run-down building, this difference might have led to different outcomes for reasons quite apart

Exhibit 1.6 Partial Evidence of Causality

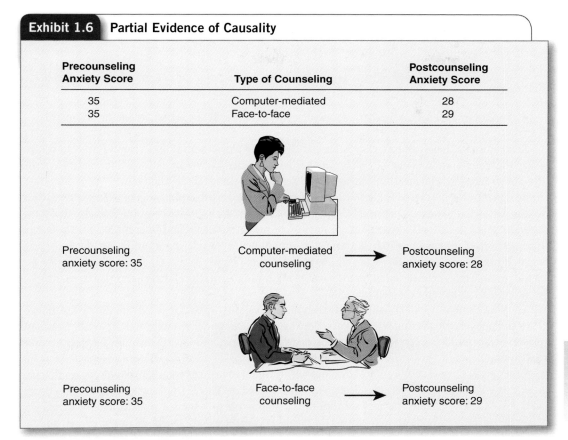

Precounseling Anxiety Score	Type of Counseling	Postcounseling Anxiety Score
35	Computer-mediated	28
35	Face-to-face	29

Precounseling anxiety score: 35 — Computer-mediated counseling → Postcounseling anxiety score: 28

Precounseling anxiety score: 35 — Face-to-face counseling → Postcounseling anxiety score: 29

Video Link 1.2
Watch more on social research methods.

from the type of counseling. Also, Cohen and Kerr didn't have a group that received no counseling. Maybe just a little quiet time or getting older would provide the same benefits as therapy.

So establishing causal validity can be quite difficult. In subsequent chapters, you will learn in more detail how experimental designs and statistics can help us evaluate causal propositions, but the solutions are neither easy nor perfect. We always have to consider critically the validity of causal statements that we hear or read.

Conclusion

This first chapter should have given you an idea of what to expect in the rest of the book. Social science provides us with a variety of methods for avoiding everyday errors in reasoning and for coming to valid conclusions about the social world. We will explore different kinds of research, using different techniques, in the chapters to come, always asking, is this answer likely to be correct? The techniques are fairly simple, but they are powerful nonetheless if properly executed. You will also learn some interesting facts about social life. We have already seen, for instance, some evidence that

- The Internet and social media may have surprising effects on our relationships with others.

- Organizational processes that build loyalty, as happens on athletic teams, can strengthen social ties.

- Neighborhoods in which social ties are weaker may result in less effective forms of parenting, but both parenting and peer group quality have stronger effects than neighborhood social ties on adolescent outcomes.

- Government programs to increase social capital in neighborhoods can increase local political participation.

- Students may benefit as much from computer-mediated counseling as from face-to-face counseling.

Remember, you must ask a direct question of each research project you examine: How valid are its conclusions? The theme of validity ties the chapters in this book together. Each technique will be evaluated in terms of its ability to help us with measurement validity, generalizability, and causal validity.

To illustrate the process of doing research, in Chapter 2, we describe studies of domestic violence, community disaster, student experience of college, and other topics. We review the types of problems that social scientists study, the role of theory, the major steps in the research process, and other sources of information that may be used in social research. We stress the importance of considering scientific standards in social research and review generally accepted ethical guidelines. In Chapter 3, we set out the general principles of ethical research that social scientists try to follow. As well, examples of ethical challenges to good research will be presented in many of the chapters that follow.

Then in Chapters 4, 5, and 6, we return to the subject of validity—the three kinds of validity and the specific techniques used to maximize the validity of our measures, our generalizations, and our causal assertions. Chapter 6 also introduces experimental studies, one of the best methods for establishing causal connections relationships.

Other methods of data collection and analysis are introduced in Chapters 7, 8, 9, and 10. Survey research is the most common method of data collection in sociology, and in Chapter 7, we devote attention to the different types of surveys. Chapter 8 is not a substitute for an entire course in statistics, but it gives you a good idea of how to use statistics honestly in reporting the results of your own studies using quantitative methods and in critically interpreting the results of research reported by others. Chapter 9 shows how qualitative methods like participant observation, intensive interviewing, and focus groups can uncover aspects of the social world that we are likely to miss in experiments and surveys, while Chapter 10, on qualitative data analysis, illustrates several approaches that researchers can take to the analysis of the data they collect in qualitative projects.

Evaluation research can use a variety of methods. Chapter 11 explains the role of evaluation research in investigating social programs and how to design evaluation research studies. Finally, Chapter 12 focuses on how to review prior research, how to propose new research, and how to report original research. We give special attention to how to formulate research proposals and how to critique, or evaluate, reports of research that you encounter.

Throughout these chapters, we will try to make the ideas interesting and useful to you, both as a consumer of research (as reported in newspapers, for instance) and as a potential producer (if, say, you do a survey in your college or neighborhood). Each chapter ends with several helpful learning tools. Lists of key terms and chapter highlights will help you review, and exercises will help you apply your knowledge. Social research isn't rocket science, but it does take some clear thinking, and these exercises should give you a chance to practice.

Here is a closing thought: Vince Lombardi, legendary coach of the Green Bay Packers of the National Football League during the 1960s, used to say that championship football was basically a matter of "four yards and a cloud of dust." Nothing too fancy, no razzle-dazzle plays, no phenomenally talented players doing it all alone—just solid, hard-working, straight-ahead fundamentals. This may sound strange, but excellent social research can be done—can "win games"—in the same way. We'll show you how to design and conduct surveys that get the right answers, interviews that discover people's true feelings, and experiments that pinpoint what causes what. And we'll show you how to avoid getting bamboozled by every "Studies Show . . . We're Committing More Crimes!" article you see in the newspaper. It takes a little effort initially, but we think you will find it worthwhile—even enjoyable.

Key Terms

Highlights

- Four common errors in everyday reasoning are overgeneralization, selective or inaccurate observation, illogical reasoning, and resistance to change. These errors result from the complexity of the social world, subjective processes that affect the reasoning of researchers and those they study, researchers' self-interestedness, and unquestioning acceptance of tradition or of those in positions of authority.

- Social science is the use of logical, systematic, documented methods to investigate individuals, societies, and social processes, as well as the knowledge these investigations produce.

- Social research can be motivated by personal interest, policy guidance and program management needs, or academic concerns.

- Social research can be descriptive, exploratory, explanatory, or evaluative—or some combination of these.

- Valid knowledge is the central concern of scientific research. The three components of validity are measurement validity, generalizability (both from the sample to the population from which it was selected and from the sample to other populations), and causal (internal) validity.

STUDENT STUDY SITE

The Student Study Site, available at **www.sagepub.com/chambliss4e,** includes useful study materials including web exercises with accompanying links, eFlashcards, videos, audio resources, journal articles, and encyclopedia articles, many of which are represented by the media links throughout the text. The site also features Interactive Exercises—represented by the green icon here—to help you understand the concepts in this book.

Exercises

Discussing Research

1. Select a social issue that interests you, such as Internet use or crime. List at least four of your beliefs about this phenomenon. Try to identify the sources of each of these beliefs.

2. Does the academic motivation to do the best possible job of understanding how the social world works conflict with policy and/or personal motivations? How could personal experiences with social isolation or with Internet use shape research motivations? In what ways might the goal of influencing policy about social relations shape how a researcher approaches this issue?

3. Pick a contemporary social issue of interest to you. List a descriptive, exploratory, explanatory, and evaluative question that you could investigate about this issue.

4. Review each of the three sets of research alternatives. Which alternatives are most appealing to you? Which combination of alternatives makes the most sense to you (one possibility, for example, is quantitative research with a basic science orientation)? Discuss the possible bases of your research preferences in terms of your academic interests, personal experiences, and policy orientations.

Finding Research

1. Read the abstracts (initial summaries) of each article in a recent issue of a major social science journal. (Ask your instructor for some good journal titles.) On the basis of the abstract only, classify each research project represented in the articles as primarily descriptive, exploratory, explanatory, or evaluative. Note any indications that the research focused on other types of research questions.

2. From the news, record statements of politicians or other leaders about some social phenomenon. Which statements do you think are likely to be in error? What evidence could the speakers provide to demonstrate the validity of these statements?

3. Check out Robert Putnam's website (www.bettertogether.org) and review survey findings about social ties in several cities. Prepare a 5- to 10-minute class presentation on what you have found about social ties and the ongoing research-based efforts to understand them.

Critiquing Research

1. Scan one of the publications about the Internet and society at the Berkman Center for Internet & Society website (http://cyber.law.harvard.edu/). Describe one of the projects discussed: its goals, methods, and major findings. What do the researchers conclude about the impact of the Internet on social life in the United States? Next, repeat this process with a report from the Pew Internet Project (www.pewinternet.org), or with the Digital Future report from the University of Southern California's Center for the Digital Future site (www.digitalcenter.org). What aspects of the methods, questions, or findings might explain differences in their conclusions? Do you think the researchers approached their studies with different perspectives at the outset? If so, what might these perspectives have been?

2. Research on social ties was publicized in a *Washington Post* article that also included comments by other sociologists (www.washingtonpost.com/wp-dyn/content/article/2006/06/22/AR2006062201763_pf.html). Read the article, and continue the commentary. Do your own experiences suggest that there is a problem with social ties in your community? Does it seem, as Barry Wellman suggests in the *Washington Post* article, that a larger number of social ties can make up for the decline in intimate social ties that McPherson et al. (2006:358) found?

Doing Research

1. What topic would you focus on if you could design a social research project without any concern for costs? What are your motives for studying this topic?

2. Develop four questions that you might investigate about the topic you just selected. Each question should reflect a different research motive: description, exploration, explanation, or evaluation. Be specific. Which question most interests you? Why?

Ethics Questions

Throughout the book, we will discuss the ethical challenges that arise in social research. At the end of each chapter, we ask you to consider some questions about ethical issues related to that chapter's focus. We introduce this critical topic formally in Chapter 3, but we begin here with some questions for you to ponder.

1. The chapter began with a brief description of research on social media and Internet use. What would you do if you were interviewing college students who spent lots of time online and found that some were very isolated and depressed or even suicidal, apparently as a result of the isolation? Do you believe that social researchers have an obligation to take action in a situation like this? What if you discovered a similar problem with a child? What guidelines would you suggest for researchers?

2. Would you encourage social researchers to announce their findings about problems such as social isolation in press conferences and to encourage relevant agencies to adopt policies encouraged to lessen social isolation? Should policies regarding attempts to garner publicity and shape policy depend on the strength of the research evidence? Do you think there is a fundamental conflict between academic and policy motivations? Do social researchers have an ethical obligation to recommend policies that their research suggests would help other people?

CHAPTER 2

The Process and Problems of Social Research

I n Chapter 1, we introduced the reasons *why* we do social research: to describe, explore, explain, and evaluate. Each type of social research can have tremendous impact. Alfred Kinsey's descriptive studies of the sex lives of Americans, conducted in the 1940s and 1950s, were at the time a shocking exposure of the wide variety of sexual practices that apparently staid, "normal" people engaged in behind closed doors—and the studies helped introduce the unprecedented sexual openness we see 60 years later (Kinsey, Pomeroy, & Martin 1948; Kinsey, Pomeroy, Martin, & Gebhard 1953). At around

Video Link 2.1
Watch some advice
for new researchers.

the same time, Gunnar Myrdal's exploratory book, *An American Dilemma* (1944/1964), forced our grandparents and great-grandparents to confront the tragedy of institutional racism. Myrdal's research was an important factor in the 1954 Supreme Court decision *Brown* v. *Topeka Board of Education,* which ended school segregation in America. The explanatory *broken windows* theory of crime, which was developed during the 1980s, dramatically changed police practices in our cities. And, evaluative social research today actively influences advertising campaigns, federal housing programs, the organization of military units (from army fire teams to navy-submarine crews), drug treatment programs, and corporate employee benefit plans.

We now introduce the *how* of social research. In this chapter, you will learn about the process of specifying a research question, developing an appropriate research strategy and design with which to investigate that question, choosing appropriate units of analysis, and conforming to scientific and ethical guidelines during the investigation. By the chapter's end, you should be ready to formulate a question, to design a strategy for answering the question, and to begin to critique previous studies that addressed the question.

🖳 What Is the Question?

A **social research question** is a question about the social world that you seek to answer through the collection and analysis of firsthand, verifiable, empirical data. Questions like this may emerge from your own experience, from research by other investigators, from social theory, or from a *request for research* issued by a government agency that needs a study of a particular problem.

> **Social research question:**
> A question about the social world that is answered through the collection and analysis of firsthand, verifiable, empirical data.

Some researchers of the health care system, for example, have had personal experiences as patients with serious diseases, as nurses or aides working in hospitals, or as family members touched directly and importantly by doctors and hospitals. They may want to learn why our health care system failed or helped them. Feminist scholars study violence against women in hopes of finding solutions to this problem as part of a broader concern with improving women's lives. One colleague of ours, Veronica Tichenor, was fascinated by a prominent theory of family relations that argues that men do less housework than women because they earn more money; Professor Tichenor did research on couples in which the woman made far more money than the man to test the theory. (She found, by the way, that the women still did more of the housework.) Some researchers working for large corporations or major polling firms conduct marketing studies simply to make money. So, a wide variety of motives can push a researcher to ask research questions.

A good research question doesn't just spring effortlessly from a researcher's mind. You have to refine and evaluate possible research questions to find one that is worthwhile. It's a good idea to develop a list of possible research questions as you think about a research area. At the appropriate time, you can narrow your list to the most interesting and feasible candidate questions.

What makes a research question "good"? Many social scientists evaluate their research questions in terms of three criteria: *feasibility* given the time and resources available, *social importance,* and *scientific relevance* (King, Keohane, & Verba 1994):

- Can you start and finish an investigation of your research question with available resources and in the time allotted? If so, your research question is feasible.

- Will an answer to your research question make a difference in the social world, even if it only helps people understand a problem they consider important? If so, your research question is socially important.

- Does your research question help to resolve some contradictory research findings or a puzzling issue in social theory? If so, your research question is scientifically relevant.

Here's a good example of a question that is feasible, socially important, and scientifically relevant: Does arresting accused spouse abusers on the spot prevent repeat incidents? Beginning in 1981, the Police Foundation and the Minneapolis Police Department began an experiment to find the answer. The Minneapolis experiment was first and foremost scientifically relevant: It built on a substantial body of contradictory theory regarding the impact of punishment on criminality (Sherman & Berk 1984). Deterrence theory predicted that arrest would deter individuals from repeat offenses, but labeling theory predicted that arrest would make repeat offenses more likely. The researchers found one prior experimental study of this issue, but it had been conducted with juveniles. Studies among adults had not yielded consistent findings. Clearly, the Minneapolis researchers had good reason for conducting a study.

As you consider research questions, you should begin the process of consulting and then reviewing the published literature. Your goal here and in subsequent stages of research should be to develop a research question and specific expectations that build on prior research and to use the experiences of prior researchers to chart the most productive directions and design the most appropriate methods. Appendix A describes how to search the literature, and Chapter 12 includes detailed advice for writing up the results of your search in a formal review of the relevant literature.

Encyclopedia Link 2.1
Read about how applied sociology helps make a difference.

🗐 What Is the Theory?

Theories have a special place in social research because they help us make connections to general social processes and large bodies of research. Building and evaluating theory is, therefore, one of the most important objectives of social science. A social **theory** is a logically interrelated set of propositions about empirical reality (i.e., the social world as it actually exists). You may know, for instance, about conflict theory, which proposes that (1) people are basically self-interested, (2) power differences between people and groups reflect the different resources available to groups, (3) ideas (religion, political ideologies, etc.) reflect the power arrangements in a society, (4) violence is always a potential resource and the one that matters most, and so on (Collins 1975). These statements are related to each other, and the sum of conflict theory is a sizable collection of such statements (entire books are devoted to it). Dissonance theory in psychology, deterrence theory in criminology, and labeling theory in sociology are other examples of social theories.

> **Theory:** A logically interrelated set of propositions about empirical reality.

Social theories suggest the areas on which we should focus and the propositions that we should consider testing. For example, Sherman and Berk's (1984) domestic violence research in the Minneapolis spouse abuse experiment was actually a test of predictions that they derived from two varying theories on the impact of punishment on crime (Exhibit 2.1).

Deterrence theory expects punishment to deter crime in two ways. General deterrence occurs when people see that crime results in undesirable punishments—that "crime doesn't pay." The persons who are punished serve as examples of what awaits those who engage in proscribed acts. Specific deterrence occurs when persons who are punished decide not to commit another offense so they can avoid further punishment (Lempert & Sanders 1986:86–87). Deterrence theory leads to the prediction that arresting spouse abusers will lessen their likelihood of reoffending.

Labeling theory distinguishes between primary deviance, the acts of individuals that lead to public sanction, and secondary deviance, the deviance that occurs in response to public sanction (Hagan 1994:33).

Journal Link 2.1
Read about understanding culture through theory.

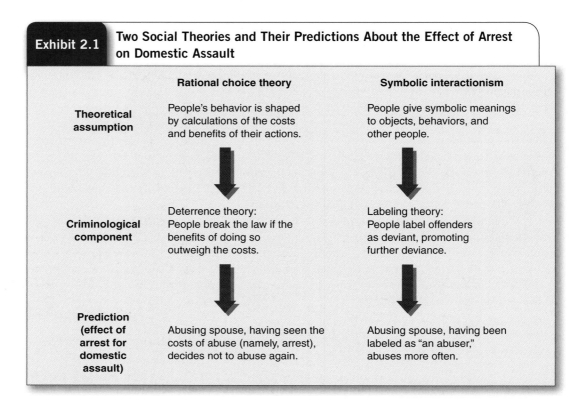

Exhibit 2.1 Two Social Theories and Their Predictions About the Effect of Arrest on Domestic Assault

Arrest or some other public sanction for misdeeds labels the offender as deviant in the eyes of others. Once the offender is labeled, others will treat the offender as a deviant, and the offender is then more likely to act in a way that is consistent with the deviant label. Ironically, the act of punishment stimulates more of the very behavior that it was intended to eliminate. This theory suggests that persons arrested for domestic assault are more likely to reoffend than those who are not punished, which is the reverse of the deterrence theory prediction.

How do we find relevant social theory and prior research? You may already have encountered some of the relevant material in courses pertaining to research questions that interest you, but that won't be enough. The social science research community is large and active, and new research results appear continually in scholarly journals and books. The World Wide Web contains reports on some research even before it is published in journals (like some of the research reviewed in Chapter 1). Conducting a thorough literature review in library sources and checking for recent results on the web are essential steps for evaluating scientific relevance. (See Appendix A for instructions on how to search the literature and the web.)

🔲 What Is the Strategy?

When conducting social research, we try to connect theory with empirical data—the evidence we obtain from the real world. Researchers may make this connection in one of two ways:

1. By starting with a social theory and then testing some of its implications with data. This is called **deductive research**; it is most often the strategy used in quantitative methods.

2. By collecting the data and then developing a theory that explains it. This **inductive research** process is typically used with qualitative methods.

A research project can use both deductive and inductive strategies. Let's examine the two different strategies in more detail. We can represent both within what is called the **research circle**.

Research in the News

INVESTIGATING CHILD ABUSE DOESN'T REDUCE IT

Congress intended the 1974 Child Abuse Prevention and Treatment Act to increase documentation of and thereby reduce the prevalence of child abuse. However, a review of records of 595 high-risk children nationwide from the ages of 4 to 8 found that those children whose families were investigated were not doing any better than those whose families were not investigated— except that mothers in investigated families had more depressive symptoms than mothers in uninvestigated families. Whatever services families were offered after being investigated failed to reduce the risk of future child abuse.

Source: Bakalar, Nicholas. 2010. Child abuse investigations didn't reduce risk, a study finds. *The New York Times,* October 12: D3.

Deductive Research

In deductive research, we start with a theory and then try to find data that will confirm or deny it. Exhibit 2.2 shows how deductive research starts with a theoretical premise and logically *deduces* a specific expectation. Let's begin with an example of a theoretical idea: When people have emotional and personal connections with coworkers, they will be more committed to their work. We could extend this idea to college life by deducing that if students know their professors well, they will be more engaged in their work. And from this, we can deduce a more specific expectation— or hypothesis—that smaller classes, which allow more student–faculty contact, will lead to higher levels of engagement. Now that we have a hypothesis, we can collect data on levels of engagement in small and large classes and compare them. We can't always directly test the general theory, but we can test specific hypotheses that are deduced from it.

A **hypothesis** states a relationship between two or more **variables**—characteristics or properties that can vary, or change. Classes can be large, like a 400-student introductory psychology course, or they can be small, like an upper-level seminar. Class size is thus a variable. And hours of homework done per week can also vary (obviously); you can do 2 hours or 20. So, too, can engagement vary, as measured in any number of ways. (Nominal designations like religion are variables, too, because they can vary among Protestant, Catholic, Jew, and so on.)

But a hypothesis doesn't just state that there is a connection between variables; it suggests that one variable actually influences another—that a change in the first one somehow propels (or predicts, influences, or causes) a change in the second. It says

Deductive research: The type of research in which a specific expectation is deduced from a general premise and is then tested.

Inductive research: The type of research in which general conclusions are drawn from specific data.

Research circle: A diagram of the elements of the research process, including theories, hypotheses, data collection, and data analysis.

Hypothesis: A tentative statement about empirical reality involving a relationship between two or more variables. *Example:* The higher the poverty rate in a community, the higher the percentage of community residents who are homeless.

Variable: A characteristic or property that can vary (take on different values or attributes). *Examples:* poverty rate, percentage of community residents who are homeless.

Dependent variable: A variable that is hypothesized to vary depending on or under the influence of another variable. *Example:* percentage of community residents who are homeless.

Independent variable: A variable that is hypothesized to cause, or lead to, variation in another variable. *Example:* poverty rate.

Direction of association: A pattern in a relationship between two variables—that is, the value of a variable tends to change consistently in relation to change in the other variable. The direction of association can be either positive or negative.

that *if* one thing happens *then* another thing is likely: *If you* stay up too late, *then* you will be tired the next day. *If you* smoke cigarettes for many years, *then* you are more likely to develop heart disease or cancer. *If* a nation loses a major war, *then* its government is more likely to collapse. And so on.

So in a hypothesis, we suggest that one variable influences another—or that the second in some ways "depends" on the first. We may believe, again, that students' reported enthusiasm for a class "depends" on the size of the class. Hence, we call enthusiasm the dependent variable—the variable that *depends* on another, at least partially, for its level. If cigarettes damage your health, then health is the dependent variable; if lost wars destabilize governments, then government stability is the dependent variable; if enthusiasm for a course depends (in some degree) on class size, then enthusiasm is the dependent variable.

The predicted result in a hypothesis, then, is called the **dependent variable**. And the hypothesized cause is called the **independent variable**, because in the stated hypothesis, it doesn't depend on any other variable.

These terms—hypothesis, variable, independent variable, and dependent variable—are used repeatedly in this book and are widely used in all fields of natural and social science, so they are worth knowing well!

You may have noticed that sometimes an increase in the independent variable leads to a corresponding increase in the dependent variable; in other cases, it leads to a decrease. An increase in your consumption of fatty foods will often lead to a corresponding increase in the cholesterol levels in your blood. But an increase in cigarette consumption leads to a decrease in health. In the first case, we say that the **direction of association** is positive; in the second, we say it is negative. Either way, you can clearly see that a change in one variable leads to a predictable change in the other.

In both explanatory and evaluative research, you should say clearly what you expect to find (your hypothesis) and design your research accordingly to test that hypothesis. Doing this strengthens the confidence we can place in the results. So the deductive researcher (to use a poker analogy) states her expectations in advance, shows her hand, and lets the chips fall where they may. The data are accepted as a fair picture of reality.

Domestic Violence and the Research Circle

The Sherman and Berk (1984) study of domestic violence is a good example of how the research circle works. Sherman and Berk's study was designed to test a hypothesis based on deterrence theory: Arrest for spouse abuse reduces the risk of repeat offenses. In this hypothesis, arrest or release is the independent variable, and variation in the risk of repeat offenses is the dependent variable (it is hypothesized to depend on arrest).

Exhibit 2.2   The Research Circle

Theory

Hypothesis

Data

Descriptive research

Empirical generalizations

Inductive research

Deductive research

Sherman and Berk (1984) tested their hypothesis by setting up an experiment in which the police responded to complaints of spouse abuse in one of three ways, one of which was to arrest the offender. When the researchers examined their data (police records for the persons in their experiment), they found that of those arrested for assaulting their spouse, only 13% repeated the offense, compared to a 26% recidivism rate for those who were separated from their spouse by the police but were not arrested. This pattern in the data, or empirical generalization, was consistent with the hypothesis that the researchers deduced from deterrence theory. The theory thus received support from the experiment (Exhibit 2.3).

Journal Link 2.2
Read about the legal actions of victims of domestic violence.

Inductive Research

In contrast to deductive research, inductive research begins with specific data, which are then used to develop (*induce*) a theory to account for the data. (Hint: When you start *in* the data, you are doing inductive research.)

One way to think of this process is in terms of the research circle. Rather than starting at the top of the circle with a theory, the inductive researcher starts at the bottom of the circle with data and then moves up to a theory. Some researchers committed to an inductive approach even resist formulating a research question before they begin to collect data. Their technique is to let the question emerge from the social situation itself (Brewer & Hunter 1989:54–58). In the research for his book *Champions: The Making of Olympic Swimmers,* Dan Chambliss (1988) spent several years living and working with world-class competitive swimmers who were training for the Olympics. Chambliss entered the research with no definite hypotheses and certainly no developed theory about how athletes became successful, what their lives were like, or how they related to their coaches and teams. He simply wanted to understand who these people were, and he decided to report on whatever struck him as most interesting in his research.

Journal Link 2.3
Read about inductive and deductive research techniques in the wake of another disaster.

As it turned out, what Chambliss learned was not how special these athletes were but actually how ordinary they were. Becoming an Olympic athlete was less about innate talent, special techniques, or inspired coaching than it was about actually paying attention to all the little things that make one perform better in one's sport. His theory was *induced* from what he learned in his studies (Chambliss 1988) while being immersed *in* the data.

Research designed using an inductive approach, as in Chambliss's study, can result in new insights and provocative questions. **Inductive reasoning** also enters into deductive research when we find unexpected patterns in data collected for testing a hypothesis. Sometimes such patterns are **anomalous**, in that they don't seem to

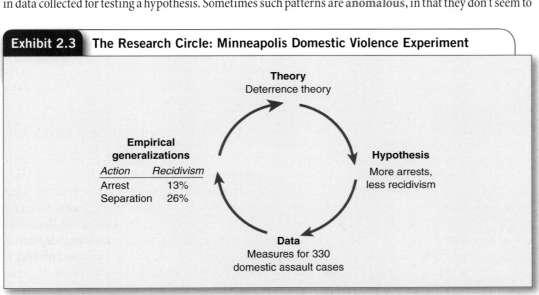

Exhibit 2.3 **The Research Circle: Minneapolis Domestic Violence Experiment**

Theory
Deterrence theory

Empirical generalizations

Action	Recidivism
Arrest	13%
Separation	26%

Hypothesis
More arrests, less recidivism

Data
Measures for 330 domestic assault cases

Inductive reasoning: The type of reasoning that moves from the specific to the general.

Anomalous: Unexpected patterns in data that do not seem to fit the theory being proposed.

Serendipitous: Unexpected patterns in data, which stimulate new ideas or theoretical approaches.

fit the theory being proposed, and they can be **serendipitous**, in that we may learn exciting, surprising new things from them. Even if we do learn inductively from such research, the adequacy of an explanation formulated after the fact is necessarily less certain than an explanation presented prior to the collection data. Every phenomenon can always be explained in some way. Inductive explanations are more trustworthy if they are tested subsequently with deductive research. Great insights and ideas can come from inductive studies, but verifiable proof comes from deductive research.

An Inductive Study of Response to a Disaster

Qualitative research is often inductive: To begin, the researcher observes social interaction or interviews social actors in depth, and then develops an explanation for what has been found. The researchers often ask questions like, What is going on here? How do people interpret these experiences? Why do people do what they do? Rather than testing a hypothesis, the researchers try to make sense of some social phenomenon.

In 1972, for example, towns along the 17-mile Buffalo Creek hollow in West Virginia were wiped out when a dam at the top of a hollow broke, sending 132 million gallons of water, mud, and garbage crashing down through the towns that bordered the creek. After the disaster, sociologist Kai Erikson went to the Buffalo Creek area and interviewed survivors. In the resulting book, *Everything in Its Path,* Erikson (1976) described the trauma suffered by those who survived the disaster. His explanation of their psychological destruction—an explanation that grew out of his interviews with the residents—was that people were traumatized not only by the violence of what had occurred but also by the "destruction of community" that ensued during the recovery efforts. Families were transplanted all over the area with no regard for placing them next to their former neighbors. Extended families were broken up in much the same way, as federal emergency housing authorities relocated people with little concern for whether they knew the people with whom they would be housed. Church congregations were scattered, lifelong friends were resettled miles apart, and entire neighborhoods simply vanished, both physically—that is, their houses were destroyed—and socially. Erikson's explanation grew out of his in-depth immersion in his data—the conversations he had with the people themselves.

Inductive explanations such as Erikson's feel authentic because we hear what people have to say "in their own words and we see the social world as they see it. These explanations are often richer and more finely textured than those in deductive research; on the other hand, they are probably based on fewer cases and drawn from a more limited area.

Descriptive Research: A Necessary Step

Research|Social Impact Link 2.1

Learn about a study that uses descriptive research.

Both deductive and inductive research move halfway around the research circle, connecting theory with data. Descriptive research does not go that far, but it is still part of the research circle shown earlier in Exhibit 2.2. Descriptive research starts with data and proceeds only to the stage of making empirical generalizations; it does not generate entire theories.

Valid description is actually critical in all research. The Minneapolis Domestic Violence Experiment was motivated in part by a growing body of descriptive research indicating that spouse abuse is very common: 572,000 reported cases of women victimized by a violent partner each year; 1.5 million women (and 500,000 men) requiring medical attention each year due to a domestic assault (Buzawa & Buzawa 1996:1–3).

Much important research for the government and private organizations is primarily descriptive: How many poor people live in this community? Is the health of the elderly improving? How frequently do convicted criminals return to crime? Description of social phenomena can stimulate more ambitious deductive and inductive research. Simply put, good description of data is the cornerstone for the scientific research process and an essential component of understanding the social world.

▣ **What Is the Design?**

Researchers usually start with a question, although some begin with a theory or a strategy. If you're very systematic, the *question* is related to a *theory,* and an appropriate *strategy* is chosen for the research. All of these, you will notice, are critical defining issues for the researcher. If your research question is trivial (How many shoes are in my closet?), or your theory sloppy (More shoes reflect better fashion sense.), or your strategy inappropriate (I'll look at lots of shoes and see what I learn.), the project is doomed from the start.

But let's say you've settled these first three elements of a sound research study. Now we must begin a more technical phase of the research: the design of a study. From this point on, we will be introducing a number of terms and definitions that may seem arcane or difficult. In every case, though, these terms will help to clarify your thinking. Like exact formulae in an algebra problem or precisely the right word in an essay, these technical terms help, or even require, scientists to be absolutely clear about what they are thinking—and to be precise in describing their work to other people.

An overall research strategy can be implemented through several different types of research design. One important distinction between research designs is whether data are collected at one point in time—a **cross-sectional research design**—or at two or more points in time—a **longitudinal research design**. Another important distinction is between research designs that focus on individuals—the **individual unit of analysis**—and those that focus on groups, or aggregates of individuals—the **group unit of analysis**.

> **Cross-sectional research design:** A study in which data are collected at only one point in time.
>
> **Longitudinal research design:** A study in which data are collected that can be ordered in time; also defined as research in which data are collected at two or more points in time.
>
> **Individual unit of analysis:** A unit of analysis in which individuals are the source of data and the focus of conclusions.
>
> **Group unit of analysis:** A unit of analysis in which groups are the source of data and the focus of conclusions.

Cross-Sectional Designs

In a cross-sectional design, all of the data are collected at one point in time. In effect, you take a *cross-section*—a slice that cuts across an entire population—and use that to see all the different parts, or sections, of that population. Imagine cutting out a slice of a tree trunk, from bark to core. In looking at this cross-section, one can see all the different parts, including the rings of the tree. In social research, you might do a cross-sectional study of a college's student body, with a sample that includes freshmen through seniors. This "slice" of the population, taken at a single point in time, would allow one to compare the different groups.

But cross-sectional studies, because they use data collected at only one time, suffer from a serious weakness: They don't directly measure the impact of time. For instance, you may see that seniors at your college write more clearly than do freshmen. You might conclude, then, that the difference is because of what transpired over time, that is, what they learned in college. But in fact, it may be because this year's seniors were recruited under a policy that favored better writers. In other words, the cross-sectional study doesn't distinguish if the seniors have learned a lot in college or if they were just better than this year's freshmen when they first enrolled.

Or let's say that in 2013, you conduct a study of the American workforce and find that older workers make more money than younger workers. You may conclude (erroneously) that as one gets older, one makes more money. But you didn't actually observe that happening because you didn't track actual people over time. It *may* be that the older generation (say, people born in 1955) have just enjoyed higher wages all along than have people born in 1985.

With a cross-sectional study, we can't be sure which explanation is correct, and that's a big weakness. Of course, we could ask workers what they made when they first started working, or we could ask college seniors what test scores they received when they were freshmen, but we are then injecting a *longitudinal* element into our cross-sectional research design. Because of the fallibility of memory and the incentives for distorting the past, taking such an approach is not a good way to study change over time.

Longitudinal Designs

In longitudinal research, data are collected over time. By measuring independent and dependent variables at each of several different times, the researcher can determine whether change in the independent variable does in fact precede change in the dependent variable—that is, whether the hypothesized cause comes before the effect, as a true cause must. In a cross-sectional study, when the data are all collected at one time, you can't really show if the hypothesized cause occurs first; in longitudinal studies, though, you can see if a cause occurs and then, later in time, an effect occurs. So if possible to do, longitudinal research is always preferable.

But collecting data more than once takes time and work. Often researchers simply cannot, or are unwilling to, delay completion of a study for even 1 year to collect follow-up data. Still, many research questions really should have a long follow-up period: What is the impact of job training on subsequent employment? How effective is a school-based program in improving parenting skills? Under what conditions do traumatic experiences in childhood result in later mental illness? The value of longitudinal data is great, so every effort should be made to develop longitudinal research designs whenever they are appropriate.

There are three basic longitudinal research designs (Exhibit 2.4). In the first, you conduct a simple *cross-sectional* study but then *repeat* that study several times; therefore, this approach is referred to as a *trend,* or *repeated cross-sectional* design. The frequency of follow-up measurement can vary, ranging from a simple before-and-after design with just one follow-up to studies in which various indicators are measured every month for many years. In such trend studies, the population from which the sample is selected may be defined broadly or narrowly, but members of the sample are rotated or completely replaced each time a measurement

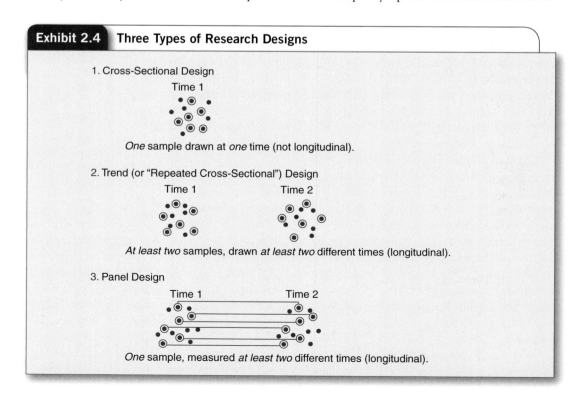

Exhibit 2.4   Three Types of Research Designs

1. Cross-Sectional Design

Time 1

One sample drawn at *one* time (not longitudinal).

2. Trend (or "Repeated Cross-Sectional") Design

Time 1   Time 2

At least two samples, drawn *at least two* different times (longitudinal).

3. Panel Design

Time 1   Time 2

One sample, measured *at least two* different times (longitudinal).

is done. In effect, you look at the population over time, drawing a new sample at each of a number of different points in time. You are looking for trends in the population.

The second major longitudinal design is called a *panel* study. A panel study uses a single group of people who are questioned or studied at multiple points across time; the same people are asked questions on multiple occasions, so how they change and develop as individuals can be studied.

Let's consider these two basic longitudinal designs first to see how they are done and their strengths and weaknesses. We'll then review the third type of longitudinal design, *cohort studies,* which is not used as often as the other two types.

Audio Link 2.1
Listen to more information on polls.

Trend Designs

Trend designs, also known as **repeated cross-sectional studies**, are conducted as follows:

1. A sample is drawn from a population at Time 1, and data are collected from the sample.

2. As time passes, some people leave the population and others enter it.

3. At Time 2, a different sample is drawn from this population.

> **Trend (repeated cross-sectional) design:** A longitudinal study in which data are collected at two or more points in time from different samples of the same population.

The Gallup polls, begun in the 1930s, are a well-known example of trend studies. One Gallup poll, for instance, asks people how well they believe the American president is doing his job (Exhibit 2.5). Every so often, the Gallup organization takes a sample of the American population (usually about 1,400 people) and asks them this question. Each time, they ask a different, though roughly demographically equivalent, group of people the question; they aren't talking to the same people every time. Then they use the results of a series of these questions to analyze trends in support for presidents. That is, they can see when support for presidents is high and when it is low, in general. This is a trend study.

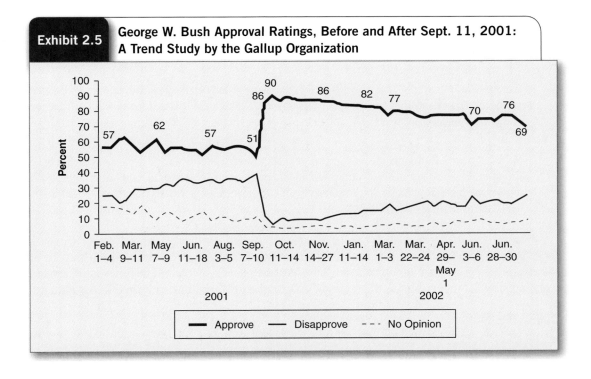

Exhibit 2.5 George W. Bush Approval Ratings, Before and After Sept. 11, 2001: A Trend Study by the Gallup Organization

When the goal is to determine whether a population has changed over time, trend (or repeated cross-sectional) designs are appropriate. Has racial tolerance increased among Americans in the past 20 years? Are employers more likely to pay maternity benefits today than they were in the 1950s? Are college students today more involved in their communities than college students were 10 years ago? These questions concern changes in populations as a whole, not changes in individuals.

Panel Designs

When we need to know whether individuals in the population changed, we must turn to a **panel design**. In the Mellon Foundation Assessment Project at Hamilton College, finishing up as this book is being written, a panel of 100 students entering college in 2001 was selected from the first-year class, each student was interviewed once a year for each of their 4 years at Hamilton; they were interviewed every two years after graduation under 2010. The goal was to determine which experiences in their college career were valuable and which were a hindrance to their education. By following the same people over a period of time, we can see how changes happen in the lives of individual students.

> **Panel design:** A longitudinal study in which data are collected from the same individuals—the panel—at two or more points in time.

Panel designs allow clear identification of changes in the units (individuals, groups, or whatever) we are studying. Here is the process for conducting fixed-sample panel studies:

1. A sample (called a panel) is drawn from a population at Time 1, and data are collected from the sample (for instance, 100 freshmen are selected and interviewed).

2. As time passes, some panel members become unavailable for follow-up, and the population changes (some students transfer to other colleges or decline to continue participating).

3. At Time 2, data are collected from the same people (the panel) as at Time 1, except for those people who cannot be located (the remaining students are reinterviewed).

A panel design allows us to determine how individuals change, as well as how the population as a whole has changed; this is a great advantage. However, panel designs are difficult to implement successfully and often are not even attempted, for two reasons:

1. *Expense and attrition*—It can be difficult and expensive to keep track of individuals over a long period, and inevitably the proportion of panel members who can be located for follow-up will decline over time. Panel studies often lose more than one quarter of their members through attrition (Miller 1991:170).

2. *Subject fatigue*—Panel members may grow weary of repeated interviews and drop out of the study, or they may become so used to answering the standard questions in the survey that they start giving stock answers rather than actually thinking about their current feelings or actions (Campbell 1992). This is called the problem of *subject fatigue*.

Because panel studies are so useful, social researchers have developed increasingly effective techniques for keeping track of individuals and overcoming subject fatigue. But if your resources do not permit use of these techniques to maintain an adequate panel and you plan to do a cross-sectional study instead, remember that it is preferable to use a repeated cross-sectional design rather than a one-time-only cross-sectional study.

Cohort Designs

Trend and panel studies can track both the results of an event (such as World War II) or the progress of a specific historical generation (for instance, people born in 1993). In this case, the historically specific group of people being studied is known as a **cohort**, and this cohort makes up the basic population for your trend or panel study. Such a study has a **cohort design**. If you were doing a trend study, the cohort would be the population from which you draw your different samples. If you were doing a panel study, the cohort provides the population from which the panel itself is drawn. Examples of cohorts include the following:

> **Cohort:** Individuals or groups with a common starting point.
>
> **Cohort design:** A longitudinal study in which data are collected at two or more points in time from individuals in a cohort.

- *Birth cohorts*—those who share a common period of birth (those born in the 1940s, 1950s, 1960s, etc.).

- *Seniority cohorts*—those who have worked at the same place for about 5 years, about 10 years, and so on.

- *Cohort*—individuals or groups with a common starting point. Examples include the college class of 2013, people who graduated from high school in the 1990s, General Motors employees who started work between 1990 and 2000, and people who were born in the late 1940s or the 1950s (the baby boom generation). Cohorts can form the initial population for either trend or panel studies.

Video Link 2.2
Watch a video about cohort designs.

We can see the value of longitudinal research in comparing two studies that estimated the impact of public and private schooling on high school students' achievement test scores. In an initial cross-sectional (not longitudinal) study, James Coleman, Thomas Hoffer, and Sally Kilgore (1982) compared standardized achievement test scores of high school sophomores and seniors in public, Catholic, and other private schools. They found that test scores were higher in the private (including Catholic) high schools than in the public high schools.

But was this difference a causal effect of private schooling? Perhaps the parents of higher-performing children were choosing to send them to private schools rather than to public ones. So James Coleman and Thomas Hoffer (1987) went back to the high schools and studied the test scores of the former sophomores 2 years later, when they were seniors; in other words, the researchers used a panel (longitudinal) design. This time, they found that the verbal and math achievement test scores of the Catholic school students had increased more over the 2 years than the scores of the public school students had. Irrespective of students' initial achievement test scores, the Catholic schools seemed to "do more" for their students than did the public schools. The researchers' causal conclusion rested on much stronger ground because they used a longitudinal panel design.

Units and Levels of Analysis

Units of analysis are the things you are studying, whose behavior you want to understand. Often these are individual people (e.g., why do certain students work harder?), but they can also be, for instance, families, groups, colleges, governments, or nations. All of these could be units of analysis for your research. Sociologist Erving Goffman, writing about face-to-face interaction, became famous in part because he realized that the interaction itself—not just the people in it—could be a unit of analysis. Goffman argued that interactions as such worked in certain ways, apart from the individuals who happened to be joining them: "Not, then, men and their moments. Rather, moments and their men" (Goffman 1967:3). Researchers must always be clear about what is the level of social life they are studying: what are their units of analysis? The units of analysis are the entities you are studying and trying to learn about.

> **Units of analysis:** The entities being studied, whose behavior is to be understood.

As the examples suggest, units exist at different *levels* of collectivity, from the most micro (small) to the most macro (large). Individual people are easily seen and talked to, and you can learn about them quite directly. A university, however, although you can certainly visit it and walk around it, is harder to visualize, and data regarding it may take longer to gather. Finally, a nation is not really a "thing" at all and can never be seen by human eyes; understanding such a unit may require many years of study. People, universities, and nations exist at different *levels* of social reality. And as probably already known, groups don't act like individuals do.

Sometimes researchers confuse levels of analysis, mistakenly using data from one level to draw conclusions about a different level. Even the best social scientists fall into this trap. In Emile Durkheim's classic (1951) study of

suicide, for example, nationwide suicide rates were compared for Catholic and Protestant countries (in an early stage of his research). Obviously, the data on suicide were collected for individual people, and religion was tallied for individuals as well. Then Durkheim used aggregated numbers to characterize entire countries as being high or low suicide countries and as Protestant (England, Germany, Norway) or Catholic (Italy, France, Spain) countries. He found that Catholic countries had lower rates of suicide than Protestant countries. His accurate finding was about countries, then, not about people; the unit of analysis was the country, and he ranked countries by their suicide rates. Yes, the data were collected from individuals and were about individuals, but it had been combined (aggregated) so as to describe entire nations. Thus, Durkheim's units of analysis were countries. So far, so good.

But Durkheim then made his big mistake. He used his findings from one level of analysis to make statements about units at a different level. He used country data to draw conclusions about individuals, claiming that Catholic individuals were less likely than Protestant individuals to commit suicide. Much of his later discussion in *Suicide* (1951) was about why Catholic individuals would be less likely to kill themselves.

Ecological fallacy: An error in reasoning in which incorrect conclusions about individual-level processes are drawn from group-level data.

Confusions about levels of analysis can take several forms (Lieberson 1985). Durkheim's mistake was to use findings from a "higher" level (countries) to draw conclusions about a "lower" level (individuals). This is called the **ecological fallacy**, because the *ecology*—the broader surrounding setting, in this case a country—is mistakenly believed to straightforwardly model how individuals will act as well. The ecological fallacy occurs when group-level data are used to draw conclusions about individual-level processes. It's a mistake, and a common one.

Try to spot the ecological fallacy in each of the following deductions. The first half of each sentence is true, but the second half doesn't logically follow from the first:

- Richer countries have higher rates of heart disease; therefore, richer people have higher rates of heart disease.

- Florida counties with the largest number of black residents have the highest rates of Ku Klux Klan membership; therefore, blacks join the Klan more than whites.

- In the fall 2010 election, Republicans won the House of Representatives, while Democrats held onto the Senate; therefore, Americans want a divided government.

In each case, a group-level finding from data is used to draw (erroneous) conclusions about individuals. In rich countries, yes, there is more heart disease, but actually it's among the poor individuals within those countries. Florida counties with more black people attract more white individuals to the Klan. And although America (as a whole) was certainly divided in the 2010 election, just as certainly many individual Americans, both Republican and Democratic, had no ambivalence whatsoever about who was their favorite candidate. *America* as a whole "may want a divided government," but relatively few *Americans* do.

A researcher who draws such hasty conclusions about individual-level processes from group-level data is committing an ecological fallacy. In August 2006, the *American Sociological Review* published a fierce exchange in which Mitchell Duneier, a well-known field researcher from Princeton University, attacked a very popular book *Heat Wave*, by Eric Klinenberg. *Heat Wave* vividly described how hundreds of poor people in Chicago died during a heat wave in July 1995. Klinenberg argued that the deaths were the result of deteriorating community conditions— for instance, that vulnerable old people, afraid to go outside and possibly be attacked or mugged, remained indoors despite literally killing temperatures in their homes. Duneier (2006) claimed that Klinenberg lacked any data on individual deaths to show that this is what happened, although it was clear that community conditions mattered. But, Duneier argued, the fact that certain features prevailed in the stricken communities did not mean that it was those conditions themselves that led to individual deaths. Klinenberg (2006) disagreed, strongly.

So, conclusions about processes at the individual level must be based on individual-level data; conclusions about group-level processes must be based on data collected about groups (Exhibit 2.6.)

Exhibit 2.6 Levels of Analysis. Data from one level of analysis should lead to conclusions only about that level of analysis.

INCORRECT

Level of Analysis	Data Findings	(Incorrect) Conclusion	Level of Analysis
NATION	Protestant countries have high suicide rates	New York State votes Republican	NATION
	Rich countries have high rates of heart disease		
GROUP	Most counties in New York State vote Republican	Platoons with high promotion rates have high morale	GROUP
INDIVIDUAL	Individual soldiers who get promoted have high morale	Individual Protestants are more likely to commit suicide	INDIVIDUAL
		Rich people are more likely to have heart disease	

Downslope line (\) indicates Ecological Fallacy; Upslope line (/) indicates Reductionism.

CORRECT

Level of Analysis	Data Findings	Conclusion	Level of Analysis
NATIONS	(Data about nations)	(Conclusion about nations)	NATIONS
STATES	(Data about states)	(Conclusion about states)	STATES
COUNTIES	(Data about counties)	(Conclusion about counties)	COUNTIES
ORGANIZATIONS	(Data about organizations)	(Conclusion about organizations)	ORGANIZATIONS
GROUPS	(Data about groups)	(Conclusion about groups)	GROUPS
INDIVIDUALS	(Data about individuals)	(Conclusion about individuals)	INDIVIDUALS

We don't want to leave you with the belief that conclusions about individual processes based on group-level data are *necessarily* wrong. We just don't know for sure. Suppose, for example, that we find that communities with higher average incomes have lower crime rates. Perhaps something about affluence improves community life such that crime is reduced; that's possible. Or, it may be that the only thing special about these communities is that they have more individuals with higher incomes, who tend to commit fewer crimes. Even though we collected data at the group level and analyzed them at the group level, they may reflect a causal process at the individual level (Sampson & Lauritsen 1994:80–83). The ecological fallacy just reminds us that we can't *know* about individuals without having individual-level information.

Confusion between levels of analysis also occurs in the other direction, when data from the individual level are used to draw conclusions about group behavior. For instance, you may know the personal preferences of everyone on a hiring committee, so you try to predict whom the committee will decide to hire; but you could easily be wrong. Or you may know two good individuals who are getting married, so you think that the marriage (the higher-level unit) will be good, too. But often, such predictions are wrong, because groups as units don't work like individuals. Nations often go to war even when most of their people (individually) don't want to. Adam Smith, in the 1700s, famously pointed out that millions of people (individuals) acting selfishly could in fact produce an economy (a group) that acted selflessly, helping everyone. You can't predict higher-level processes or outcomes from lower-level ones. You can't, in short, always reduce group behavior to individual behavior added up; doing so is called the **reductionist fallacy**, or **reductionism**, (since it *reduces* group behavior to that of individuals), and it's basically the reverse of the ecological fallacy.

> **Reductionist fallacy (reductionism):** An error in reasoning that occurs when incorrect conclusions about group-level processes are based on individual-level data.

Both involve confusion of levels of analysis.

▣ But Is It Ethical?

Thus far, we've only described how one conducts research, not whether the research project is morally justifiable. But every scientific investigation, whether in the natural or social sciences, has an ethical dimension to it, so we address such issues throughout this book. The next chapter, sections in many other chapters, and questions at the end of each chapter specifically deal with formal ethical principles in research, but here we give a quick introduction to the topic and mention some common areas of ethical difficulty.

Honesty and Openness

Research distorted by political or personal pressures to find particular outcomes or to achieve the most marketable results is unlikely to be carried out in an honest and open fashion. And what about the ethics of concealing from your subjects that you're even doing research? Carolyn Ellis (1986) spent several years living in and studying two small fishing communities on Chesapeake Bay in Massachusetts. Living with these "fisher folk," as she called them, she learned quite a few fairly intimate details about their lives, including their less-than-perfect hygiene habits (many simply smelled bad from not bathing). When the book was published, many townspeople were enraged that Ellis had lived among them and then, in effect, betrayed their innermost secrets without having told them that she was planning to write a book. There was enough detail in the book, in fact, that some of the fisher folk could be identified, and Ellis had never fully disclosed to the fisher folk that she was doing research. The episode stirred quite a debate among professional sociologists as well.

Here's another example of hiding one's motives from one's subjects. In the early 1980s, Professor Erich Goode spent three and a half years doing research on the National Association to Aid Fat Americans. Professor Goode was interested primarily in how overweight people managed their identity and enhanced their own self-esteem by forming support groups. Twenty years after the research, in 2002, Goode published an article in which he revealed that in doing the research, he met and engaged in romantic and sexual relationships with more than a dozen women in that organization. There was a heated discussion among the editors and board members of the journal in which the article was published, not only about the ethics of the researcher doing such a thing but also about the ethics of the journal then publishing an article that seemed to take inappropriate advantage of the unusual subject matter.

Openness about research procedures and results goes hand in hand with honesty in research design. Openness is also essential if researchers are to learn from the work of others. In the 1980s, there was a long legal battle between a U.S. researcher, Robert Gallo, and a French researcher, Luc Montagnier, about which of them first discovered the virus that causes AIDS. In 2008, Montagnier received the Nobel Prize in recognition of his work (Altman 2008). (Scientists are like other people in their desire to be first.) Enforcing standards of honesty and encouraging openness about research is the best solution for these problems.

The Uses of Science

Scientists must also consider the uses to which their research is put. For example, during the 1980s, Murray Straus, a prominent researcher of family violence (wife battering, child abuse, corporal punishment, and the like) found in his research that in physical altercations between husband and wife, the wife was just as likely as the husband to throw the first punch. This is a startling finding when taken by itself. But Straus also learned that regardless of who actually hit first, the wife nearly always wound up being physically injured far more severely than the man. Whoever started the fight, she lost it (Straus & Gelles 1988). In this respect (as well as in certain others), Straus's finding that "women hit first as often as men" is quite misleading when taken by itself. When Straus published his findings, a host of social scientists and feminists protested loudly on the grounds that his research was likely to be misused by those who believe that wife battering is not, in fact, a serious problem. It seemed to suggest that, really, men are no worse in their use of violence than are women. Do researchers have an obligation to try to correct what seem to be misinterpretations of their findings?

Research|Social Impact Link 2.2
Read more on ethics.

Social scientists who conduct research on behalf of organizations and agencies may face additional difficulties when the organization, not the researcher, controls the final report and the publicity it receives. If organizational leaders decide that particular research results are unwelcome, the researcher's desire to have findings used appropriately and reported fully can conflict with contractual obligations. This possibility cannot be avoided entirely, but because of it, researchers should acknowledge their funding sources in reports.

Research on People

The chapter on ethics deals fully with the general history and topic of human subjects research, but here we suggest a couple of the problems that arise.

Maintaining **confidentiality**, for instance, is a key ethical obligation; it should be reflected in a statement in the informed consent agreement about how each subject's privacy will be protected (Sieber 1992). Procedures, such as locking records and creating special identifying codes, must be created to minimize the risk of unauthorized persons' access. However, statements about confidentiality should be realistic. In 1993, sociologist Rik Scarce was jailed for 5 months for contempt of court after refusing to testify to a grand jury about so-called eco-terrorists. Scarce, a PhD candidate at Washington State University at the time, was researching radical

Confidentiality: Provided by research in which identifying information that could be used to link respondents to their responses is available only to designated research personnel for specific research needs.

environmentalists and may have had information about a 1991 "liberation" raid on an animal research lab at Washington State. Scarce was eventually released from jail, but he never did violate the confidentiality he claimed to have promised his informants (Scarce 2005). Laws allow research records to be subpoenaed and may require reporting child abuse. A researcher also may feel compelled to release information if a health- or life-threatening situation arises and participants need to be alerted.

The potential of withholding a beneficial treatment from some subjects is another cause for ethical concern. Sometimes, in an ethically debatable practice, researchers will actually withhold treatments from some subjects, knowing that those treatments would probably help the people, to accurately measure *how much* they helped. For example, in some recent studies of AIDS drugs conducted in Africa, researchers provided different levels of AIDS-combating drugs to different groups of patients with the disease. Some patients received no drug therapy at all, despite the fact that all indications were that the drug treatments would help them. From the point of view of pure science, this makes sense: You can't really know how effective the drugs are unless you try different treatments on different people who start from the same situation (e.g., having AIDS). But the research has provoked a tremendous outcry across the world because many people find the practice of deliberately not treating people—in particular, impoverished black people living in Third World countries—to be morally repugnant.

The extent to which ethical issues are a problem for researchers and their subjects varies dramatically with research design. Most survey research, in particular, creates few ethical problems (Reynolds 1979:56–57). On the other hand, some experimental studies in the social sciences that have put people in uncomfortable or embarrassing situations have generated vociferous complaints and years of debate about ethics (Reynolds 1979; Sjoberg 1967). Moreover, adherence to ethical guidelines must take into account each aspect of the research procedures. For example, full disclosure of "what is really going on" in an experimental study is unnecessary if subjects are unlikely to be harmed.

What it comes down to is that the researcher must think through in advance the potential for ethical problems, must make every effort to foresee all possible risks and to weigh the possible benefits of the research against these risks, must establish clear procedures that minimize the risks and maximize the benefits, and must inform research subjects in advance about the potential risks.

Institutional review board (IRB): A group of organizational and community representatives required by federal law to review the ethical issues in all proposed research that is federally funded, involves human subjects, or has any potential for harm to subjects.

Ultimately, these decisions about ethical procedures are not just up to you, as a researcher, to make. Your university's human subjects protection committee—**institutional review board**, or **IRB**—sets the human subjects protection standards for your institution and may even require that you submit your research proposal to it for review. Before submitting a project for review, you should also consult with individuals with different perspectives to develop a realistic risk–benefit assessment (Sieber 1992:75–108).

▣ Conclusion

Social researchers can find many questions to study, but not all questions are equally worthy. The ones that warrant the expense and effort of social research are feasible, socially important, and scientifically relevant.

Selecting a worthy research question does not guarantee a worthwhile research project. The simplicity of the research circle presented in this chapter belies the complexity of the social research process. In the following chapters, we will focus on particular aspects of that process. Chapter 4 examines the interrelated processes of conceptualization and measurement, arguably the most important parts of research. Measurement validity is the foundation for the other two aspects of validity, which are discussed in Chapters 5 and 6. Chapter 5 reviews the meaning of generalizability and the sampling strategies that help us to achieve this goal. Chapter 6 introduces the third aspect of validity—causal validity—and illustrates different methods for achieving causal validity and explains basic experimental data collection. The next two chapters introduce approaches to data collection—surveys and qualitative research—that help us, in different ways, to achieve validity.

Ethical issues also should be considered in the evaluation of research proposals and completed research studies. As the preceding examples show, ethical issues in social research are no less complex than the other issues that researchers confront. And it is inexcusable to jump into research on people without any attention to ethical considerations. Chapter 3 continues our discussion of research ethics.

You are now forewarned about the difficulties that all scientists, but social scientists in particular, face in their work. We hope that you will return often to this chapter as you read the subsequent chapters, when you criticize the research literature, and when you design your own research projects. To be conscientious, thoughtful, and responsible—this is the mandate of every social scientist. If you formulate a feasible research problem, ask the right questions in advance, try to adhere to the research guidelines, and steer clear of the most common difficulties, you will be well along the road to fulfilling this mandate.

Key Terms

Anomalous   23
Cohort   29
Cohort design   29
Confidentiality   33
Cross-sectional research design   25
Deductive research   20
Dependent variable   22
Direction of association   22
Ecological fallacy   30

Group unit of analysis   25
Hypothesis   21
Independent variable   22
Individual unit of analysis   25
Inductive reasoning   23
Inductive research   21
Institutional review board (IRB)   34
Longitudinal research design   25
Panel design   28

Reductionist fallacy (reductionism)   32
Research circle   21
Serendipitous   24
Social research question   18
Theory   19
Trend (repeated cross-sectional)
   design   27
Units of analysis   29
Variable   21

Highlights

- Research questions should be feasible (within the time and resources available), socially important, and scientifically relevant.

- Building social theory is a major objective of social science research. Investigate relevant theories before starting social research projects and draw out the theoretical implications of research findings.

- The type of reasoning in most research can be described as primarily deductive or inductive. Research based on deductive reasoning proceeds from general ideas, deduces specific expectations from these ideas, and then tests the ideas with empirical data. Research based on inductive reasoning begins with (*in*) specific data and then develops (*induces*) general ideas or theories to explain patterns in the data.

- It may be possible to explain unanticipated research findings after the fact, but such explanations have less credibility than those that have been tested with data collected for the purpose of the study.

- The scientific process can be represented as circular, with connections from theory, to hypotheses, to data, and to empirical generalizations. Research investigations may begin at different points along the research circle and traverse different portions of it. Deductive research begins at the point of theory; inductive research begins with data but ends with theory. Descriptive research begins with data and ends with empirical generalizations.

- Scientific research should be conducted and reported in an honest and open fashion. Contemporary ethical standards also require that social research cause no harm to subjects, that participation be voluntary as expressed in informed consent, that researchers fully disclose their identity, that benefits to subjects outweigh any foreseeable risks, and that anonymity or confidentiality be maintained for participants unless it is voluntarily and explicitly waived.

STUDENT STUDY SITE

The Student Study Site, available at **www.sagepub.com/chambliss4e**, includes useful study materials including web exercises with accompanying links, eFlashcards, videos, audio resources, journal articles, and encyclopedia articles, many of which are represented by the media links throughout the text. The site also features Interactive Exercises—represented by the green icon here—to help you understand the concepts in this book.

Exercises

Discussing Research

1. Pick a social issue about which you think research is needed. Draft three research questions about this issue. Refine one of the questions and evaluate it in terms of the three criteria for good research questions.

2. Identify variables that are relevant to your three research questions. Now formulate three related hypotheses. Which are the independent and which are the dependent variables in these hypotheses?

3. If you were to design research about domestic violence, would you prefer an inductive approach or a deductive approach? Explain your preference. What would be the advantages and disadvantages of each approach? Consider in your answer the role of social theory, the value of searching the literature, and the goals of your research.

4. Sherman and Berk's (1984) study of the police response to domestic violence tested a prediction derived from deterrence theory. Propose hypotheses about the response to domestic violence that are consistent with labeling theory. Which theory seems to you to provide the best framework for understanding domestic violence and how to respond to it?

5. Review our description of the research projects in the section "Social Research in Practice" in Chapter 1. Can you identify the stages of each project corresponding to the points on the research circle? Did each project include each of the four stages? Which theory (or theories) seem applicable to each of these projects?

Finding Research

1. State a problem for research—some feature of social life that interests you. If you have not already identified a problem for study, or if you need to evaluate whether your research problem is doable, a few suggestions should help to get the ball rolling and keep you on course.

   a. Jot down several questions that have puzzled you about people and social relations, perhaps questions that have come to mind while reading textbooks or research articles, talking with friends, or hearing news stories.

   b. Now take stock of your interests, your opportunities, and the work of others. Which of your research questions no longer seem feasible or interesting? What additional research questions come to mind? Pick out one question that is of interest and seems feasible and that has probably been studied before.

c. Do you think your motives for doing the research would affect how the research is done? How? Imagine several different motives for doing the research. Might any of them affect the quality of your research? How?

d. Write out your research question in one sentence; then elaborate on it in one paragraph. List at least three reasons why it is a good research question for you to investigate. Then present your question to your classmates and instructor for discussion and feedback.

2. Review Appendix A, Finding Information, and then search the literature (and the Internet) on the research question you identified. Copy down at least five citations for articles (with abstracts from CSA Sociological Abstracts) and two websites reporting research that seems highly relevant to your research question. Look up at least two of these articles and one of the websites. Inspect the article bibliographies and the links at the website and identify at least one more relevant article and website from each source.

Write a brief description of each article and website you consulted and evaluate its relevance to your research question. What additions or changes to your thoughts about the research question do the sources suggest?

3. To brush up on a range of social theorists, visit the following site (www.sociologyprofessor.com), pick a theorist, and read some of what you find. What social phenomena does this theorist focus on? What hypotheses seem consistent with his or her theorizing? Describe a hypothetical research project to test one of these hypotheses.

4. You've been assigned to write a paper on domestic violence and the law. To start, you can review relevant research on the American Bar Association's website (www.americanbar.org/groups/domestic_violence/resources/statistics.html). What does the research summarized at this site suggest about the prevalence of domestic violence, its distribution about social groups, and its causes and effects? Write your answers in a one- to two-page report.

Critiquing Research

1. Using recent newspapers or magazines, find three articles that report on large interview or survey research studies. Describe each study briefly. Then say (a) whether the study design was longitudinal or cross-sectional and (b) if that mattered—that is, if the study's findings would possibly have been different using the alternative design.

2. Search the journal literature for three studies concerning some social program or organizational policy after you review the procedures in Appendix A. Several possibilities are research on Head Start, on the effects of welfare payments, on boot camps for offenders, and on standardized statewide

testing in the public schools. Would you characterize the findings as largely consistent or inconsistent? How would you explain discrepant findings?

3. Criticize one of the studies described in this chapter in terms of its adherence to each of the ethics guidelines for social research. List each guideline and indicate what problem or problems might have occurred as a result of deviation from it. How would you weigh the study's contribution to knowledge and social policy against its potential risks to human subjects?

Doing Research

1. Formulate four research questions about support for capital punishment. Provide one question for each research purpose: descriptive, exploratory, explanatory, and evaluative.

2. State four hypotheses in which support for capital punishment is the dependent variable and some other variable is the independent variable.

   a. Justify each hypothesis in a sentence or two.

   b. Propose a design to test each hypothesis. Design the studies to use different longitudinal designs and different

units of analysis. What difficulties can you anticipate with each design?

3. Write a statement for one of your proposed research designs that states how you will ensure adherence to each ethical guideline for the protection of human subjects. Which standards for the protection of human subjects might pose the most difficulty for researchers on your proposed topic? Explain your answers and suggest appropriate protection procedures for human subjects.

Ethics Questions

1. Sherman and Berk (1984) and those who replicated their research on the police response to domestic violence assigned persons accused of domestic violence by chance (randomly) to be arrested or not. Their goal was to ensure that the people who were arrested were similar to those who were not arrested. Based on what you now know, do you feel that this random assignment procedure was ethical? Why or why not?

2. Concern with how research results are used is one of the hallmarks of ethical researchers, but deciding what form that concern should take is often difficult. You learned in this chapter about the controversy that occurred after Sherman and Berk (1984) encouraged police departments to adopt a pro-arrest policy in domestic abuse cases, based on findings from their Minneapolis study. Do you agree with the researchers' decision, in an effort to minimize domestic abuse, to suggest policy changes to police departments based on their study? Several replication studies failed to confirm the Minneapolis findings. Does this influence your evaluation of what the researchers should have done after the Minneapolis study was completed? What about Larry Sherman's (1992) argument that failure to publicize the Omaha study's finding of the effectiveness of arrest warrants resulted in some cases of abuse that could have been prevented?

Ethics in Research

I magine this: One spring morning as you are drinking coffee and reading the newspaper, you notice a small ad for a psychology experiment at the local university.

Video Link 3.1
Watch excerpts from Milgram's experiment.

> **WE WILL PAY YOU $45 FOR ONE HOUR OF YOUR TIME**
>
> *Persons Needed for a Study of Memory*

"Earn money and learn about yourself," it continues. Feeling a bit bored, you call and schedule an evening visit to the lab.

You are about to enter one of the most ethically controversial experiments in the history of social science.

You arrive at the assigned room at the university and are immediately impressed by the elegance of the building and the professional appearance of the personnel. In the waiting room, you see a man dressed in a lab technician's coat talking to another visitor, a middle-aged fellow dressed in casual attire. The man in the lab coat turns, introduces himself, and explains that, as a psychologist, he is interested in whether people learn better when they are punished for making mistakes. He quickly convinces you that this is an important question; he then explains that his experiment on punishment and learning will discover the answer. Then he announces, "I'm going to ask one of you to be the teacher here tonight and the other one to be the learner."

The experimenter [as we'll refer to him from now on] says he will write either *teacher* or *learner* on small identical slips of paper and then asks both of you to draw one. Yours says teacher.

The experimenter now says, in a matter-of-fact way, "All right. Now the first thing we'll have to do is to set the learner up so that he can get some type of punishment."

He leads you both behind a curtain, sits the learner in the chair, straps down both of his arms, and attaches an electric wire to his left wrist (Exhibit 3.1). The wire is connected to a console with 30 switches and a large dial, on the other side of the curtain. When you ask what the wire is for, the experimenter demonstrates. He asks you to take hold of the end of the wire, walks back to the control console, and flips several switches. You hear a clicking noise, see the dial move, and then feel an electric shock in your hand. When the experimenter flips the next switch, the shock increases.

"Ouch!" you say. "So that's the punishment. Couldn't it cause injury?" The experimenter explains that the machine is calibrated so that it will not cause permanent injury but admits that when turned up all the way, it is very, very painful.

Exhibit 3.1   Learner Strapped in Chair With Electrodes

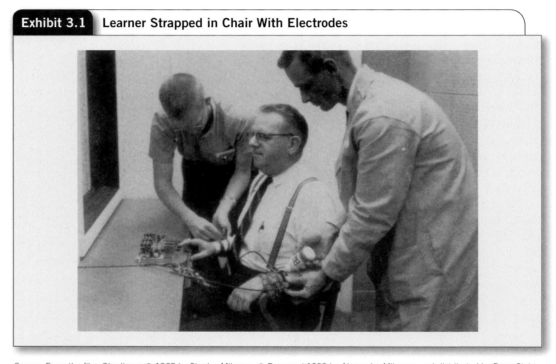

Source: From the film *Obedience* © 1968 by Stanley Milgram, © Renewed 1993 by Alexandra Milgram, and distributed by Penn State, Media Sales.

Now you walk back to the other side of the room (so that the learner is behind the curtain) and sit before the console (Exhibit 3.2). The experimental procedure has four simple steps:

1. You read aloud a series of word pairs, like *blue box, nice day, wild duck,* and so on.

2. You read one of the first words from those pairs and a set of four words, one of which is the original paired word. For example, you might say, "blue: sky-ink-box-lamp."

3. The learner states the word that he thinks was paired with the first word you read ("blue"). If he gives a correct response, you compliment him and move on to the next word. If he makes a mistake, you flip a switch on the console. This causes the Learner to feel a shock on his wrist.

4. After each mistake, you are to flip the next switch on the console, progressing from left to right. You note that a label corresponds to every fifth mark on the dial, with the first mark labeled *slight shock*, the fifth mark labeled *moderate shock*, the tenth *strong shock*, and so on through *very strong shock, intense shock, extreme intensity shock,* and *danger: severe shock*.

Exhibit 3.2   Milgram's "Shock Generator"

Source: From the film *Obedience* © 1968 by Stanley Milgram, © Renewed 1993 by Alexandra Milgram, and distributed by Penn State, Media Sales.

You begin. The learner at first gives some correct answers, but then he makes a few errors. Soon you are beyond the fifth mark (slight shock) and are moving in the direction of more and more severe shocks. As you turn the dial, the learner's reactions increase in intensity: from a grunt at the fifteenth mark (strong shock) to painful groans at higher levels, to anguished cries of "get me out of here" at the extreme intensity shock levels, to a deathly silence at the highest level. When you protest at administering the stronger shocks, the experimenter tells you, "The experiment requires that you continue." Occasionally he says, "It is absolutely essential that you continue."

This is a simplified version of the famous **Stanley Milgram's obedience experiments**, begun at Yale University in 1960. Outside the laboratory, Milgram surveyed Yale undergraduates and asked them to indicate at what level they would terminate their "shocks" if they were in the study. Now, please mark on the console below the most severe shock that you would agree to give the learner (Exhibit 3.3).

The average (mean) maximum shock level predicted by the Yale undergraduates was 9.35, corresponding to a strong shock. Only one student predicted that he would provide a stimulus above that level, at the very strong level. Responses were similar from nonstudent groups.

But the actual average level of shock the 40 adults who volunteered for the experiment administered was 24.53—higher than extreme intensity shock and just short of danger: severe shock. Of Milgram's original 40 subjects, 25 complied entirely with the experimenter's demands, going all the way to the top of the scale (labeled simply as *XXX*). Judging from the subjects' visibly high stress, and from their subsequent reports, they believed that the learner was receiving physically painful shocks. **(In fact, no electric shocks were actually delivered.)**

We introduce the Milgram experiment not to discuss obedience to authority but instead to introduce research ethics. We refer to Milgram's obedience studies throughout this chapter, since they ultimately had as profound an influence on scientists' thinking about ethics as on how we understand obedience to authority.

Obedience experiments (Milgram's): A series of famous experiments conducted during the 1960s by Stanley Milgram, a psychologist from Yale University, testing subjects' willingness to cause pain to another person if instructed to do so.

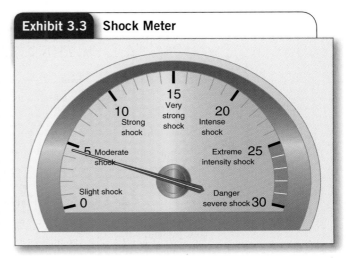

Exhibit 3.3   Shock Meter

Throughout this book, we discuss ethical problems common to various research methods; in this particular chapter, we present in more detail some of the general ethical principles that professional social scientists use in monitoring their work.

Historical Background

Formal procedures for the protection of participants in research grew out of some widely publicized abuses. A defining event occurred in 1946, when the **Nuremberg war crime trials** exposed horrific medical experiments conducted during World War II by Nazi doctors in the name of "science." During the 1950s and 1960s, American military personnel and Pacific Islanders were sometimes unknowingly exposed to radiation during atomic bomb tests. And in the 1970s, Americans were shocked to learn that researchers funded by the U.S. Public Health Service had, for decades, studied 399 low-income African American men diagnosed with syphilis in the 1930s to follow the "natural" course of the illness (Exhibit 3.4). In the **Tuskegee syphilis study**, many participants were not informed of their illness and were denied treatment until 1972, even though a cure (penicillin) was developed in the 1950s (Jones 1993).

Such egregious violations of human rights resulted, in the United States, in the creation of a National Commission for the Protection of Human Subjects of

Nuremberg war crime trials: Trials held in Nuremberg, Germany, in the years following World War II, in which the former leaders of Nazi Germany were charged with war crimes and crimes against humanity; frequently considered the first trials for people accused of genocide.

Audio Link 3.1
Listen to a case of unethical human treatment.

Exhibit 3.4 **Tuskegee Syphilis Experiment**

Biomedical and Behavioral Research. The Commission's 1979 *Belmont Report* (U.S. Department of Health, Education, and Welfare 1979) established three basic ethical principles for the protection of human subjects (Exhibit 3.5):

1. **Respect for persons**—treating persons as autonomous agents and protecting those with diminished autonomy

2. **Beneficence**—minimizing possible harms and maximizing benefits

3. **Justice**—distributing benefits and risks of research fairly

The Department of Health and Human Services and the Food and Drug Administration then translated these principles into specific regulations, which were adopted in 1991 as the Federal Policy for the Protection of Human Subjects. This policy has shaped the course of social science research ever since, and you will have to take it into account as you design your own research investigations. Some professional associations—such as the American Psychological Association, the American Political Science Association, the American Sociological Association, university review boards, and ethics committees in other organizations—set standards for the treatment of human subjects by their members, employees, and students; these standards are designed to comply with the federal policy.

Federal regulations require that every institution that seeks federal funding for biomedical or behavioral research on human subjects have an institutional review board (IRB) that reviews research proposals. If you do research for a class assignment, you may need to prepare a brief IRB proposal, so they can be sure that your project meets all ethical standards. IRBs at universities and other agencies apply ethics standards that are set by federal regulations but can be expanded or specified by the IRB itself (Sieber 1992:5, 10). To promote adequate review of ethical issues, the regulations require that IRBs include members with diverse backgrounds. The **Office for Protection From Research Risks** in the **National Institutes of Health** monitors IRBs, with the exception of research involving drugs (which is the responsibility of the federal Food and Drug Administration).

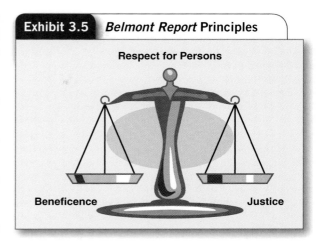

Exhibit 3.5 **Belmont Report Principles**

Respect for Persons

Beneficence   Justice

Tuskegee syphilis study: Research study conducted by a branch of the U.S. government, lasting for roughly 50 years (ending in the 1970s), in which a sample of African American men diagnosed with syphilis were deliberately left untreated, without their knowledge, to learn about the lifetime course of the disease.

Belmont Report: Report in 1979 of the National Commission for the Protection of Human Subjects of Biomedical and Behavioral Research stipulating three basic ethical principles for the protection of human subjects: respect for persons, beneficence, and justice.

🔲 Ethical Principles

The American Sociological Association (ASA), like other professional social science organizations, has adopted, for practicing sociologists, ethical guidelines that are more specific than the federal regulations. Professional organizations may also review complaints of unethical practices when asked.

The *Code of Ethics* of the ASA (1997) is summarized at the ASA website (www .asanet.org); the complete text of the code is also available at this site.

Respect for persons: In human subjects ethics discussions, treating persons as autonomous agents and protecting those with diminished autonomy.

Beneficence: Minimizing possible harms and maximizing benefits.

Justice: As used in human research ethics discussions, distributing benefits and risks of research fairly.

Federal Policy for the Protection of Human Subjects: Federal regulations codifying basic principles for conducting research on human subjects; used as the basis for professional organizations' guidelines.

Office for Protection From Research Risks, National Institutes of Health: Federal agency that monitors institutional review boards (IRBs).

Mostly, ethical issues in research are covered by four guidelines:

1. To protect research subjects
2. To maintain honesty and openness
3. To achieve valid results
4. To encourage appropriate application

Each of these guidelines became a focus of the debate about Milgram's experiments, to which we will refer frequently. Did Stanley Milgram respect the spirit expressed in these principles? You will find that there is no simple answer to the question of what is (or isn't) ethical research practice.

Protecting Research Subjects

This guideline, our most important, can be divided into four specific directions:

1. Avoid harming research participants.
2. Obtain informed consent.
3. Avoid deception in research, except in limited circumstances.
4. Maintain privacy and confidentiality.

Avoid Harming Research Participants

Encyclopedia Link 3.1
Read about the ethics of the researcher-participant relationship.

This standard may seem straightforward, but can be difficult to interpret in specific cases. Does it mean that subjects should not be harmed even mentally or emotionally? That they should feel no anxiety or distress?

The most serious charge leveled against the ethics of Milgram's study was that he had harmed his subjects. A verbatim transcript of one session will give you an idea of what participants experienced as the "shock generator," which made it appear they were delivering increasingly severe shocks to the learner (Milgram 1965:67):

150 volts delivered. You want me to keep going?

165 volts delivered. That guy is hollering in there . . . He's liable to have a heart condition. You want me to go on?

180 volts delivered. He can't stand it! I'm not going to kill that man in there! You hear him hollering? He's hollering. He can't stand it . . . I mean who is going to take responsibility if anything happens to that gentleman? *[The experimenter accepts responsibility.]* All right.

195 volts delivered. You see he's hollering. Hear that. Gee, I don't know. *[The experimenter says: "The experiment requires that you go on."]* I know it does, sir, but I mean—phew—he don't know what he's in for. He's up to 195 volts.

210 volts delivered. *225 volts delivered. 240 volts delivered.*

The experimental manipulation generated "extraordinary tension" (Milgram 1963:377):

Subjects were observed to sweat, tremble, stutter, bite their lips, groan and dig their fingernails into their flesh . . . Full-blown, uncontrollable seizures were observed for 3 subjects. One . . . seizure so violently convulsive that it was necessary to call a halt to the experiment [for that individual]. (p. 375)

An observer (behind a one-way mirror) reported, "I observed a mature and initially poised business-man enter the laboratory smiling and confident. Within 20 minutes he was reduced to a twitching, stuttering wreck, who was rapidly approaching a point of nervous collapse" (Milgram 1963:377).

Milgram's "Behavioral Study of Obedience" was published in 1963 in the *Journal of Abnormal and Social Psychology*. In the next year, the *American Psychologist* published a critique of the experiment's ethics by psychologist Diana Baumrind (1964:421). From Baumrind's perspective, the emotional disturbance in subjects was "potentially harmful because it could easily effect an alteration in the subject's self-image or ability to trust adult authorities in the future" (p. 422). Stanley Milgram (1964) quickly countered that

> momentary excitement is not the same as harm. As the experiment progressed there was no indica-tion of injurious effects in the subjects; and as the subjects themselves strongly endorsed the experi-ment, the judgment I made was to continue the experiment. (p. 849)

Milgram (1963) also attempted to minimize harm to subjects with postexperiment procedures "to assure that the subject would leave the laboratory in a state of well being" (p. 374). A friendly reconciliation was arranged between the subject and the victim, and an effort was made to reduce any tensions that arose as a result of the experiment.

In some cases, the "dehoaxing"—or **debriefing**—discussion was extensive, and all subjects were promised (and later received) a comprehensive report (Milgram, 1964:849). But Baumrind (1964) was uncon-vinced: "It would be interesting to know what sort of procedures could dissipate the type of emotional disturbance just described" (p. 422).

When Milgram (1964:849) surveyed subjects in a follow-up, 83.7% endorsed the statement that they were "very glad" or "glad" "to have been in the experiment," 15.1% were "neither sorry nor glad," and just 1.3% were "sorry" or "very sorry" to have participated. Interviews by a psychiatrist a year later found no evidence "of any traumatic reactions" (Milgram, 1974:197). Subsequently, Milgram argued that "the central moral justification for allowing my experiment is that it was judged accept-able by those who took part in it" (Milgram as cited in Cave & Holm 2003:32).

> **Debriefing:** A researcher's informing subjects after an experiment about the experiment's purposes and methods and evaluating subjects' personal reactions to the experiment.

In a later article, Baumrind (1985:168) dismissed the value of the self-reported "lack of harm" of subjects who had been willing to participate in the experiment and noted that 16% did *not* endorse the statement that they were "glad" they had participated in the experiment. Many social scientists, ethicists, and others con-cluded that Milgram's procedures had not harmed subjects and so were justified by the knowledge they pro-duced; others sided with Baumrind's criticisms (Miller 1986:88–138).

Or, consider the possible harm to subjects in the famous prison simulation study at Stanford University (Haney, Banks, & Zimbardo 1973). **Zimbardo's prison simulation study** was designed to investigate the impact of being either a guard or a prisoner in a prison, a "total institution." The researchers selected apparently stable and mature young male volunteers and asked them to sign a contract to work for 2 weeks as a guard or a prisoner in a simulated prison. Within the first 2 days after the prisoners were incarcerated in a makeshift basement prison, the prisoners began to be passive and disorganized, while the guards became "sadistic"—verbally and physically aggressive (Exhibit 3.6). Five "prisoners" were soon released for depression, uncontrollable crying, fits of rage, and, in one case, a psychosomatic rash. Instead of letting things continue for 2 weeks as planned, Zimbardo and his colleagues terminated the experiment after 6 days to avoid harming subjects.

Video Link 3.2
Watch excerpts from Zimbardo's Stanford prison experiment.

> **Prison simulation study (Zimbardo's):** Famous study from the early 1970s, organized by Stanford psychologist Philip Zimbardo, demonstrating the willingness of average college students quickly to become harsh disciplinarians when put in the role of (simulated) prison guards over other students; usually interpreted as demonstrating an easy human readiness to become cruel.

Participants playing the prisoner role certainly felt some stress, but postexperi-ment discussion sessions seemed relieve this; follow-up during the next year indi-cated no lasting negative effects on the participants and some benefits in the form of

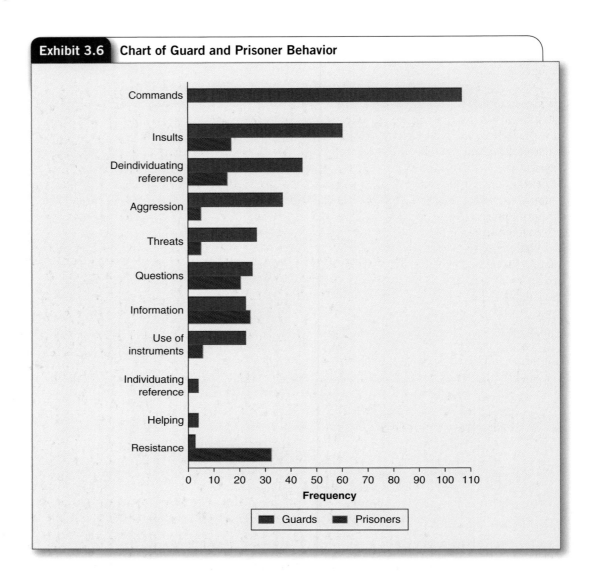

Exhibit 3.6   Chart of Guard and Prisoner Behavior

greater insight. And besides, Zimbardo and his colleagues had no way of predicting the bad outcome; indeed, they were themselves surprised (Haney et al. 1973).

Even well-intentioned researchers may fail to foresee potential ethical problems. Milgram (1974:27–31) reported that he and his colleagues were surprised by the subjects' willingness to administer such severe shocks. In Zimbardo's prison simulation, all the participants signed consent forms, but even the researchers did not realize that participants would fall apart so quickly, that some prisoners would have to be released within a few days, or that others would soon be begging to be released from the mock prison. Since some risks cannot be foreseen, they cannot be consented to.

Obtain Informed Consent

Journal Link 3.1
Read about the written consent needed for a youth smoking prevention trial.

Just defining informed consent may also be more difficult than it first appears. To be informed, consent must be given by persons who are competent to consent, have consented voluntarily, are fully informed about the research, and have comprehended what they have been told (Reynolds 1979). Yet you probably realize, as did

Diana Baumrind (1985), that due to the inability to communicate perfectly, "Full disclosure of everything that could possibly affect a given subject's decision to participate is not possible, and therefore cannot be ethically required" (p. 165).

Obtaining informed consent creates additional challenges for researchers. For instance, the language of the consent form must be clear and understandable yet sufficiently long and detailed to explain what will actually happen in the research. Examples A (Exhibit 3.7) and B (Exhibit 3.8) illustrate two different approaches to these tradeoffs. Consent form A was approved by a university for a substance abuse survey with undergraduate students. It is brief and to the point but leaves quite a bit to the imagination of the prospective participants. Consent form B reflects the requirements of an academic hospital's IRB. Because the hospital is used to reviewing research proposals involving drugs and other treatment interventions with hospital patients, it requires a very detailed and lengthy explanation of procedures and related issues, even for a simple survey. Requiring prospective participants to sign such lengthy forms can reduce their willingness to participate in research and perhaps influence their responses if they do agree to participate (Larson 1993:114).

When an experimental design requires subject deception, researchers may withhold information before the experiment but then debrief subjects after the experiment ends (Milgram did this). In the debriefing, the researcher explains what really happened in the experiment, and why, and responds to subjects' questions. A carefully designed debriefing procedure can often help research participants deal with their anger or embarrassment at having been deceived (Sieber 1992:39–41), thus substituting for fully informed consent prior to the experiment.

Finally, some participants can't truly give informed consent. College students, for instance, may feel unable to refuse if their professor asks them to be in an experiment. Legally speaking, children cannot give consent to participate in research; a child's legal guardian must give written informed consent to have the

Research|Social Impact Link 3.1
Read about research ethics.

Exhibit 3.7 Consent Form A

University of Massachusetts at Boston

Department of Sociology

October 28, 1996

Dear _____ :

    The health of students and their use of alcohol and drugs are important concerns for every college and university. The enclosed survey is about these issues at UMass/Boston. It is sponsored by University Health Services and the PRIDE Program (Prevention, Resources, Information, and Drug Education). The questionnaire was developed by graduate students in Applied Sociology, Nursing, and Gerontology.

    You were selected for the survey with a scientific, random procedure. Now it is important that you return the questionnaire so that we can obtain an unbiased description of the undergraduate student body. Health Services can then use the results to guide campus education and prevention programs.

    The survey requires only about 20 minutes to complete. Participation is completely voluntary and anonymous. No one will be able to link your survey responses to you. In any case, your standing at the University will not be affected whether or not you choose to participate. Just be sure to return the enclosed postcard after you mail the questionnaire so that we know we do not have to contact you again.

    Please return the survey by November 15th. If you have any questions or comments, call the PRIDE program at 287–5680 or Professor Schutt at 287–6250. Also call the PRIDE program if you would like a summary of our final report.

    Thank you in advance for your assistance.

Russell K. Schutt, PhD
Professor and Chair

Exhibit 3.8 Consent Form B

Research Consent Form for Social and Behavioral Research

Dana-Farber/Harvard Cancer Center

BIDMC/BWH/CH/DFCI/MGH/Partners Network Affiliates

OPRS 11-05

Protocol Title: <u>ASSESSING COMMUNITY HEALTH WORKERS' ATTITUDES AND KNOWLEDGE ABOUT EDUCATING COMMUNITIES ABOUT CANCER CLINICAL TRIALS</u>

DF/HCC Principal Research Investigator / Institution: Dr. Russell Schutt, PhD/ Beth Israel Deaconess Medical Center and Univ. of Massachusetts, Boston

DF/HCC Site-Responsible Research Investigator(s) / Institution(s): Lidia Schapira, MD/ Massachusetts General Hospital

Interview Consent Form

A. <u>INTRODUCTION</u>

We are inviting you to take part in a research study. Research is a way of gaining new knowledge. A person who participates in a research study is called a "subject." This research study is evaluating whether community health workers might be willing and able to educate communities about the pros and cons of participating in research studies.

It is expected that about 10 people will take part in this research study.

An institution that is supporting a research study either by giving money or supplying something that is important for the research is called the "sponsor." The sponsor of this protocol is National Cancer Institute and is providing money for the research study.

This research consent form explains why this research study is being done, what is involved in participating in the research study, the possible risks and benefits of the research study, alternatives to participation, and your rights as a research subject. The decision to participate is yours. If you decide to participate, please sign and date at the end of the form. We will give you a copy so that you can refer to it while you are involved in this research study.

If you decide to participate in this research study, certain questions will be asked of you to see if you are eligible to be in the research study. The research study has certain requirements that must be met. If the questions show that you can be in the research study, you will be able to answer the interview questions.

If the questions show that you cannot be in the research study, you will not be able to participate in this research study.

Page 1 of 6

DFCI Protocol Number: <u>06-085</u>	Date DFCI IRB Approved This Consent Form: <u>January 16, 2007</u>
Date Posted for Use: <u>January 16, 2007</u>	Date DFCI IRB Approval Expires: <u>August 13, 2007</u>

Exhibit 3.8 (Continued)

**Research Consent Form for Social and
Behavioral Research**

Dana-Farber/Harvard Cancer Center

BIDMC/BWH/CH/DFCI/MGH/Partners Network Affiliates

OPRS 11-05

We encourage you to take some time to think this over and to discuss it with other people and to ask questions now and at any time in the future.

B. WHY IS THIS RESEARCH STUDY BEING DONE?

Deaths from cancer in general and for some specific cancers are higher for black people compared to white people, for poor persons compared to nonpoor persons, and for rural residents compared to nonrural residents. There are many reasons for higher death rates between different subpopulations. One important area for changing this is to have more persons from minority groups participate in research about cancer. The process of enrolling minority populations into clinical trials is difficult and does not generally address the needs of their communities. One potential way to increase participation in research is to use community health workers to help educate communities about research and about how to make sure that researchers are ethical. We want to know whether community health workers think this is a good strategy and how to best carry it out.

C. WHAT OTHER OPTIONS ARE THERE?

Taking part in this research study is voluntary. Instead of being in this research study, you have the following option:

- Decide not to participate in this research study.

D. WHAT IS INVOLVED IN THE RESEARCH STUDY?

Before the research starts (screening): After signing this consent form, you will be asked to answer some questions about where you work and the type of community health work you do to find out if you can be in the research study.

If the answers show that you are eligible to participate in the research study, you will be eligible to participate in the research study. If you do not meet the eligibility criteria, you will not be able to participate in this research study.

After the screening procedures confirm that you are eligible to participate in the research study:
You will participate in an interview by answering questions from a questionnaire. The interview will take about 90 minutes. If there are questions you prefer not to answer we can skip those questions. The questions are about the type of work you do and your opinions about participating in research. If you agree, the interview will be taped and then transcribed. Your name and no other information about you will be associated:

. . .

Page 2 of 6

| DFCI Protocol Number: 06-085 | Date DFCI IRB Approved this Consent Form: January 16, 2007 |
| Date Posted for Use: January 16, 2007 | Date DFCI IRB Approval Expires: August 13, 2007 |

(Continued)

Exhibit 3.8 (Continued)

Research Consent Form for Social and Behavioral Research

Dana-Farber/Harvard Cancer Center

BIDMC/BWH/CH/DFCI/MGH/Partners Network Affiliates   OPRS 11-05

with the tape or the transcript. Only the research team will be able to listen to the tapes. Immediately following the interview, you will have the opportunity to have the tape erased if you wish to withdraw your consent to taping or participation in this study. You will receive $30.00 for completing this interview.

After the interview is completed: Once you finish the interview there are no additional interventions.

...

N. DOCUMENTATION OF CONSENT

My signature below indicates my willingness to participate in this research study and my understanding that I can withdraw at any time.

_____   _____
Signature of Subject   Date
or Legally Authorized Representative

_____   _____
Person obtaining consent   Date

To be completed by person obtaining consent:

The consent discussion was initiated on _____ (date) at _____ (time).

☐ A copy of this signed consent form was given to the subject or legally authorized representative.

For Adult Subjects

☐ The subject is an adult and provided consent to participate.

☐ The subject is an adult who lacks capacity to provide consent, and his/her legally authorized representative:

    ☐ gave permission for the adult subject to participate

    ☐ did not give permission for the adult subject to participate

Page 6 of 6

DFCI Protocol Number: 06-085	Date DFCI IRB Approved this Consent Form: January 16, 2007
Date Posted for Use: January 16, 2007	Date DFCI IRB Approval Expires: August 13, 2007

child participate in research (Sieber 1992). Then, the child must in most circumstances be given the opportunity to give or withhold *assent* to participate in research, usually by a verbal response to an explanation of the research. Special protections exist for other vulnerable populations—prisoners, pregnant women, mentally disabled persons, and educationally or economically disadvantaged persons. And in a sense, anyone deliberately deceived in an experiment cannot be said to really have given "informed" consent, since the person wasn't honestly told what would happen.

Social media and digital technologies have in recent years opened the doors to new kinds of ethical problems in research, by blurring the lines between public and private behavior. If you have a Facebook or MySpace page with 600 "friends," is that your private page, or a public document? In Chapter 4, we'll see how social researchers are eagerly mining such data for information on people's social networks; "Employers are looking at people's online postings and Googling information about them, and I think researchers are right behind them," said Professor Nicholas Christakis (as cited in Rosenbloom 2007:2), a Harvard sociologist in a *New York Times* article in 2007. But the federal guidelines under which Institutional Review Boards are set up didn't anticipate the Internet. "The [human subject] rules were made for a different world, a pre-Facebook world," said Samuel D. Gosling, a psychology professor at the University of Texas who uses Facebook as a data source. "There is a rule that you are allowed to observe public behavior, but it's not clear if online behavior is public or not" (as cited in Rosenbloom 2007:2).

In truth, though, the public vs. private debate is a long-standing issue in social science. Laud Humphreys (1970) decided that truly informed consent would be impossible to obtain for his study of the social background of men who engage in homosexual behavior in public facilities. Humphreys served as a lookout—a "watch queen"—for men who were entering a public bathroom in a city park with the intention of having sex. In a number of cases, he then left the bathroom and copied the license plate numbers of the cars driven by the men. One year later, he visited the homes of the men and interviewed them as part of a larger study of social issues. Humphreys changed his appearance so that the men did not recognize him. In his book *Tearoom Trade,* Humphreys concluded that the men who engaged in what were widely viewed as deviant acts were, for the most part, married, suburban men whose families were unaware of their sexual practices. But debate has continued ever since about Humphreys's failure to tell the men what he was really doing in the bathroom or why he had come to their homes for the interview. He was criticized by many, including some faculty members at the University of Washington who urged that his doctoral degree be withheld. However, many other professors and some members of the gay community praised Humphreys for helping to normalize conceptions of homosexuality (Miller 1986:135).

If you served on your university's IRB, would you allow research such as Humphreys's to be conducted?

Research|Social Impact Link 3.2
Read more about informed consent.

Tearoom Trade: Book by Laud Humphreys investigating the social background of men who engage in homosexual behavior in public facilities; controversially, he did not obtain informed consent from his subjects.

Avoid Deception in Research, Except in Limited Circumstances

Deception occurs when subjects are misled about research procedures. Frequently, this is done to simulate real-world conditions in the lab. The goal is to get subjects "to accept as true what is false or to give a false impression" (Korn 1997:4). In Milgram's (1964) experiment, for example, deception seemed necessary because actually giving electric shocks to the "stooge" would be cruel. Yet to test obedience, the task had to be troubling for the subjects. Milgram (1974:187–188) insisted that the deception was absolutely essential. Many other psychological and social psychological experiments would be worthless if subjects understood what was really happening to them while the experiment was in progress. But is this sufficient justification to allow the use of deception?

Some important topics have been cleverly studied using deception. Gary Marshall and Philip Zimbardo (of prison study fame), in a 1979 study, told the student volunteers that they were being injected with a vitamin supplement to test its effect on visual acuity (Korn 1997:2–3). But to determine the physiological basis of

Researcher Interview Link 3.1
Watch a researcher describe how an IRB works.

emotion, they actually injected them with adrenaline, so that their heart rate and sweating would increase, and then placed them in a room with a student stooge who acted silly. Piliavin and Piliavin, in a 1972 study, staged fake seizures on subway trains to study helpfulness (Korn:3–4). Again, would you allow such deceptive practices if you were a member of your university's IRB? Giving people stimulating drugs, apart from the physical dangers, is using their very bodies for research without their knowledge. Faking an emergency may lessen one's willingness to help in the future or may, in effect, punish the research subjects—through embarrassment—for their reaction to what is really "just an experiment."

But perhaps risk, not deception per se, is the real problem. Aronson and Mills's (1959) study of severity of initiation to groups is a good example of experimental research that does not pose greater-than-everyday risks to subjects but still uses deception. This study was conducted at an all-women's college in the 1950s. The student volunteers who were randomly assigned to the "severe initiation" experimental condition had to read a list of embarrassing words. Even in the 1950s, reading a list of potentially embarrassing words in a laboratory setting, then listening to a taped discussion, was unlikely to increase the risks to which students were exposed in their everyday lives. Moreover, the researchers informed subjects that they would be expected to talk about sex and could decline to participate in the experiment if this requirement would bother them. None dropped out. To further ensure that no psychological harm was caused, Aronson and Mills explained the true nature of the experiment to subjects after the experiment. The subjects did not seem perturbed: "None of the Ss expressed any resentment or annoyance at having been misled. In fact, the majority were intrigued by the experiment, and several returned at the end of the academic quarter to ascertain the result" (p. 179).

Are you satisfied that this procedure caused no harm? The minimal deception in the Aronson and Mills experiment, coupled with the lack of any ascertainable risk to subjects and a debriefing, satisfies the ethical standards for research of most psychologists and IRBs, even today.

Maintain Privacy and Confidentiality

Journal Link 3.2
Read a qualitative study that protects subjects' identities by using pseudonyms.

Maintaining privacy and confidentiality after a study is completed is another way to protect subjects, and the researcher's commitment to that standard should be included in the informed consent agreement (Sieber 1992). Procedures to protect each subject's privacy, such as locking records and creating special identifying codes, must be created to minimize the risk of access by unauthorized persons. For the protection of health care data, the **Health Insurance Portability and Accountability Act (HIPAA)**, passed by Congress in 1996, created much more stringent regulations. As implemented by the U.S. Department of Health and Human Services in 2000 (and revised in 2002), the HIPAA Final Privacy Rule applies to oral, written, and electronic information that "relates to the past, present, or future physical or mental health or condition of an individual" (Legal Information Institute, 2006. § 1320d[6][B]). The HIPAA Rule requires that researchers have valid authorization for any use or disclosure of "protected health information" (PHI) from a health care provider. Waivers of authorization can be granted in special circumstances (Cava, Cushman, & Goodman 2007).

> **Health Insurance Portability and Accountability Act (HIPAA):** A U.S. federal law passed in 1996 that guarantees, among other things, specified privacy rights for medical patients, in particular those in research settings.

However, statements about confidentiality also need to be realistic. The law allows even confidential research records to be subpoenaed and may require reporting child abuse. A researcher may feel compelled to release information if a health- or life-threatening situation arises and participants need to be alerted. The National Institutes of Health can issue a **Certificate of Confidentiality** to protect researchers from being legally required to disclose confidential information. Researchers who focus on high-risk populations or behaviors or sensitive topics, such as crime, substance abuse, sexual activity, or genetic information, can request such a certificate. Suspicions of child abuse or neglect must still be reported, and in some states, researchers may still be required to report such crimes as elder abuse (Arwood & Panicker 2007).

> **Certificate of Confidentiality:** Document issued by the National Institutes of Health to protect researchers from being legally required to disclose confidential information.

Maintaining Honesty and Openness

Protecting subjects, then, is the primary focus of research ethics. But researchers have obligations to other groups, including the scientific community, whose concern with validity requires that scientists be open in disclosing their methods and honest in presenting their findings. To assess the validity of a researcher's conclusions and the ethics of this researcher's procedures, you need to know how the research was conducted. This means that articles or other reports must include a detailed methodology section, perhaps supplemented by appendixes containing the research instruments or websites or other contact information where more information can be obtained. Biases or political motives should be acknowledged, since research distorted by political or personal pressures to find particular outcomes is unlikely to be carried out in an honest and open fashion.

Stanley Milgram's research exemplifies adherence to the goal of honesty and openness. His initial 1963 article included a description of study procedures, including details about the procedures involved in the learning task, administration of the "sample shock," the shock instructions and the preliminary practice run, the standardized feedback from the "victim" and from the experimenter, and the measures used. Many more details, including pictures, were provided in Milgram's (1974) subsequent book.

The act of publication itself is a vital element in maintaining openness and honesty, since then others can review procedures and debate with the researcher. Although Milgram disagreed sharply with Diana Baumrind's criticisms of his experiments, their mutual commitment to public discourse in journals widely available to psychologists resulted in more comprehensive presentation of study procedures and more thoughtful conversation about research ethics. Almost 50 years later, this commentary continues to inform debates about research ethics (Cave & Holm 2003).

In spite of this need for openness, researchers may hesitate to disclose their procedures or results to prevent others from "stealing" their ideas and taking the credit. However, failure to be open about procedures can result in difficult disputes. In the 1980s, for instance, as mentioned in Chapter 2, there was a long legal battle between a U.S. researcher, Robert Gallo, and a French researcher, Luc Montagnier, both of whom claimed credit for discovering the AIDS virus. Eventually the dispute was settled at the highest levels of government, through an agreement announced by American president Ronald Reagan and French prime minister Jacques Chirac (Altman 1987). Gallo and Montagnier jointly developed a chronology of discovery as part of the agreement. Enforcing standards of honesty and encouraging openness about research are often the best solutions to such problems.

Encyclopedia Link 3.2
Read about how to retain ethical communication with research participants.

Journal Link 3.3
Read about a study where only a portion of participants received a beneficial treatment.

Research in the News

HONESTY IN RESEARCH

The number of scientific articles increased by 44% between 2001 and 2010, but during that same period, the number of articles that were retracted due to the exposure of fraudulent or otherwise misleading findings multiplied more than 15 times, to a total of 339. Misleading findings have led to retractions in psychological research about animals, in biological research about cloning, and in immunological research about vaccines and autism. In response, researchers have been fired, funding has been terminated, and scientific journals have become more diligent.

Source: Cook, Gareth. 2011. Fraud in a lab coat: Science is finally getting serious about misconduct. *The Boston Sunday Globe,* August 21: K9.

Achieving Valid Results

It is the pursuit of objective knowledge—the goal of validity—that justifies our investigations and our claims to the use of human subjects. We have no business asking people to answer questions, submit to observations, or participate in experiments if we are simply trying to trumpet our own prejudices or pursue our personal interests. If, on the other hand, we approach our research projects objectively, setting aside our predilections in the service of learning a bit more about human behavior, we can honestly represent our actions as potentially contributing to the advancement of knowledge.

Audio Link 3.2
Listen to more about
Milgram's experiment.

The details in Milgram's 1963 article and 1974 book on the obedience experiments make a compelling case for his commitment to achieving valid results—to learning how obedience influences behavior. In Milgram's (1963) own words,

> It has been reliably established that from 1933–45 millions of innocent persons were systematically slaughtered on command . . . Obedience is the psychological mechanism that links individual action to political purpose. It is the dispositional cement that binds men to systems of authority . . . for many persons obedience may be a deeply ingrained behavior tendency . . . Obedience may [also] be ennobling and educative and refer to acts of charity and kindness, as well as to destruction. (p. 371)

Milgram (1963) then explains how he devised experiments to study the process of obedience in a way that would seem realistic to the subjects and still allow "important variables to be manipulated at several points in the experiment" (p. 372). Every step in the experiment was carefully designed to ensure that subjects received identical stimuli and that their responses were measured carefully.

Milgram's (1963) attention to validity is also apparent in his reflections on "the particular conditions" of his experiment, for, he notes, "Understanding of the phenomenon of obedience must rest on an analysis of [these conditions]" (p. 377). These particular conditions included the setting for the experiment at Yale University, its purported "worthy purpose" to advance knowledge about learning and memory, and the voluntary participation of the subject as well as of the learner—as far as the subject knew. The importance of some of these "particular conditions" (such as the location at Yale) was then tested in subsequent replications of the basic experiment (Milgram 1965).

However, not all psychologists agreed that Milgram's approach could achieve valid results. Baumrind's (1964) critique begins with a rejection of the external validity—the generalizability—of the experiment. "The laboratory is unfamiliar as a setting and the rules of behavior ambiguous . . . Therefore, the laboratory is not the place to study degree of obedience or suggestibility, as a function of a particular experimental condition" (p. 423). And so, "the parallel between authority-subordinate relationships in Hitler's Germany and in Milgram's laboratory is unclear" (p. 423).

Video Link 3.3
Watch an interview about
naturally occurring data.

Stanley Milgram (1964) quickly published a rejoinder in which he disagreed with (among other things) the notion that it is inappropriate to study obedience in a laboratory setting: "A subject's obedience is no less problematical because it occurs within a social institution called the psychological experiment" (p. 850).

Milgram (1974:169–178) also pointed out that his experiment had been replicated in other places and settings with the same results, that there was considerable evidence that subjects had believed that they actually were administering shocks, and that the "essence" of his experimental manipulation—the request that subjects comply with a legitimate authority—was shared with the dilemma faced by people in Nazi Germany and soldiers at the My Lai massacre in Vietnam (Miller 1986:182–183).

But Baumrind (1985) was still not convinced. In a follow-up article in the *American Psychologist*, she argued that "far from illuminating real life, as he claimed, Milgram in fact appeared to have constructed a set of conditions so internally inconsistent that they could not occur in real life" (p. 171).

Milgram assumed that obedience could fruitfully be studied in the laboratory; Baumrind disagreed. Both, however, buttressed their ethical arguments with assertions about the external validity (or invalidity) of the experimental results. They agreed, in other words, that a research study is in part justified by its valid findings—the knowledge to be gained. If the findings aren't valid, they can't justify the research at all. It is hard to justify any risk for human subjects, or even any expenditure of time and resources, if our findings tell us nothing about human behavior.

Encouraging Appropriate Application

Finally, scientists must consider the uses to which their research is put. Although many scientists believe that personal values should be left outside the laboratory, some feel that it is proper—even necessary—for scientists to concern themselves with the way their research is used.

Stanley Milgram made it clear that he was concerned about the phenomenon of obedience precisely because of its implications for people's welfare. As you have already learned, his first article (1963) highlighted the atrocities committed under the Nazis by citizens and soldiers who were "just following orders." In his more comprehensive book on the obedience experiments (1974), he also used his findings to shed light on the atrocities committed in the Vietnam War at My Lai, slavery, the destruction of the American Indian population, and the internment of Japanese Americans during World War II. Milgram makes no explicit attempt to "tell us what to do" about this problem. In fact, as a dispassionate psychological researcher, Milgram (1974) tells us, "What the present study [did was] to give the dilemma [of obedience to authority] contemporary format by treating it as subject matter for experimental inquiry, and with the aim of understanding rather than judging it from a moral standpoint" (p. xi).

Yet it is impossible to ignore the very practical implications of Milgram's investigations. His research highlighted the extent of obedience to authority and identified multiple factors that could be manipulated to lessen blind obedience (such as encouraging dissent by just one group member, removing the subject from direct contact with the authority figure, and increasing the contact between the subject and the victim).

A widely publicized experiment on the police response to domestic violence, mentioned earlier, provides an interesting cautionary tale about the uses of science. Lawrence Sherman and Richard Berk (1984) arranged with the Minneapolis police department for the random assignment of persons accused of domestic violence to be either arrested or simply given a warning. The results of this field experiment indicated that those who were arrested were less likely subsequently to commit violent acts against their partners. Sherman (1993) explicitly cautioned police departments not to adopt mandatory arrest policies based solely on the results of the Minneapolis experiment, but the results were publicized in the mass media and encouraged many jurisdictions to change their policies (Binder & Meeker 1993; Lempert 1989). Although we now know that the original finding of a deterrent effect of arrest did not hold up in many other cities where the experiment was repeated, Sherman (1992:150–153) later suggested that implementing mandatory arrest policies might have prevented some subsequent cases of spouse abuse. In particular, in a follow-up study in Omaha, arrest warrants reduced repeat offenses among spouse abusers who had already left the scene when police arrived. However, this Omaha finding was not publicized, so it could not be used to improve police policies. So how much publicity is warranted, and at what point in the research should it occur?

Social scientists who conduct research on behalf of specific organizations may face additional difficulties when the organization, instead of the researcher, controls the final report and the publicity it receives. If organizational leaders decide that particular research results are unwelcome, the researcher's desire to have findings used appropriately and reported fully can conflict with contractual obligations. Researchers can anticipate such dilemmas and resolve them when the contract is negotiated—or simply decline a particular research opportunity altogether. But often such problems arise only after a report has been drafted, or a researcher who needs the job or to maintain a personal relationship ignores the problems. These possibilities cannot be avoided entirely, but because of them, it is always important to acknowledge the source of research funding in reports and to consider carefully the sources of funding for research reports written by others.

Encyclopedia Link 3.3
Read a brief overview about research ethics.

▣ Conclusion

Different kinds of research produce different kinds of ethical problems. Most survey research, for instance, creates few if any ethical problems and can even be enjoyable for participants. In fact, researchers from Michigan's Institute for Survey Research interviewed a representative national sample of adults and found that 68% of those who had participated in a survey were somewhat or very interested in participating in another; the more times respondents had been interviewed, the more willing they were to participate again (Reynolds 1979:56–57). On the other hand, some experimental studies in the social sciences that have put people in uncomfortable or embarrassing situations have generated vociferous complaints and years of debate about ethics (Reynolds 1979; Sjoberg 1967).

Research ethics should be based on a realistic assessment of the overall potential for harm and benefit to research subjects. In this chapter, we have presented some basic guidelines, and examples in other chapters suggest applications, but answers aren't always obvious. For example, full disclosure of "what is really going on" in an experimental study is unnecessary if subjects are unlikely to be harmed. In one student observation study on cafeteria workers, for instance, the IRB didn't require consent forms to be signed. The legalistic forms and signatures, they felt, would be more intrusive or upsetting to workers than the very benign and confidential research itself. The committee put the feelings of subjects above the strict requirement for consent.

Ultimately, then, these decisions about ethical procedures are not just up to you, as a researcher, to make. Your university's IRB sets the human subjects protection standards for your institution and will require that researchers—even, in most cases, students—submit their research proposal to the IRB for review. So an institutional committee, following professional codes and guidelines, will guard the ethical propriety of your research; but still, that is an uncertain substitute for your own conscience.

Key Terms

Belmont Report   43
Beneficence   43
Certificate of Confidentiality   52
Debriefing   45
Federal Policy for the Protection of
    Human Subjects   44

Health Insurance Portability and
    Accountability Act (HIPAA)   52
Justice   43
Nuremberg war crime trials   42
Obedience experiments
    (Milgram's)   41

Office for Protection from Research Risks,
    National Institutes of Health   43
Prison simulation study (Zimbardo's)   45
Respect for persons   43
Tearoom Trade   51
Tuskegee syphilis study   42

Highlights

- Stanley Milgram's obedience experiments led to intensive debate about the extent to which deception could be tolerated in psychological research and how harm to subjects should be evaluated.

- Egregious violations of human rights by researchers, including scientists in Nazi Germany and researchers in the Tuskegee

syphilis study, led to the adoption of federal ethical standards for research on human subjects.

- The 1979 *Belmont Report* developed by a national commission established three basic ethical standards for the protection of human subjects: (1) respect for persons, (2) beneficence, and

(3) justice. The Department of Health and Human Services adopted in 1991 the Federal Policy for the Protection of Human Subjects. The policy requires that every institution seeking federal funding for biomedical or behavioral research on human subjects have an institutional review board to exercise oversight.

- Standards for the protection of human subjects require avoiding harm, obtaining informed consent, avoiding deception

except in limited circumstances, and maintaining privacy and confidentiality. Scientific research should maintain high standards for validity and be conducted and reported in an honest and open fashion.

- Effective debriefing of subjects after an experiment can help to reduce the risk of harm due to the use of deception in the experiment.

STUDENT STUDY SITE

The Student Study Site, available at **www.sagepub.com/chambliss4e,** includes useful study materials including web exercises with accompanying links, eFlashcards, videos, audio resources, journal articles, and encyclopedia articles, many of which are represented by the media links throughout the text. The site also features Interactive Exercises—represented by the green icon here—to help you understand the concepts in this book.

Exercises

Discussing Research

1. Should social scientists be permitted to conduct replications of Milgram's obedience experiments? Zimbardo's prison simulation? Can you justify such research as permissible within the current ASA ethical standards? If not, do you believe that these standards should be altered so as to permit Milgram-type research?

2. Why does unethical research occur? Is it inherent in science? Does it reflect "human nature"? What makes ethical research more or less likely?

3. Does debriefing solve the problem of subject deception? How much must researchers reveal after the experiment is over, as well as before it begins?

Finding Research

1. The Collaborative Institutional Training Initiative (CITI) offers an extensive online training course in the basics of human subjects protections issues. Go to the public access CITI site (www .citiprogram.org/rcrpage.asp?affiliation=100) and complete the course in social and behavioral research. Write a short summary of what you have learned.

2. The U.S. Department of Health and Human Services maintains extensive resources concerning the protection of human subjects in research. Read several documents that you find on its website (www.hhs.gov/ohrp/) and share your findings in a short report.

Critiquing Research

1. Pair up with one other student and select one of the research articles you have reviewed for other exercises. Criticize the research in terms of its adherence to each of the ethical principles for research on human subjects, as well as for the authors' apparent honesty, openness, and consideration of social consequences. Try to be critical but fair. The student with whom you are working should critique the article in the

same way but from a generally positive standpoint, defending its adherence to the four guidelines but without ignoring the study's weak points. Together, write a summary of the study's strong and weak points or conduct a debate in class.

2. How do you evaluate the current American Sociological Association ethical code? Is it too strict, too lenient, or just

about right? Are the enforcement provisions adequate? What provisions could be strengthened?

3. IRB members and the researchers who submit proposals to them must be familiar with a number of key concepts about ethical principles. The interactive exercises "Ethics" lesson at the text's study site will help you learn how to do this.

   To use these lessons, choose one of the four "Ethics" exercises from the opening menu for the Interactive Exercises.

Follow the instructions for entering your answers and responding to the program's comments.

4. Now go to the book's Study Site (www.sagepub.com/chambliss4e) and choose the Learning from Journal Articles option. Read one article based on research involving human subjects. What ethical issues did the research pose, and how were they resolved? Does it seem that subjects were appropriately protected?

Doing Research

1. List elements in a research plan for the project you envisioned for the "Doing Research" section in Chapter 2 that an IRB might consider to be relevant to the protection of human subjects. Rate each element from 1 to 5, where 1 indicates no more than a minor ethical issue and 5 indicates a major ethical problem that probably cannot be resolved.

2. Write one page for the application to the IRB that explains how you will ensure that your research adheres to each relevant standard.

Ethics Questions

1. Read the entire American Sociological Association *Code of Ethics* at the ASA website (www.asanet.org/about/ethics.cfm).

2. Discuss the potential challenges in adhering to the ASA's ethical standards in research.

CHAPTER 4

Conceptualization and Measurement

E very time you begin to review or design a research study, you will have to answer two questions: (1) What do the main concepts mean in this research? (2) How are the main concepts measured? Both questions must be answered to evaluate the validity of any research. For instance, to study a hypothesized link between religious fundamentalism and terrorism, you may conceptualize terrorism as *nongovernmental political violence* and measure incidents of terrorism by counting, over a 5-year period, the number of violent attacks that have explicit political aims. You will also need to define and measure *religious fundamentalism*—no easy task. What counts? And how should you decide what counts? We cannot make sense of a researcher's study until we know how the concepts were *defined* and *measured.* Nor can we begin our own research until we have defined our concepts clearly and constructed valid measures of them.

In this chapter, we briefly address the issue of conceptualization, or defining your main terms. We then describe measurement sources such as available archive data; questions; observations; and less direct, or unobtrusive, measures. We then discuss the level of measurement reflected in different measures. The final topic is to assess the validity and reliability of these measures. By the chapter's end, you should have a good understanding of measurement, the first of the three legs (measurement, generalizability, and causality) on which a research project's validity rests.

What Do We Have In Mind?

A May 2000 *New York Times* article (Stille 2000) announced that the "social health" of the United States had risen a bit, after a precipitous decline in the 1970s and 1980s. Should we be relieved? Concerned? What, after all, does *social health* mean? To social scientist Marc Miringoff, it had to do with social and economic inequalities. To political pundit William J. Bennett, it was more a matter of moral values. The concept of social health means different things to different people. Most agree that it has to do with "things that are not measured in the gross national product" and is supposed to be "a more subtle and more meaningful way of measuring what's important to [people]" (Stille:A19). But until we agree on a definition of social health, we can't decide whether it has to do with child poverty, trust in government, out-of-wedlock births, alcohol-related traffic deaths, or some combination of these or other phenomena.

Conceptualization

A continuing challenge for social scientists, then, rests on the fact that many of our important topics of study (social health, for instance) are not clearly defined things or objects (like trees or rocks) but are abstract concepts or ideas. A **concept** is an image or idea, not a simple object. Some concepts are relatively simple, such as a person's age or sex: Almost everyone would agree what it means to be 14-years-old or female. But other concepts are more ambiguous. For instance, if you want to count the number of families in Chicago, what counts

Concept: A mental image that summarizes a set of similar observations, feelings, or ideas.

as a family? A husband and wife with two biological children living in one house—yes, that's a family. Do cousins living next door count? Cousins living in California? Or maybe the parents are divorced, the children are adopted, or the children are grown. Maybe two women live together with one adopted child and one biological child fathered by a now-absent man. So perhaps "living together" is what defines a family—or is it biology? Or is it a crossing of generations—that is, the presence of adults and children? The particular definition you develop will affect your research findings, and some people probably won't like it whatever you do, but how you define *family* obviously affects your results.

Often social concepts can be used sloppily or even misleadingly. In some years, you may hear that "the economy" is doing well, but even then, many people may be faring badly. Typically in news reports, *the economy* refers to the gross domestic product (GDP)—the total amount of economic activity (value of goods and services, precisely) in the country in a given year. When the GDP goes up, reporters say, "the economy is improving." But that's very different from saying that the average working person makes more money than this person would have 30 years ago—in fact, the average American man makes a little less than 30 years ago, and for women it's close. We could use the concept of *the economy* to refer to the economic well-being of actual people, but that's not typically how it's used.

Defining concepts clearly can be quite difficult because many concepts have several meanings and can be measured in many ways. What is meant, for instance, by the idea of *power*? The classic definition, provided by German sociologist Max Weber (1947/1997:152), is that power is the ability to meet your goals over the objections of other people. That definition implies that unknown people can be quite powerful, whereas certain presidents of the United States, very well known, have been relatively powerless. A different definition might equate power to one's official position; in that case, the president of the United States would always be powerful. Or perhaps power is equated with prestige, so famous intellectuals like Albert Einstein would be considered powerful. Or maybe power is defined as having wealth, so that rich people are seen as powerful.

And even if we can settle on a definition, how then do we actually measure power? Should we ask a variety of people if a certain person is powerful? Should we review that person's acts over the last 10 years and see when the person exerted his or her will over others? Should we try to uncover the true extent of the individual's wealth and use that? How about power at a lower level, say, as a member of student government? The most visible and vocal people in your student assembly may be, in fact, quite unpopular and perhaps not very powerful at all—just loud. At the same time, there may be students who are members of no official body whatsoever, but somehow they always get what they want. Isn't that power? From these varied cases, you can see that power can be quite difficult to conceptualize.

Likewise, describing what causes *crime,* or even what causes *theft,* is inherently problematic, since the very definition of these terms is spectacularly flexible and indeed forms part of their interest for us. What counts as theft varies dramatically, depending on who is the thief—a next-door neighbor, a sister, or a total stranger wandering through town—and what item is taken: a bottle of water, your watch, a lawn mower, a skirt, your reputation, or $5. Indeed, part of what makes social science interesting is the debates over, for instance, what is a theft or what is crime.

So **conceptualization**—working out what your key terms will mean in your research—is a crucial part of the research process. Definitions need to be explicit. Sometimes conceptualization is easy: "Older men are more likely to suffer myocardial infarction than younger men," or "Career military officers mostly vote for Republican candidates in national elections." Most of the concepts used in those statements are easily understood and easy to measure (gender, age, military status, voting). In other cases, conceptualization is quite difficult: "As people's moral standards deteriorate, the family unit starts to die," or "Intelligence makes you more likely to succeed."

Conceptualization, then, is the process of matching up terms (family, sex, happiness, power) to clarified definitions for them—really, figuring out what are the social "things" you'll be talking about.

It is especially important to define clearly concepts that are abstract or unfamiliar. When we refer to concepts like *social capital, whiteness,* or *dissonance,* we cannot count on others knowing exactly what we mean. Even experts may disagree about the meaning of frequently used concepts if they base their conceptualizations on different theories. That's okay. The point is not that there can be only one definition of a concept; rather, we have to specify clearly what we mean when we use a concept, and we should expect others to do the same.

Conceptualization: The process of specifying what we mean by a term. In deductive research, conceptualization helps to translate portions of an abstract theory into testable hypotheses involving specific variables. In inductive research, conceptualization is an important part of the process used to make sense of related observations.

Conceptualization also involves creating concepts, or thinking about how to conceive of the world: What things go together? How do we slice up reality? Cell phones, for instance, may be seen as communication devices, like telephones, radios, telegraphs, or two tin cans connected by a string. But they can also be conceived in another way: a college administrator we know, seeing students leaving class outside her building, said, "Cell phones have replaced cigarettes." She reconceptualized cell phones, seeing them not as communication tools but as something to fiddle with, like cigarettes, chewing gum wrappers, keys on a lanyard, or the split ends of long hair. In conceptualizing the world, we create the lenses through which we see it.

Our point is not that conceptualization problems are insurmountable, but that (1) you need to develop and clearly state what you *mean* by your key concepts and (2) your measurements will need to be clear and consistent with the definitions you've settled on (more on that topic shortly).

Research in the News

HOW POVERTY IS MEASURED MATTERS

Calculation of the federal poverty rate does not take into account variations in the local cost of living, expenses for health care, commuting and day care, or the value of such benefits as food stamps, housing allowances, or tax credits. According to the federal poverty rate, the level of poverty in New York City remained constant at 17.3% from 2008 to 2009. However, the city calculates the poverty rate after taking into account all these other factors. Using this way of operationalizing poverty, the poverty rate increased from 17.3% to 19.9% from 2008 to 2009. Housing and rental subsidies reduced the poverty rate by 6 percentage points, while unreimbursed medical expenses increased the rate by 3.1 percentage points.

Source: Roberts, Sam. 2011. Food stamps and tax aid kept poverty rate in check. *The New York Times,* March 21:A19.

Variables and Constants

**Research|Social Impact
Link 4.1**
Read more about variables.

After we define the concepts for a study, we must identify variables that correspond to those concepts. For example, we might be interested in what affects students' engagement in their academic work—when they are excited about their studies, when they become eager to learn more, and so on. Our main concept, then, would be *engagement.* We could use any number of variables to measure engagement: the student's reported interest in classes, teacher evaluations of student engagement, the number of hours spent on homework, or an index summarizing a number of different questions. Any of these variables could show a high or low level of student engagement. If we are to study variation in engagement, we must identify variables to measure that are most pertinent to our theoretical concerns.

You should be aware that not every concept in a particular study is represented by a variable. In our student engagement study, all of the students *are* students—there is no variation in that. So student, in this study, is a **constant** (it's always the same), not a variable.

Many variables could measure student engagement. Which variables should we select? It's very tempting, and all too common, to simply try to "measure everything" by including in a study every variable we can think of. We could collect self-reports of engagement, teacher ratings, hours studied per week, pages of essays written for class, number of visits to the library per week, frequency of participation in discussion, times met with professors, and on and on. This haphazard approach will inevitably result in the collection of some useless data and the failure to collect some important data. Instead, we should take four steps:

Constant: A number that has a fixed value in a given situation; a characteristic or value that does not change.

1. Examine the theories that are relevant to our research question to identify those concepts that would be expected to have some bearing on the phenomenon we are investigating.

2. Review the relevant research literature and assess the utility of variables used in prior research.

3. Consider the constraints and opportunities for measurement that are associated with the specific setting(s) we will study. Distinguish constants from variables in this setting.

4. Look ahead to our analysis of the data. What role will each variable play in our analysis?

Remember: A few well-chosen variables are better than a barrel full of useless ones.

🔲 How Will We Know When We've Found It?

Once we have defined our concepts in the abstract—that is, after conceptualizing—and we have identified the variables that we want to measure, we must develop our measurement procedures. The goal is to devise **operations** that actually measure the concepts we intend to measure—in other words, to achieve measurement validity.

> **Operation:** A procedure for identifying or indicating the value of cases on a variable.

Exhibit 4.1 represents the **operationalization** process in three studies. The first researcher defines his concept, binge drinking, and chooses one variable—frequency of heavy episodic drinking—to represent it. This variable is then measured with responses to a single question, or *indicator:* "How often within the last 2 weeks did you drink five or more drinks containing alcohol in a row?" Because "heavy" drinking is defined differently for men and women (relative to their differ-

> **Operationalization:** The process of specifying the operations that will indicate the value of cases on a variable.

ent metabolisms), the question is phrased in terms of "four or more drinks" for women. The second researcher defines her concept—poverty—as having two aspects or dimensions, subjective poverty and absolute poverty. Subjective poverty is measured with responses to a survey question: "Would you say that you are poor?" Absolute poverty is measured by comparing family income to the poverty threshold. The third researcher decides that his concept—social class—is defined by the sum of measurements of three variables: (1) income, (2) education, and (3) occupational prestige.

Measures can be based on activities as diverse as asking people questions, reading judicial opinions, observing social interactions, coding words in books, checking census data tapes, enumerating the contents of trash receptacles, or drawing urine and blood samples. Experimental researchers may operationalize a concept by manipulating its value; for example, to operationalize the concept of exposure to antidrinking messages, some subjects may listen to a talk about binge drinking while others do not. We will focus here on the operations of using published data, asking questions, observing behavior, and using unobtrusive means of measuring people's behavior and attitudes.

The variables and measurement operations chosen for a study should be consistent with the purpose of the research question. Suppose we hypothesize that college students who go abroad for the junior year have a more valuable experience than those who remain at the college. If our purpose is *evaluation* of different junior-year options, we can operationalize *junior-year programs* by comparing (1) traditional coursework at home, (2) study in a foreign country, and (3) internships at home that are not traditional college courses. A simple question—for example, asking students in each program, "How valuable do you feel your experience was?"—would help to provide the basis for determining the relative value of these programs. But if our purpose is

Video Link 4.1
Watch operationalization in action.

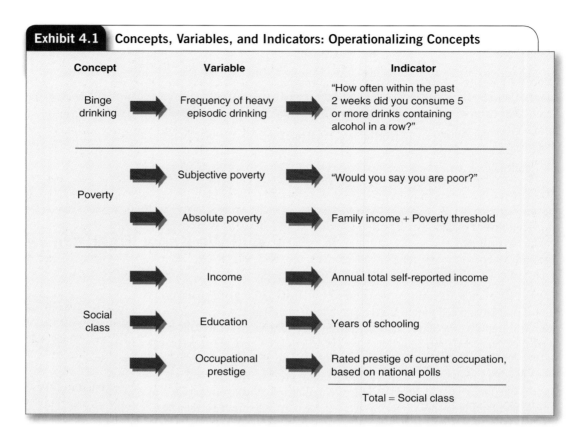

Exhibit 4.1 Concepts, Variables, and Indicators: Operationalizing Concepts

Concept	Variable	Indicator
Binge drinking	Frequency of heavy episodic drinking	"How often within the past 2 weeks did you consume 5 or more drinks containing alcohol in a row?"
Poverty	Subjective poverty	"Would you say you are poor?"
	Absolute poverty	Family income + Poverty threshold
Social class	Income	Annual total self-reported income
	Education	Years of schooling
	Occupational prestige	Rated prestige of current occupation, based on national polls

Total = Social class

explanation, we would probably want to interview students to learn what features of the different programs made them valuable to find out the underlying dynamics of educational growth.

Time and resource limitations also must be taken into account when we select variables and devise measurement operations. For many sociohistorical questions (such as "How has the poverty rate varied since 1950?"), census data or other published counts must be used.

On the other hand, a historical question about the types of social bonds among combat troops in wars since 1940 probably requires retrospective interviews with surviving veterans. The validity of the data is lessened by the unavailability of many veterans from World War II and by problems of recall, but direct observation of their behavior during the war is certainly not an option.

Using Available Data

Data can be collected in a wide variety of ways; indeed, much of this book describes different technologies for data collection. But some data are already gathered and ready for analysis. Government reports, for instance, are rich, accessible sources of social science data. Organizations ranging from nonprofit service groups to private businesses also compile a wealth of figures that may be available to some social scientists. Data from many social science surveys are archived and made available for researchers who were not involved in the original survey project.

Before we assume that available data will be useful, we must consider how appropriate they are for our concepts of interest, whether other measures would work better, or whether our concepts can be measured at all with these data. For example, many organizations informally (and sometimes formally) use turnover—that is,

Journal Link 4.1
Read an article which uses available data to look at the relationship between social capital and democracy.

how many employees quit each year—as a measure of employee morale (or satisfaction). If turnover is high (or retention rates are low), morale must be bad and needs to be raised. Or so the thinking goes.

But obviously, factors other than morale affect whether people quit their jobs. When a single chicken-processing plant is the only employer in a small town, other jobs are hard to find, and people live on low wages, then turnover may be very low even among miserable workers. In the dot-com companies of the late 1990s, turnover was high—despite amazingly good conditions, salary, and morale—because the industry was so hungry for good workers that companies competed ferociously to attract them. Maybe the concepts *morale* and *satisfaction,* then, can't be measured adequately by the most easily available data (that is, turnover rates).

We also cannot assume that available data are accurate, even when they appear to measure the concept. "Official" counts of homeless persons have been notoriously unreliable because of the difficulty of locating homeless persons on the streets, and government agencies have at times resorted to "guesstimates" by service providers. Even available data for such seemingly straightforward measures as counts of organizations can contain a surprising amount of error. For example, a 1990 national church directory reported 128 churches in a Midwestern county; an intensive search in that county in 1992 located 172 churches (Hadaway, Marler, & Chaves 1993:744). Still, when legal standards, enforcement practices, and measurement procedures have been taken into account, comparisons among communities become more credible.

However, such adjustments may be less necessary when the operationalization of a concept is seemingly unambiguous, as with the homicide rate: after all, dead is dead, right? And when a central authority imposes a common data collection standard, as with the FBI's *Uniform Crime Reports,* data become more comparable across communities. But even here careful review of measurement operations is still important, because (for instance) procedures for classifying a death as a homicide can vary between jurisdictions and over time.

Another rich source of already-collected data is survey datasets archived and made available to university researchers by the Inter-University Consortium for Political and Social Research (1996). One of its most popular survey datasets is the General Social Survey (GSS). The GSS is administered regularly by the National Opinion Research Center (NORC) at the University of Chicago to a sample of more than 1,500 Americans (annually until 1994; biennially since then). GSS questions vary from year to year, but an unchanging core of questions includes measures of political attitudes, occupation and income, social activities, substance abuse, and many other variables of interest to social scientists. College students can easily use this dataset to explore a wide range of interesting topics. However, when surveys are used in this way, after the fact, researchers must carefully evaluate the survey questions. Are the available measures sufficiently close to the measures needed that they can be used to answer the new research question?

Researcher Interview Link 4.1
Watch a researcher explain the GSS and how we use it.

Content Analysis

One particular method for using available data is **content analysis**, a method for systematically analyzing and making inferences from text (Weber 1985:9). You can think of a content analysis as a survey of "documents," ranging from newspapers, books, or TV shows to persons referred to in other communications, themes expressed in government documents, or propositions made in tape-recorded debates. Words or other features of these units are then coded to measure the variables involved in the research question (Weber). As a simple example of content analysis, you might look at a variety of women's magazines over the past 25 years and count the number of articles in each year devoted to various topics, such as makeup, weight loss, relationships, sex, and so on. You might count the number of articles on different subjects as a measure of the media's emphasis on women's anxiety about these issues and see how that emphasis (i.e., the number of articles) has increased or decreased over the past quarter century. At the simplest level, you could code articles by whether key words (*fat, weight, pounds,* etc.) appeared in the titles.

Journal Link 4.2
Read an example of content analysis design which focuses on media coverage of youth voting.

Content analysis: A research method for systematically analyzing and making inferences from text.

After coding procedures are developed, their reliability should be assessed by comparing different coders' results for the same variables. Computer programs for content analysis can be used to enhance reliability (Weitzman & Miles 1994). The computer is programmed with certain rules for coding text so that these rules will be applied consistently. We describe content analysis in detail in Chapter 10.

Constructing Questions

Asking people questions is the most common, and probably most versatile, operation for measuring social variables. Do you play on a varsity team? What is your major? How often, in a week, do you go out with friends? How much time do you spend on schoolwork? Most concepts about individuals can be measured with such simple questions. In this section, we introduce some options for writing questions, explain why single questions can sometimes be inadequate measures, and then examine the use of multiple questions to measure a concept.

In principle, questions, asked perhaps as part of a survey, can be a straightforward and efficient means by which to measure individual characteristics, facts about events, level of knowledge, and opinions of any sort. In practice, though, survey questions can easily result in misleading or inappropriate answers. All questions proposed for a survey must be screened carefully for their adherence to basic guidelines and then tested and revised until the researcher feels some confidence that they will be clear to the intended respondents (Fowler 1995). Some variables may prove to be inappropriate for measurement with any type of question. We have to recognize that memories and perceptions of the events about which we might like to ask can be limited.

Specific guidelines for reviewing questions are presented in Chapter 7; here, our focus is on the different types of survey questions.

Single Questions

Measuring variables with single questions is very popular. Public opinion polls based on answers to single questions are reported frequently in newspaper articles and TV newscasts: Do you favor or oppose U.S. policy in Iraq? If you had to vote today, for which candidate would you vote? Social science surveys also rely on single questions to measure many variables: Overall, how satisfied are you with your job? How would you rate your current health?

**Researcher Interview
Link 4.2**
Watch a researcher describe how to construct questions.

Single questions can be designed with or without explicit response choices. The question that follows is a **closed-ended**, or **fixed-choice**, **question**, because respondents are offered explicit responses from which to choose. It has been selected from the Core Alcohol and Drug Survey distributed by the Core Institute, Southern Illinois University, for the FIPSE Core Analysis Grantee Group (Presley, Meilman, & Lyerla 1994).

> **Closed-ended (fixed-choice) question:** A survey question that provides preformatted response choices for the respondent to circle or check.

Compared with other campuses with which you are familiar, this campus's use of alcohol is . . . (Mark one)

_____*Greater than other campuses*

_____*Less than other campuses*

_____*About the same as other campuses*

Most surveys of a large number of people contain primarily fixed-choice questions, which are easy to process with computers and analyze with statistics. However, fixed-response choices can obscure what people really think, unless the choices are designed carefully to match the range of possible responses to the question.

Most important, response choices should be **mutually exclusive** and **exhaustive**, so that respondents can each find *one and only one* choice that applies to them (unless the question is of the "Check all that apply" variety). To make response choices exhaustive, researchers may need to offer at least one option with room for ambiguity. For example, a questionnaire asking college students to indicate their school status should not use freshman, sophomore, junior, senior, and graduate student as the only response choices. Most campuses also have students in a "special" category, so you might add "Other (please specify)" to the five fixed responses to this question. If respondents do not find a response option that corresponds to their answer to the question, they may skip the question entirely or choose a response option that does not indicate what they are really thinking.

> **Mutually exclusive:** A variable's attributes (or values) are mutually exclusive when every case can be classified as having only one attribute (or value).
>
> **Exhaustive:** Every case can be classified as having at least one attribute (or value) for the variable.

Researchers who study small numbers of people often use **open-ended questions**, which don't have explicit response choices and allow respondents to write in their answers. The next question is an open-ended version of the earlier fixed-choice question:

Encyclopedia Link 4.1
Read about why and how open-ended questions should be implemented.

How would you say alcohol use on this campus compares to that on other campuses?

An open-ended format is preferable when the full range of responses cannot be anticipated, especially when questions have not been used previously in surveys or when questions are asked of new groups. Open-ended questions also can allow clear answers when questions involve complex concepts. In the previous question, for instance, "alcohol use" may cover how many students drink, how heavily they drink, if the drinking is public or not, if it affects levels of violence on campus, and so on.

> **Open-ended question:** A survey question to which respondents reply in their own words, either by writing or by talking.

Just like fixed-choice questions, open-ended questions should be reviewed carefully for clarity before they are used. For example, if respondents are asked, "When did you move to Boston?" they might respond with a wide range of answers: "In 1987." "After I had my first child." "When I was 10." "20 years ago." Such answers would be very hard to compile. To avoid such ambiguity, rephrase the question to clarify the form of the answer; for instance, "In what year did you move to Boston?" Or, provide explicit response choices (Center for Survey Research 1987).

Indexes and Scales

When several questions are used to measure one concept, the responses may be combined by taking the sum or average of responses. A composite measure based on this type of sum or average is termed an **index**. The idea is that idiosyncratic variation in response to particular questions will average out, so that the main influence on the combined measure will be the concept that all the questions focus on. In addition, the index can be considered a more complete measure of the concept than can any one of the component questions.

Creating an index is not just a matter of writing a few questions that seem to focus on a concept. Questions that seem to you to measure a common concept might seem to respondents to concern several different issues. The only way to know that a given set of questions does, in fact, form an index is to administer the questions to people like those you plan to study. If a common concept is being measured, people's responses to the different questions should display some consistency.

> **Index:** A composite measure based on summing, averaging, or otherwise combining the responses to multiple questions that are intended to measure the same concept.

Because of the popularity of survey research, indexes already have been developed to measure many concepts, and some of these indexes have proved to be reliable in a range of studies. Usually it is much better to use such an index than it is to try to form a new one. Use of a preexisting index both simplifies the work of designing a study and facilitates the comparison of findings from other studies.

The questions in Exhibit 4.2 represent a short form of an index used to measure depression; it is called the Center for Epidemiologic Studies Depression Index (CES-D). Many researchers in different studies have found

Exhibit 4.2	Example of an Index: Excerpt From the Center for Epidemiologic Studies Depression Index (CES-D)

At any time during the past week . . . (Circle one response on each line)	Never	Some of the time	Most of the time
a. Was your appetite so poor that you did not feel like eating?	1	2	3
b. Did you feel so tired and worn out that you could not enjoy anything?	1	2	3
c. Did you feel depressed?	1	2	3
d. Did you feel unhappy about the way your life is going?	1	2	3
e. Did you feel discouraged and worried about your future?	1	2	3
f. Did you feel lonely?	1	2	3

that these questions form a reliable index. Note that each question concerns a symptom of depression. People may well have one particular symptom without being depressed; for example, persons who have been suffering from a physical ailment may say that they have a poor appetite. By combining the answers to questions about several symptoms, the index reduces the impact of this idiosyncratic variation. (This set of questions uses what is termed a *matrix* format, in which a series of questions that concern a common theme are presented together with the same response choices.)

Usually an index is calculated by simply averaging responses to the questions, so that every question counts equally. But sometimes, either intentionally by the researcher or by happenstance, questions on an index arrange themselves in a kind of hierarchy in which an answer to one question effectively provides answers to others. For instance, a person who supports abortion on demand almost certainly supports it in cases of rape and incest as well. Such questions form a **scale**. In a scale, we give different weights to the responses to different questions before summing or averaging the responses. Responses to one question might be counted two or three times as much as responses to another. For example, based on Christopher Mooney and Mei Hsien Lee's (1995) research on abortion law reform, a scale to indicate support for abortion might give a 1 to agreement that abortion should be allowed "when the pregnancy results from rape or incest" and a 4 to agreement with the statement that abortion should be allowed "whenever a woman decides she wants one." A 4 rating is much stronger, in that anyone who gets a 4 would probably agree to all lower-number questions as well.

Scale: A composite measure based on combining the responses to multiple questions pertaining to a common concept after these questions are differentially weighted, such that questions judged on some basis to be more important for the underlying concept contribute more to the composite score.

Making Observations

Asking questions, then, is one way to operationalize, or measure, a variable. *Observations* can also be used to measure characteristics of individuals, events, and places. The observations may be the primary form of measurement in a study, or they may supplement measures obtained through questioning.

Direct observations can be used as indicators of some concepts. For example, Albert J. Reiss Jr. (1971) studied police interaction with the public by riding in police squad cars, observing police–citizen interactions, and recording the characteristics of the interactions on a form. Notations on the form indicated such variables as how many police–citizen contacts occurred, who initiated the contacts, how compliant citizens were with police directives, and whether or not police expressed hostility toward the citizens.

Often, observations can supplement what is initially learned from interviews or survey questions, putting flesh on the bones of what is otherwise just a verbal self-report. In Chambliss's (1996) book, *Beyond Caring,* a theory of the nature of moral problems in hospital nursing that was originally developed through interviews was expanded with lessons learned from observations. Chambliss found, for instance, that in interviews, nurses described their daily work as exciting, challenging, dramatic, and often even heroic. But when Chambliss himself sat for many hours and watched nurses work, he found that their daily lives were rather humdrum and ordinary, even to them. Occasionally there were bursts of energetic activity and even heroism—but the reality of day-to-day nursing was far less exciting than interviews would lead one to believe. Indeed, Chambliss modified his original theory to include a much broader role for routine in hospital life.

Direct observation is often the method of choice for measuring behavior in natural settings, as long as it is possible to make the requisite observations. Direct observation avoids the problems of poor recall and self-serving distortions that can occur with answers to survey questions. It also allows measurement in a context that is more natural than an interview. But observations can be distorted, too. Observers do not see or hear everything, and their own senses and perspectives filter what they do see. Moreover, in some situations, the presence of an observer may cause people to act differently from the way they would otherwise (Emerson 1983). If you set up a video camera in an obvious spot on campus to monitor traffic flows, you may well change the flow—just because people will see the camera and avoid it (or come over to make faces). We will discuss these issues in more depth in Chapter 9, but it is important to begin to consider them whenever you read about observational measures.

Collecting Unobtrusive Measures

Unobtrusive measures allow us to collect data about individuals or groups without their direct knowledge or participation, hence they are *nonreactive* measures—they can be gathered without causing a reaction among the people being studied. In their recently revised classic book, Eugene Webb and his colleagues (Webb, Campbell, Schwartz, & Sechrest) identified four types of unobtrusive measures: (1) physical trace evidence, (2) archives (available data), (3) simple observation, and (4) contrived observation (using hidden recording hardware or manipulation to elicit a response). These measures provide valuable supplements or alternatives to more standard survey-based measures, because they are not affected by an interviewer's appearance or how the interviewer asks questions. We have already considered some types of archival and observational data.

> **Unobtrusive measure:** A measurement based on physical traces or other data that are collected without the knowledge or participation of the individuals or groups that generated the data.

Potential unobtrusive measures are everywhere. Webb and his colleagues (2000:37) suggested measuring viewers' interest in museum exhibits by the frequency with which tiles in front of the exhibits needed to be replaced. If auto mechanics note the radio dial settings in cars brought in for repairs, they can target their advertising to those stations to which their customers listen most. A quick glance at the hands of patrons in a neighborhood bar could help you see if they do heavy manual work (calluses).

The physical traces of past behavior are one type of unobtrusive measure that provides creative opportunities; for instance, "community urinalysis." As reported in the *New York Times* "Year in Ideas" issue, "since all drug users urinate, and since the urine eventually winds up in the sewers, [Oregon State chemist Jennifer] Field and her fellow researchers figured that sewer water would contain traces of whatever drugs the citizens were using." Samples detected varying usage by city of cocaine, methamphetamine, and—most popular of all—caffeine. Cocaine use, interestingly, peaked on weekends, while methamphetamine use tended to be steady across the week (Thompson 2007).

Unobtrusive measures can also be created from such diverse forms of media as newspaper archives, magazine articles, TV or radio talk shows, legal opinions, historical documents, personal letters, or e-mail messages. Researchers may read and evaluate the text of Internet LISTSERVs, as Fox and Roberts (1999) did in a study of British physicians, using the techniques of content analysis. We could even learn about cities by comparing their "yellow book" business telephone directories. For example, we find that Sarasota, Florida, has many pages devoted to nursing homes and hospital appliances; Chattanooga, Tennessee, which has approximately the same number of people, rather than having pages devoted to medical care, has many pages listing churches. Admittedly, that's a rough way to compare cities, but it may alert us to key differences between them.

A treasure trove for unobtrusive research lies in new social media such as Facebook and Myspace. Even as recently at 2005, no such data existed. "We're on the cusp of a new way of doing social science," said Nicholas Christakis, a sociology professor at Harvard, in another *New York Times* article from 2007 (as cited in Rosenbloom, p. 1). Christakis and colleagues from Harvard and UCLA studied an entire class of students at a college via Facebook pages, to see how personal tastes, habits, and values affect social relationships—and vice versa. Meanwhile, other scholars at various universities were looking at Facebook to learn about (for instance) how details from the Virginia Tech mass shootings were disseminated; how young adults treat privacy issues online; and how people meet and learn about potential romantic partners, among other issues. "For studying young adults," said Vincent Roscigno, a member of the editorial board at the American Sociological Review, "Facebook is the key site of the moment" (as cited in Rosenbloom 2007, p. 1).

But there are some obvious drawbacks to using such sites. Their users are self-selecting, so they are not at all representative of the public at large, in terms of age, ethnicity, education, and income. Hispanic students, for instance, are more likely to use Myspace, whereas white and Asian students prefer Facebook (Hargittai 2007). Social media can be a wonderful data source so long as you remain aware of their limitations and the potential ethical problems (as mentioned in Chapter 3).

Combining Measurement Operations

Encyclopedia Link 4.2
Read more about using triangulation in social research.

The choice of a particular measurement method—questions, observations, archives, and the like—is often determined by available resources and opportunities, but measurement is improved if this choice also takes into account the particular concept or concepts to be measured. Responses to questions such as "How socially adept were you at the party?" or "How many days did you use sick leave last year?" are unlikely to provide valid information on shyness or illness. Direct observation or company records may work better. On the other hand, observations at cocktail parties may not fully answer our questions about why some people are shy; we may just have to ask people. Or if a company keeps no record of sick leave, we may have to ask direct questions and hope for accurate memories. Every choice of a measurement method entails some compromise between the perfect and the possible.

Triangulation—the use of two or more different measures of the same variable—can strengthen measurement considerably (Brewer & Hunter 1989:17). When we achieve similar results with different measures of the same variable, particularly when they are based on such different methods as survey questions and field-based observations, we can be more confident of the validity of each measure. In surveys, for instance, people may say that they would return a lost wallet they found on the street. But field observation may prove that in practice, many succumb to the temptation to keep the wallet. The two methods produce different results. In a contrasting example, postcombat interviews of American soldiers in World War II found that most GIs never fired their weapons in battle, and the written, archival records of ammunition resupply patterns confirmed this interview finding (Marshall 1947/1978). If results diverge when using different measures, it may indicate that we are sustaining more measurement error than we can tolerate.

Triangulation: The use of multiple methods to study one research question.

Divergence between measures could also indicate that each measure actually operationalizes a different concept. An interesting example of this interpretation of divergent results comes from research on crime. Crime statistics are often inaccurate measures of actual crime; what gets reported to the police and shows up in official statistics is not at all the same thing as what happens according to victimization surveys (in which random people are asked if they have been a crime victim). Social scientists generally regard victim surveys as a more valid measure of crime than police-reported crime. We know, for instance, that rape is a dramatically *underreported* crime, with something like 4 to 10 times the number of rapes occurring as are reported to police. But auto theft is an *overreported* crime: More auto thefts are reported to police than actually occur. This may strike you as odd, but remember that almost everyone who owns a car also owns car insurance; if the car is stolen, the victim will definitely report it to the police to claim the insurance. Plus, some other people might report cars stolen when they haven't been because of the financial incentive. (By the way, insurance companies are quite good at discovering this scam, so it's a bad way to make money.)

Murder, however, is generally reported to police at roughly the same rate at which it actually occurs (i.e., official police reports generally match victim surveys). When someone is killed, it's very difficult to hide the fact: a body is missing, a human being doesn't show up for work, people find out. At the same time, it's very hard to pretend that someone was murdered when they weren't. There they are, still alive, in the flesh. Unlike rape or auto theft, there are no obvious incentives for either underreporting or overreporting murders. The official rate is generally valid.

So if you can, it's best to use multiple measures of the same variable; that way, each measure helps to check the validity of the others.

Researcher Interview Link 4.3
Watch a researcher discuss how to receive adequate measurements.

🔲 How Much Information Do We Really Have?

There are many ways of collecting information, or different *operations* for gathering data: asking questions, using previously gathered data, analyzing texts, and so on. Some of this data contains mathematically detailed information; it represents a higher level of measurement. There are four **levels of measurement**: (1) nominal, (2) ordinal, (3) interval, and (4) ratio. Exhibit 4.3 depicts the differences among these four levels.

Level of measurement: The mathematical precision with which the values of a variable can be expressed. The nominal level of measurement, which is qualitative, has no mathematical interpretation; the quantitative levels of measurement—ordinal, interval, and ratio—are progressively more precise mathematically.

Nominal level of measurement: Variables whose values have no mathematical interpretation; they vary in kind or quality but not in amount.

Nominal Level of Measurement

The **nominal level of measurement** identifies variables whose values have no mathematical interpretation; they vary in kind or quality but not in amount. *State* (referring to the United States) is one example. The variable has 50 attributes (or categories or qualities), but none of them is more *state* than another. They're just different. *Religious affiliation* is another nominal variable, measured in categories: Christian, Muslim, Hindu, Jewish, and so on. *Nationality, occupation,* and *region of the country* are also measured at the nominal level. A person may be Spanish or Portuguese, but one nationality does not represent more nationality than another—just a different nationality (Exhibit 4.3). A person may be a doctor or a truck driver, but one does not represent three units "more occupation" than the other. Of course, more people may identify themselves as being of one nationality than of another, or one occupation may have a higher average income than another occupation, but these are comparisons involving variables other than *nationality* or *occupation* themselves.

Journal Link 4.3
Read an article that uses nominal variables to help assess community networks.

Exhibit 4.3 Levels of Measurement

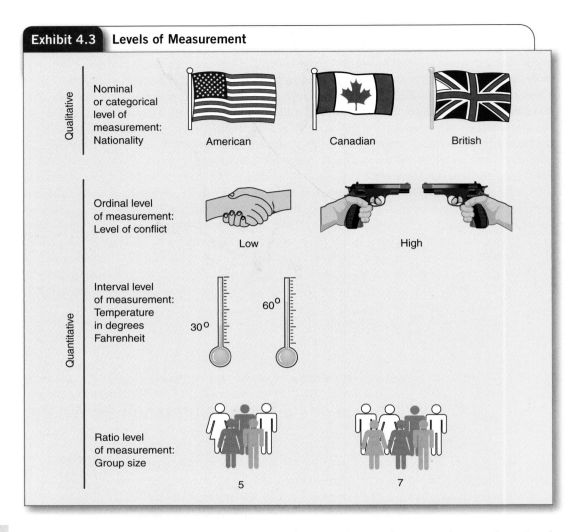

Video Link 4.2
Watch more about levels
of measurement.

Although the attributes of nominal variables do not have a mathematical meaning, they must be assigned to cases with great care. The attributes we use to measure, or categorize, cases must be mutually exclusive and exhaustive:

- A variable's attributes or values are mutually exclusive if every case can have only one attribute.

- A variable's attributes or values are exhaustive when every case can be classified into one of the categories.

When a variable's attributes are mutually exclusive and exhaustive, every case corresponds to one—and only one—attribute.

Ordinal Level of Measurement

The first of the three quantitative levels is the **ordinal level of measurement**. At this level, you specify only the order of the cases in *greater than* and *less than* distinctions. At the coffee shop, for example, you might choose between a small, medium, or large cup of decaf—that's ordinal measurement.

The properties of variables measured at the ordinal level are illustrated in Exhibit 4.3 by the contrast between the levels of conflict in two groups. The first group, symbolized by two people shaking hands, has a low level of conflict. The second group, symbolized by two people pointing guns at each other, has a high level of conflict. To measure conflict, we could put the groups "in order" by assigning 1 to the low-conflict group and 2 to the high-conflict group, but the numbers would indicate only the relative position, or order, of the cases.

As with nominal variables, the different values of a variable measured at the ordinal level must be mutually exclusive and exhaustive. They must cover the range of observed values and allow each case to be assigned no more than one value.

> **Ordinal level of measurement:**
> A measurement of a variable in which the numbers indicating a variable's values specify only the order of the cases, permitting *greater than* and *less than* distinctions.

Interval Level of Measurement

At the **interval level of measurement**, numbers represent fixed measurement units but have no absolute zero point. This level of measurement is represented in Exhibit 4.3 by the difference between two Fahrenheit temperatures. Note, for example, that 60 degrees is 30 degrees higher than 30 degrees, but 60 is not "twice as hot" as 30. Why not? Because heat does not "begin" at 0 degrees on the Fahrenheit scale. The numbers can therefore be added and subtracted, but ratios of them (2 to 1 or "twice as much") are not meaningful. There are thus few true interval-level measures in the social sciences; most are ratio-level, because they have zero points.

> **Interval level of measurement:**
> A measurement of a variable in which the numbers indicating a variable's values represent fixed measurement units but have no absolute, or fixed, zero point.

Sometimes, though, social scientists will create indexes by combining responses to a series of variables measured at the ordinal level and then treat these indexes as interval-level measures. An index of this sort could be created with responses to the Core Institute's questions about friends' disapproval of substance use (Exhibit 4.4). The survey has 13 questions on the topic, each of

---

Exhibit 4.4 Ordinal Measures: Core Alcohol and Drug Survey. Responses could be combined to create an interval scale (see text).

26. **How do you think your close friends feel (or would feel) about you . . .** *(mark one for each line)*

	Don't disapprove	Disapprove	Strongly disapprove
a. Trying marijuana once or twice	○	○	○
b. Smoking marijuana occasionally	○	○	○
c. Smoking marijuana regularly	○	○	○
d. Trying cocaine once or twice	○	○	○
e. Taking cocaine regularly	○	○	○
f. Trying LSD once or twice	○	○	○
g. Taking LSD regularly	○	○	○
h. Trying amphetamines once or twice	○	○	○
i. Taking amphetamines regularly	○	○	○
j. Taking one or two drinks of an alcoholic beverage (beer, wine, liquor) nearly every day	○	○	○
k. Taking four or five drinks nearly every day	○	○	○
l. Having five or more drinks in one sitting	○	○	○
m. Taking steroids for bodybuilding or improved athletic performance	○	○	○

which has the same three response choices. If "Don't disapprove" is valued at 1, "Disapprove" is valued at 2, and "Strongly disapprove" is valued at 3, the summed index of disapproval would range from 13 to 39. A score of 20 could be treated as if it were 4 more units than a score of 16. Or the responses could be averaged to retain the original 1 to 3 range.

Ratio Level of Measurement

Ratio level of measurement: A measurement of a variable in which the numbers indicating the variable's values represent fixed measuring units *and* an absolute zero point.

A **ratio level of measurement** represents fixed measuring units with an absolute zero point. Zero, in this situation, means absolutely no amount of whatever the variable indicates. On a ratio scale, 10 is 2 points higher than 8 and is also 2 times as great as 5. Ratio numbers can be added and subtracted, and because the numbers begin at an absolute zero point, they can also be multiplied and divided (so ratios can be formed between the numbers).

For example, people's ages can be represented by values ranging from 0 years (or some fraction of a year) to 120 or more. A person who is 30-years-old is 15 years older than someone who is 15-years old ($30 - 15 = 15$) and is also twice as old as that person ($30/15 = 2$). Of course, the numbers also are mutually exclusive and exhaustive, so that every case can be assigned one and only one value. Age (in years) is clearly a ratio-level measure.

Exhibit 4.3 displays an example of a variable measured at the ratio level. The number of people in the first group is 5, and the number in the second group is 7. The ratio of the two groups' sizes is then 1.4, a number that mirrors the relationship between the sizes of the groups. Note that there does not actually have to be any "group" with a size of zero; what is important is that the numbering scheme begins at an absolute zero—in this case, the absence of any people.

Comparison of Levels of Measurement

Exhibit 4.5 summarizes the types of comparisons that can be made with different levels of measurement, as well as the mathematical operations that are legitimate with each. All four levels of measurement allow researchers to assign different values to different cases. All three quantitative measures allow researchers to rank cases in order.

Researchers choose levels of measurement in the process of operationalizing variables; the level of measurement is not inherent in the variable itself. Many variables can be measured at different levels with different procedures. Age can be measured as *young* or *old*; as 0 to 10, 11 to 20, 21 to 30, and so on; or as 1-, 2-, or 3-years-old. We could gather the data by asking people their age, by having an observer guess ("Now *there's* an

Exhibit 4.5	Properties of Measurement Levels					
Examples of Comparison Statements	**Appropriate Math Operations**	**Relevant Level of Measurement**				
		Nominal	**Ordinal**	**Interval**	**Ratio**	
A is equal to (not equal to) B	$= (\neq)$	✓	✓	✓	✓	
A is greater than (less than) B	$> (<)$		✓	✓	✓	
A is three more than (less than) B	$+ (-)$			✓	✓	
A is twice (half) as large as B	$\times (/)$				✓	

old guy!"), or by searching through hospital records for exact dates and times of birth. Any of these approaches could work, depending on our research goals.

Usually, though, it is a good idea to measure variables at the highest level of measurement possible. The more information available, the more ways we have to compare cases. We also have more possibilities for statistical analysis with quantitative than with qualitative variables. Even if your primary concern is only to compare teenagers to young adults, you should measure age in years rather than in categories; you can always combine the ages later into categories corresponding to *teenager* and *young adult.*

Be aware, however, that other considerations may preclude measurement at a high level. For example, many people are very reluctant to report their exact incomes, even in anonymous questionnaires. So asking respondents to report their income in categories (such as less than $10,000, $10,000–$19,999, $20,000–$29,999, and so on) will elicit more responses, and thus more valid data, than asking respondents for their income in dollars.

Did We Measure What We Wanted To Measure?

Do the operations developed to measure our variables actually do so—are they valid? If we have weighed our measurement options, carefully constructed our questions and observational procedures, and selected sensibly from the available data indicators, we should be on the right track. But we cannot have much confidence in a measure until we have empirically evaluated its validity. We must also evaluate the reliability of our measures. The reliability of a measure is the degree to which it produces a consistent answer; such reliability (consistency) is a prerequisite for measurement validity.

Audio Link 4.1
Listen to an issue regarding measurement validity.

Measurement Validity

In Chapter 1, you learned that measurement validity refers to how well your indicators measure what they are intended to measure. For instance, a good measure of a person's age is the current year minus the year given on that person's birth certificate. Very probably, the resulting number accurately represents the person's age. A less valid measure would be for the researcher to ask the person (who may lie or forget) or for the researcher to simply guess. Measurement validity can be assessed with four different approaches: (1) face validation, (2) content validation, (3) criterion validation, and (4) construct validation.

Face Validity

Researchers apply the term **face validity** to the confidence gained from careful inspection of a concept to see if it is appropriate "on its face." More precisely, we can say that a measure has face validity if it obviously pertains to the meaning of the concept being measured more than to other concepts (Brewer & Hunter 1989:131). For example, a count of the number of drinks people have consumed in the past week would be a face valid measure of their alcohol consumption.

Although every measure should be inspected in this way, face validation in itself does not provide convincing evidence of measurement validity. Face validity has some plausibility but often not much. For instance, let's say that Sara is having some worries about her boyfriend, Jeremy. She wants to know if he loves her. So she asks him, "Jeremy, do you really love me?" He replies, "Sure, baby, you know I do." And yet he routinely goes out with other women, only calls Sara once every 3 weeks, and

Face validity: The type of validity that exists when an inspection of items used to measure a concept suggests that they are appropriate "on their face."

isn't particularly nice to her when they do go out. His answer that he loves her has a certain face validity, but Sara should probably look for other validating measures. (At the least, she might ask her friends if they think his claim has validity.)

Content Validity

Content validity establishes that the measure covers the full range of the concept's meaning. To determine that range of meaning, the researcher may solicit the opinions of experts and review literature that identifies the different aspects, or dimensions, of the concept. A measure of student engagement based on how much one talks in class won't count the quiet, but attentive and hardworking, person in the front row. Or, if you measure power by listing only elected government officials, you may miss the most important people altogether.

Content validity: The type of validity that exists when the full range of a concept's meaning is covered by the measure.

Criterion Validity

Criterion validity is established when the results from one measure match those obtained with a more direct or an already validated measure of the same phenomenon (the *criterion*). A measure of blood-alcohol concentration, for instance, could be the criterion for validating a self-report measure of drinking. In other words, if Jason says he hasn't been drinking, we establish criterion validity by giving him a Breathalyzer test. Observations of drinking by friends or relatives could also, in some limited circumstances, serve as a criterion for validating a self-report.

Criterion validity: The type of validity that is established by comparing the scores obtained on the measure being validated to those obtained with a more direct or already validated measure of the same phenomenon (the criterion).

The criterion that researchers select can be measured either at the same time as the variable to be validated or after that time. **Concurrent validity** exists when a criterion conducted at the same time yields scores that are closely related to scores on a measure. A store might validate a test of sales ability by administering the test to its current salespeople and then comparing their test scores to their actual sales performance. Or, a measure of walking speed based on mental counting might be validated concurrently with a stopwatch. With **predictive validity**, predicting scores on a criterion measured in the future validates a measure—for instance, predicting a student's college grades validates SAT scores.

Concurrent validity: The type of validity that exists when scores on a measure are closely related to scores on a criterion measured at the same time.

Predictive validity: The type of validity that exists when a measure predicts scores on a criterion measured in the future.

Criterion validation greatly increases our confidence that a measure works, but for many concepts of interest to social scientists, it's difficult to find a criterion. Yes, if you and your roommate are together every evening, you can count the beers he drinks. You definitely know about his drinking. But if we are measuring feelings or beliefs or other subjective states, such as feelings of loneliness, what direct indicator could serve as a criterion? How do you know he's lonely? Even with variables for which a reasonable criterion exists, the researcher may not be able to gain access to the criterion—as would be the case with a tax return or employer document that we might wish we could use as a criterion for self-reported income.

Construct Validity

Construct validity: The type of validity that is established by showing that a measure is related to other measures as specified in a theory.

Measurement validity also can be established by relating a measure to other measures specified in a theory. This validation approach, known as **construct validity**, is commonly used in social research when no clear criterion exists for validation purposes.

A historically famous example of construct validity is provided by the work of Theodor W. Adorno, Nevitt Sanford, Else Frenkel-Brunswik, and Daniel

Levinson (1950) in their book *The Authoritarian Personality.* Adorno and his colleagues, working in the United States and Germany immediately after World War II, were interested in a question that troubled much of the world during the 1930s and 1940s: Why were so many people attracted to Nazism and to its Italian and Japanese fascist allies? Hitler was not an unpopular leader in Germany. In fact, in January 1933, he came to power by being elected chancellor (something like president) of Germany, although some details of the election were a bit suspicious. Millions of people supported him enthusiastically. Why did so many Germans during the 1930s come to nearly worship Adolf Hitler and believe strongly in his program—which proved, of course, to be so disastrous for Europe and the rest of the world? The Adorno research group proposed the existence of what they called an "authoritarian personality," a type of person who would be drawn to a dictatorial leader of the Hitler type. Their key concept, then, was *authoritarianism.*

But of course there's no such "thing" as authoritarianism; it's not like a tree, something you can look at. It's a *construct,* an idea that we use to help make sense of the world. To establish construct validity of this idea, the researchers created a number of different scales made up of interview questions. One scale was called the "anti-Semitism" scale, in which hatred of Jews was measured. Another measure was a "fascism" scale, measuring a tendency toward favoring a militaristic, nationalist government. Another was the "political and economic conservatism" scale, and so on. Adorno and his colleagues interviewed lots of Germans and found that high scores on these different scales tended to correlate; a person who scored high on one tended to score high on the others. Hence, they determined that the authoritarian personality was a legitimate construct. The idea of authoritarianism, then, was validated through construct validity.

In short, a construct (authoritarianism) was validated through the use of a number of other measures that all tended to be high or low at the same time. Simultaneous high scores on them validated the idea of authoritarianism.

Construct and criterion validation, then, compare scores on one measure to scores on other measures that are predicted to be related. Distinguishing the two forms (construct and criterion) matters less than thinking clearly about the comparison measures and whether they actually represent different views of the same phenomenon. For example, correspondence between scores on two different self-report measures of alcohol use is a weak indicator of measurement validity. A person just reports in two different ways how much she drinks; of course the two will be related. But the correspondence of a self-report measure with an observer-based measure of substance use is a much stronger demonstration of validity. The subject (1) reports how much she drinks, and then (2) an observer reports on the subject's drinking. If the results match up, it's strong evidence of validity.

Reliability

Reliability means that a measurement procedure yields consistent scores (or that the scores change only to reflect actual changes in the phenomenon). If a measure is *reliable,* it is affected less by random error, or chance variation, than if it is unreliable. Reliability is a prerequisite for measurement validity: We cannot really measure a phenomenon if the measure we are using gives inconsistent results. Let's say, for example, that you would like to know your weight and have decided on two different measures: the scales in the bathroom and your mother's estimate. Clearly, the scales are more reliable, in the sense that they will show pretty much the same thing from one day to the next unless your weight actually changes. But your mother, bless her, may say, "You're so skinny!" on Sunday; but on Monday, when she's not happy, she may say, "You look terrible! Have you gained weight?" Her estimates may bounce around quite a bit. The bathroom scales are not so fickle; they are *reliable.*

This doesn't mean that the scales are *valid*—in fact, if they are spring-operated and old, they may be off by quite a few pounds. But they will be off by the same amount every day—hence not being valid but *reliable* nonetheless.

Audio Link 4.2
Listen to what reliability is and how it affects social research.

Reliability: A measurement procedure yields consistent scores when the phenomenon being measured is not changing.

There are four possible indications of unreliability. For example, a test of your knowledge of research methods would be unreliable if every time you took it, you received a different score, even though your knowledge of research methods had not changed in the interim, not even as a result of taking the test more than once. This is **test-retest reliability**. Similarly, an index composed of questions to measure knowledge of research methods would be unreliable if respondents' answers to each question were totally independent of their answers to the others. The index has **interitem reliability** if the component items are closely related. A measure also would be unreliable if slightly different versions of it resulted in markedly different responses (it would not achieve **alternate-forms reliability**). Finally, an assessment of the level of conflict in social groups would be unreliable if ratings of the level of conflict by two observers were not related to each other—it would then lack **interobserver reliability**.

Test-Retest Reliability

When researchers measure an unchanging phenomenon at two different times, the degree to which the two measurements are related is the test-retest reliability of the measure. If you take a test of your math ability and then retake the test 2 months later, the test is reliable if you receive a similar score both times, presuming that your math ability stayed constant. Of course, if events between the test and the retest have changed the variable being measured, then the difference between the test and retest scores should reflect that change.

Interitem Reliability (Internal Consistency)

When researchers use multiple items to measure a single concept, they must be concerned with interitem reliability (or internal consistency). For example, if the questions in Exhibit 4.2 reliably measure depression, the answers to the different questions should be highly associated with one another. The stronger the association among the individual items and the more items that are included, the higher the reliability of the index.

Alternate-Forms Reliability

When researchers compare subjects' answers to slightly different versions of survey questions, they are testing alternate-forms reliability (Litwin 1995:13–21). A researcher may reverse the order of the response choices in an index or may modify the question wording in minor ways and then readminister the index to subjects. If the two sets of responses are not too different, alternate-forms reliability is established.

A related test of reliability is the **split-halves reliability** approach. A survey sample is divided in two by flipping a coin or using some other random assignment method. The two forms of the questions are then administered to the two halves of the sample. If the responses of the two halves of the sample are about the same, the reliability of the measure is established.

Interobserver Reliability

When researchers use more than one observer to rate the same people, events, or places, interobserver reliability is their goal. If observers are using the same instrument to rate the same thing, their ratings should be very similar. If they are similar, we can have much more confidence that the ratings reflect the phenomenon being assessed rather than the orientations of the observers.

Test-retest reliability: A measurement showing that measures of a phenomenon at two points in time are highly correlated, if the phenomenon has not changed or has changed only as much as the phenomenon itself.

Interitem reliability (internal consistency): An approach that calculates reliability based on the correlation among multiple items used to measure a single concept.

Alternate-forms reliability: A procedure for testing the reliability of responses to survey questions in which subjects' answers are compared after the subjects have been asked slightly different versions of the questions or when randomly selected halves of the sample have been administered slightly different versions of the questions.

Interobserver reliability: When similar measurements are obtained by different observers rating the same persons, events, or places.

Split-halves reliability: Reliability achieved when responses to the same questions by two randomly selected halves of a sample are about the same.

Assessing interobserver reliability is most important when the rating task is complex. Consider a commonly used measure of mental health, the Global Assessment of Functioning Scale (GAFS). The rating task seems straightforward, with clear descriptions of the subject characteristics that are supposed to lead to high or low GAFS scores. But in fact, the judgments that the rater must make while using this scale are very complex. They are affected by a wide range of subject characteristics, attitudes, and behaviors as well as by the rater's reactions. As a result, interobserver agreement is often low on the GAFS, unless the raters are trained carefully.

Can We Achieve Both Reliability and Validity?

The reliability and validity of measures in any study must be tested after the fact to assess the quality of the information obtained. But then, if it turns out that a measure cannot be considered reliable and valid, little can be done to save the study. Hence, it is supremely important to select in the first place measures that are likely to be both reliable and valid. The Dow Jones Industrials Index is a perfectly *reliable* measure of the state of the American economy—any two observers of it will see the same numbers—but its validity is shaky: There's more to the economy than the rise and fall of stock prices. In contrast, a good therapist's interview of a married couple may produce a *valid* understanding of their relationship, but such interviews are often not reliable, because another interviewer could easily reach different conclusions.

Finding measures that are both reliable and valid can be challenging. Don't just choose the first measure you find or can think of. Consider the different strengths of different measures and their appropriateness to your study. Conduct a pretest in which you use the measure with a small sample and check its reliability. Provide careful training to ensure a consistent approach if interviewers or observers will administer the measures. In most cases, however, the best strategy is to use measures that have been used before and whose reliability and validity have been established in other contexts. But even the selection of "tried and true" measures does not absolve researchers from the responsibility of testing the reliability and validity of the measure in their own studies.

Remember that a reliable measure is not necessarily a valid measure, as Exhibit 4.6 illustrates. The discrepancy shown is a common flaw of self-report measures of substance abuse. People's answers to the questions are consistent, but they are consistently misleading: A number of respondents will not admit to drinking, even though they drink a lot. Most respondents answer the multiple questions in self-report indexes of substance abuse in a consistent way, so the indexes are reliable. As a result, some indexes based on self-report are reliable but invalid. Such indexes are not useful and should be improved or discarded.

Research|Social Impact Link 4.2
Read more about reliability and validity.

🔲 Conclusion

Remember always that measurement validity is a necessary foundation for social research. Gathering data without careful conceptualization or conscientious efforts to operationalize key concepts often is a wasted effort.

The difficulties of achieving valid measurement vary with the concept being operationalized and the circumstances of the particular study. The examples in this chapter of difficulties in achieving valid measures should sensitize you to the need for caution.

Planning ahead is the key to achieving valid measurement in your own research; careful evaluation is the key to sound decisions about the validity of measures in others' research. Statistical tests can help to determine

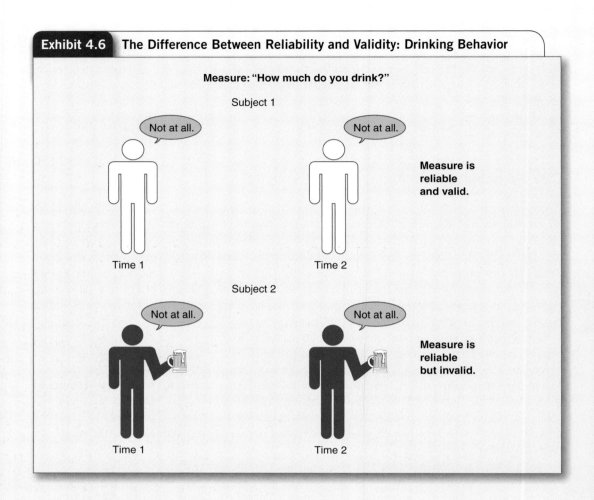

Exhibit 4.6   The Difference Between Reliability and Validity: Drinking Behavior

whether a given measure is valid after data have been collected, but if it appears after the fact that a measure is invalid, little can be done to correct the situation. If you cannot tell how key concepts were operationalized when you read a research report, don't trust the findings. And if a researcher does not indicate the results of tests used to establish the reliability and validity of key measures, remain skeptical.

Key Terms

Highlights

- Conceptualization plays a critical role in research. In deductive research, conceptualization guides the operationalization of specific variables; in inductive research, it guides efforts to make sense of related observations.

- Concepts may refer to either constant or variable phenomena. Concepts that refer to variable phenomena may be very similar to the actual variables used in a study, or they may be much more abstract.

- Concepts are operationalized in research by one or more indicators, or measures, which may derive from observation, self-report, available records or statistics, books and other written documents, clinical indicators, discarded materials, or some combination.

- Indexes and scales measure a concept by combining answers to several questions and so reducing idiosyncratic variation. Several issues should be explored with every intended index: Does each question actually measure the same concept? Does combining items in an index obscure important relationships between individual questions and other variables? Is the index multidimensional?

- If differential weighting, based on differential information captured by questions, is used in the calculation of index scores, then we say that the questions constitute a scale.

- Level of measurement indicates the type of information obtained about a variable and the type of statistics that can be used to describe its variation. The four levels of measurement can be ordered by complexity of the mathematical operations they permit: nominal (or qualitative), ordinal, interval, and ratio (most complex). The measurement level of a variable is determined by how the variable is operationalized.

- The validity of measures should always be tested. There are four basic approaches: face validation, content validation, criterion validation (either predictive or concurrent), and construct validation. Criterion validation provides the strongest evidence of measurement validity, but often there is no criterion to use in validating social science measures.

- Measurement reliability is a prerequisite for measurement validity, although reliable measures are not necessarily valid. Reliability can be assessed through a test-retest procedure, an interitem comparison of responses to alternate forms of the test, or the consistency of findings among observers.

STUDENT STUDY SITE

The Student Study Site, available at **www.sagepub.com/chambliss4e,** includes useful study materials including web exercises with accompanying links, eFlashcards, videos, audio resources, journal articles, and encyclopedia articles, many of which are represented by the media links throughout the text. The site also features Interactive Exercises—represented by the green icon here—to help you understand the concepts in this book

Exercises

Discussing Research

1. What does *trust* mean to you? Identify two examples of "trust in action" and explain how they represent your concept of trust. Now develop a short definition of *trust* (without checking a dictionary). Compare your definition to those of your classmates and what you find in a dictionary. Can you improve your definition based on some feedback?

2. What questions would you ask to measure the level of trust among students? How about feelings of being "in" or "out" with regard to a group? Write five questions for an index and suggest response choices for each. How would you validate this measure using a construct validation approach? Can you think of a criterion validation procedure for your measure?

3. If you were given a questionnaire right now that asked you about your use of alcohol and illicit drugs in the past year, would you disclose the details fully? How do you think others would respond? What if the questionnaire was anonymous? What if there was a confidential ID number on the questionnaire so that the researcher could keep track of who responded? What criterion validation procedure would you suggest for assessing measurement validity?

Finding Research

1. What are some of the research questions you could attempt to answer with available statistical data? Visit your library and ask for an introduction to the government documents collection. Inspect the U.S. Census Bureau website (www.census .gov) and find the population figures broken down by city and state. List five questions that you could explore with such data. Identify six variables implied by these research questions that you could operationalize with the available data. What are three factors that might influence variation in these measures other than the phenomenon of interest? (Hint: Consider how the data are collected.)

2. How would you define *alcoholism?* Write a brief definition. Based on this conceptualization, describe a method of measurement that would be valid for a study of alcoholism (as you define it).

   Now go to the American Council for Drug Education and read some their facts about alcohol (www.acde.org/common/ alcohol2.pdf). Is this information consistent with your definition?

Critiquing Research

1. Shortly before the year 2000 national census of the United States, a heated debate arose in Congress over whether instead of a census—a total headcount—a sample should be used to estimate the number and composition of the U.S. population. As a practical matter, might a sample be more accurate in this case than a census? Why?

2. Develop a plan for evaluating the validity of a measure. Your instructor will give you a copy of a questionnaire actually used in a study. Pick out one question and define the concept that you believe it is intended to measure. Then develop a construct validation strategy involving other measures in the questionnaire that you think should be related to the question of interest—if it measures what you think it measures.

3. The questions in Exhibit 4.7 are selected from a survey of homeless shelter staff (Schutt & Fennell 1992). First, identify the level of measurement for each question. Then rewrite each question so that it measures the same variable but at a different level. For example, you might change a question that measures age at the ratio level, in years, to one that measures age at the ordinal level, in categories. Or you might change a variable measured at the ordinal level to one measured at the ratio level. For the categorical variables, those measured at the nominal level, try to identify at least two underlying quantitative dimensions of variation and write questions to measure variation along these dimensions. For example, you might change a question asking which of several factors the respondent thinks is responsible for homelessness to a series of questions that ask how important each factor is in generating homelessness.

4. What are the advantages and disadvantages of phrasing each question at one level of measurement rather than another? Do you see any limitations on the types of questions for which levels of measurement can be changed?

Doing Research

1. Some people have said in discussions of international politics that "democratic governments don't start wars." How could you test this hypothesis? Clearly state how you would operationalize (1) *democratic* and (2) *start.*

2. Now it's time to try your hand at operationalization with survey-based measures. Formulate a few fixed-choice questions to measure variables pertaining to the concepts you researched for Exercise 1 under "Discussing Research." Arrange to interview

---

Exhibit 4.7   Selected Shelter Staff Survey Questions

1. What is your current job title? _____

2. What is your current employment status?

   Paid, full-time 1
   Paid, part-time (less than 30 hours per week) 2

3. When did you start your current position? _____ / _____ / _____
   　　　　　　　　　　　　　　　　　　　　　　Month 　　　Day 　　　Year

4. In the past month, how often did you help guests deal with each of the following types of problems?
   (Circle one response on each line.)

	Very often						Never
Job training/placement...................	1	2	3	4	5	6	7
Lack of food or bed........................	1	2	3	4	5	6	7
Drinking problems..........................	1	2	3	4	5	6	7

5. How likely is it that you will leave this shelter within the next year?

   Very likely.. 1
   Moderately.. 2
   Not very likely... 3
   Not likely at all.. 4

6. What is the highest grade in school you have completed at this time?

   First through eighth grade... 1
   Some high school.. 2
   High school diploma.. 3
   Some college.. 4
   College degree.. 5
   Some graduate work... 6
   Graduate degree... 7

7. Are you a veteran?

   Yes.. 1
   No.. 2

---

one or two other students with the questions you have developed. Ask one fixed-choice question at a time, record your interviewee's answer, and then probe for additional comments and clarifications. Your goal is to discover what respondents take to be the meaning of the concept you used in the question and what additional issues shape their response to it.

When you have finished the interviews, analyze your experience: Did the interviewees interpret the fixed-choice questions and response choices as you intended? Did you learn about the concepts you were working on? Should your conceptual definition be refined? Should the questions be rewritten, or would more fixed-choice questions be necessary to capture adequately the variation among respondents?

3. Now try index construction. You might begin with some of the questions you wrote for Exercise 2. Write four or five fixed-choice questions that each measure the same concept.

(For instance, you could ask questions to determine whether someone is alienated.) Write each question so it has the same response choices (a matrix design). Now conduct a literature search to identify an index that another researcher used to measure your concept or a similar concept. Compare your index to the published index. Which seems preferable to you? Why?

4. List three attitudinal variables.

   a. Write a conceptual definition for each variable. Whenever possible, this definition should come from the existing literature—either a book you have read for a course or the research literature that you have searched. Ask two class members for feedback on your definitions.

   b. Develop measurement procedures for each variable: Two measures should be single questions, and one should be an index used in prior research (search the Internet and the journal literature in Soc Abstracts or Psych Abstracts). Ask classmates to answer these questions and give you feedback on their clarity.

c. Propose tests of reliability and validity for the measures.

5. Exercise your cleverness on this question: For each of the following, suggest two unobtrusive measures that might help you discover (a) how much of the required reading for this course students actually complete, (b) where are the popular spots to sit in a local park, and (c) which major U.S. cities have the highest local taxes.

Ethics Questions

1. The ethical guidelines for social research require that subjects give their "informed consent" prior to participating in an interview. How "informed" do you think subjects have to be? If you are interviewing people to learn about substance abuse and its impact on other aspects of health, is it okay just to tell respondents in advance that you are conducting a study of health issues? What if you plan to inquire about victimization experiences? Explain your reasoning.

2. Both some Homeland Security practices and inadvertent releases of web searching records have raised new concerns about the use of unobtrusive measures of behavior and attitudes. If all identifying information is removed, do you think social scientists should be able to study the extent of prostitution in different cities by analyzing police records? How about how much alcohol different types of people use by linking credit card records to store purchases?

CHAPTER 5

Sampling

A n old history professor was renowned for his ability, at semester's end, to finish grading large piles of student papers (many of them undistinguished) in a matter of a few short hours. When asked by a younger colleague how he accomplished this feat, the codger replied with a snort. "You don't have to eat the whole tub of butter to know if it's rancid." Harsh, but true.

That is the essence of sampling: A small portion, carefully chosen, can reveal the quality of a much larger whole. A survey of 1,400 Americans telephoned one Saturday afternoon can tell us very accurately how 40 million will vote for president on the following Tuesday morning. A quick check of reports from a few selected banks can tell the Federal Reserve how strong inflation is. And, when you go to the health clinic with a possible case of mononucleosis and a blood test is done, the phlebotomist needn't take all of your blood to see if you have too many atypical lymphocytes. Sampling techniques tell us how to select cases that can lead to valid generalizations about a **population**, or the entire group you wish to learn about. In this chapter, we define the key components of sampling strategy and then present the types of sampling one may use in a research study along with the strengths and weaknesses of each.

> **Population:** The entire set of individuals or other entities to which study findings are to be generalized.

How Do We Prepare to Sample?

Define Sample Components and the Population

To understand how sampling works, you'll first need a few useful definitions. A **sample** is a subset of the population that we want to learn about. For instance, suppose the human resources (HR) offices at a large retail clothing chain wants to understand the career aspirations of their employees. The population would be all current employees of the company. The sample could be, say, 200 individuals whom HR will select to interview. The individual members of this sample are called **elements**, or **elementary units**—that is, the specific people selected. These are the cases that we actually study. To select these elements, we often rely on some list of all elements in the population—**a sampling frame**. In our example, this would be a list of all current employees. In some cases, a sampling frame may be quite difficult to produce: all homeless people in Chicago, all drug users at your universities, or all professional comedians in San Francisco.

> **Sample:** A subset of a population used to study the population as a whole.
>
> **Elements:** The individual members of the population whose characteristics are to be measured.
>
> **Sampling frame:** A list of all elements or other units containing the elements in a population.

It is important to remember that a sample can only represent the population from which it was drawn. So if we sample students in one high school, the population for our study is the student body of that school, not all high school students in the nation. Some populations, such as frequent moviegoers, are not identified by a simple criterion, such as a geographic boundary or an organizational membership. Clear definition of such a population is difficult but quite necessary. Anyone should be able to determine just what population was actually studied, so we would have to define clearly the concept of *frequent moviegoers* and specify how we determined their status.

Oftentimes researchers make fundamental sampling mistakes even before they start examining their data, for instance, by selecting the wrong sampling frame—one that does not adequately represent the population. Perhaps the most common version of this error is called *sampling on the dependent variable,* in which cases are chosen not to represent the population but because they represent a (usually) interesting outcome—that is, only one value of the dependent variable. Even the best social scientists sometimes fall into this trap. In their fascinating and important book *Rampage: The Social Roots of School Shootings,* Katherine S. Newman and her coauthors studied in detail the case histories of 27 different teenagers who had gone into their schools and killed (mostly random) fellow students—the Columbine attack of April 20, 1999, may be the most famous case, where the shooters killed 13 and wounded 21 others, then killed themselves (Newman, Fox, Harding, Mehta, & Roth 2004). Based on their study of shooters, Newman and colleagues concluded that there were five "necessary but not sufficient" factors in school shootings: (1) a self-perception of shooters as socially marginal, (2) psychosocial problems, (3) cultured scripts linking masculinity and violence, (4) failure of surveillance systems (so troubled kids are "under the radar"), and (5) availability of guns. Virtually all school shooters fit this description; they have all these characteristics. *Rampage* is a valuable piece of serious exploratory social science.

But this model still does not explain shootings, or even tell us much about who will commit them. The fact is, all of the shooters were also boys, they were all teenagers, and they all attended high school. Were these also important factors in explaining their participation in the school shootings? And, were there other students who perceived themselves as socially marginal, or who had psychosocial problems, and so on? Why didn't these other students turn into school shooters? The problem, in other words, is that Newman and her

Video Link 5.1
Watch a guide to sampling.

colleagues (2004) only looked at shooters, instead of comparing shooters with nonshooters to see what made the difference. Their sampling frame (a list of school shooters) allowed them to generalize to other school shooters, but not to tell you how shooters differ from other teenagers.

Sometimes our sources of information are not actually the elements in our study. For example, for a survey about educational practices, a researcher might first sample schools and then, within sampled schools, interview a sample of teachers. The schools and the teachers are both termed **sampling units**, because the researcher sampled from both (Levy & Lemeshow 1999:22). The schools are selected in the first stage of the sample, so they are the *primary sampling units* (and in this case, the elements in the study). The teachers are *secondary sampling units* (but they are not elements, because they are used to provide information about the entire school) (Exhibit 5.1).

> **Sampling units:** Units listed at each stage of a multistage sampling design.

Exhibit 5.1   Sample Components in a Two-Stage Study

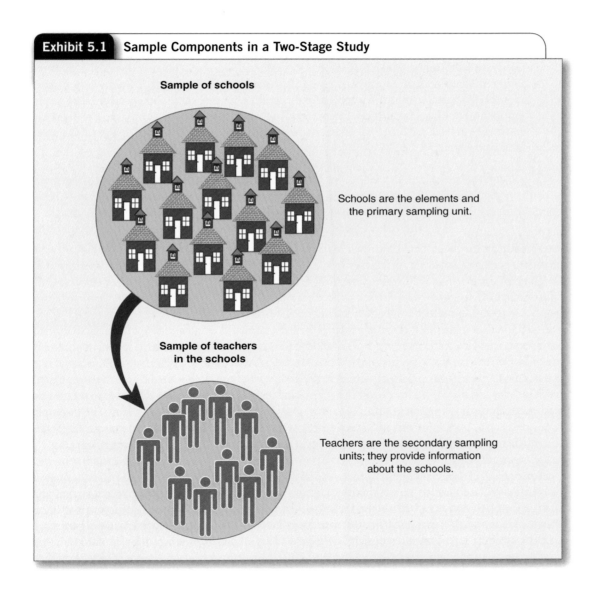

Sample of schools

Schools are the elements and the primary sampling unit.

Sample of teachers in the schools

Teachers are the secondary sampling units; they provide information about the schools.

Evaluate Generalizability

Once we have defined clearly the population from which we will sample, we need to determine the scope of the generalizations we will seek to make from our sample. Do you recall the two different meanings of *generalizability* from Chapter 1?

Journal Link 5.1
Read an article that takes multiple individual countries' data to create a generalizable finding.

- *Can the findings from a sample of the population be generalized to the population from which the sample was selected?* This issue was defined in Chapter 1. Again, when the Gallup polls ask some Americans for their political opinions, can those answers be generalized to the U.S. population? Probably so. But if Gallup's sampling was haphazard—say, if the pollsters just talked to some people in the office—they probably couldn't make the same accurate generalizations.

- *Can the findings from a study of one population be generalized to another, somewhat different population?* Are residents of three impoverished communities in the city of Enschede, the Netherlands, similar to those in other communities? In other cities? In other nations? The problem here was defined in Chapter 1 as *cross-population generalizability.* For example, many psychology studies are run using (easily available) college students as subjects. Because such research is often on tasks that require no advanced education, such as memorizing lists of nonsense syllables or spotting patterns in an array of dots, college students may in this respect be like most other human beings, so the generalization seems legitimate. But when psychoanalyst Sigmund Freud talked with a very narrow sample of Viennese housewives in 1900, could his findings be accurately generalized (as he attempted) to the entire human race? Probably not.

This chapter focuses attention primarily on the problem of sample generalizability: Can findings from a sample be generalized to the population from which the sample was drawn? This is really the most basic question to ask about a sample, and social research methods provide many tools with which to address it.

Target population: A set of elements larger than or different from the population sampled and to which the researcher would like to generalize study findings.

But researchers often project their theories onto groups or populations much larger than, or simply different from, those they have actually studied. The population to which generalizations are made in this way can be termed the **target population**—a set of elements larger than or different from the population that was sampled and to which the researcher would like to generalize any study findings. Because the validity of cross-population generalizations cannot be tested empirically, except by conducting more research in other settings, we will not focus much attention on this problem here.

Assess the Diversity of the Population

Sampling is unnecessary if all the units in the population are identical. The blood in one person is constantly being mixed and stirred, so it's very homogeneous—any pint is the same as any other. Nuclear physicists don't need a representative sample of all atomic particles to learn about basic atomic processes, because in crucial respects all such particles are alike.

Journal Link 5.2
Read an article and assess generalizability based on its sample of 19–24 year olds.

What about people? Certainly all people are not identical, but if we are studying fundamental physical or psychological processes that are the same among all people, sampling is not needed to achieve generalizable findings. Psychologists and social psychologists often conduct experiments on college students to learn about such processes (basic cognitive functioning, for instance). But we must always bear in mind that we don't really know how generalizable our findings are to populations that we haven't actually studied.

So, we usually must study the larger population in which we are interested if we want to make generalizations about it. For this purpose, we must obtain a **representative sample** of the population to which generalizations are sought (Exhibit 5.2).

Consider a Census

In some circumstances, it may be feasible to establish generalizability by simply conducting a **census**—studying an entire population—rather than drawing a sample. This is what the federal government tries to do every 10 years with the U.S. Census. Censuses could also include, for instance, studies of all the employees in a small business, studies comparing all 50 states, or studies of all the museums in some region.

> **Representative sample:** A sample that "looks like" the population from which it was selected in all respects that are potentially relevant to the study. The distribution of characteristics among the elements of a representative sample is the same as the distribution of those characteristics among the total population. In an unrepresentative sample, some characteristics are overrepresented or underrepresented.

Exhibit 5.2   Representative and Unrepresentative Samples

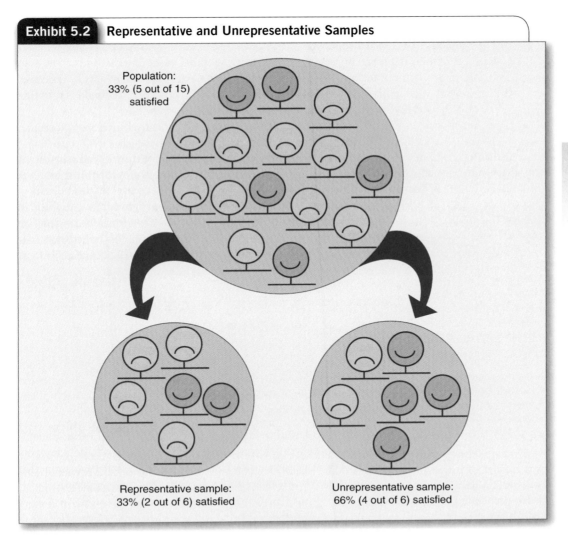

Population:
33% (5 out of 15) satisfied

Representative sample:
33% (2 out of 6) satisfied

Unrepresentative sample:
66% (4 out of 6) satisfied

Video Link 5.2
Watch a clip about samples.

Census: Research in which information is obtained through responses from or information about all available members of an entire population.

Social scientists don't often attempt to collect data from all the members of some large population because doing so would be too expensive and time consuming. The 2010 U.S. Census, for example, is estimated to have cost around $15 billion, or about $48 per person counted. But fortunately, a well-designed sampling strategy can result in a representative sample of the same population at far less cost.

▣ What Sampling Method Should We Use?

Probability sampling method: A sampling method that relies on a random, or chance, selection method so that the probability of selection of population elements is known.

Nonprobability sampling methods: Sampling methods in which the probability of selection of population elements is unknown.

Probability of selection: The likelihood that an element will be selected from the population for inclusion in the sample. In a census of all the elements of a population, the probability that any particular element will be selected is 1.0. If half the elements in the population are sampled on the basis of chance (say, by tossing a coin), the probability of selection for each element is one half, or 0.5. As the size of the sample as a proportion of the population decreases, so does the probability of selection.

Random sampling: A method of sampling that relies on a random, or chance, selection method so that every element of the sampling frame has a known probability of being selected.

Certain features of samples make them more or less likely to represent the population from which they are selected; the more representative the sample, the better. The crucial distinction about samples is whether they are based on a probability or a nonprobability sampling method. **Probability sampling methods** allow us to know in advance how likely it is that any element of a population will be selected. Sampling methods that do not let us know in advance the likelihood of selecting each element are termed **nonprobability sampling methods**.

Probability sampling methods rely on a random, or chance, selection procedure, which is in principle the same as flipping a coin to decide which of two people "wins" and which one "loses." Heads and tails are equally likely to turn up in a coin toss, so both persons have an equal chance to win. That chance, their **probability of selection**, is 1 out of 2, or 0.5.

There is a natural tendency to confuse the concept of **random sampling**, in which cases are selected only on the basis of chance, with *haphazard* sampling. On first impression, "leaving things up to chance" seems to imply not exerting any control over the sampling method. But to achieve true randomness, the researcher must in fact proceed very methodically, following careful procedures.

Two common problems can bias even what appear to be random samples:

1. If the sampling frame is incomplete, a random sample from that list will not really be a random sample of the population. You should always consider the adequacy of the sampling frame. Even for a fairly small population like a university's student body, the registrar's list is likely to be at least somewhat out-of-date at any given time—and the missing students are probably different from those in the list.

2. Nonresponse is a major hazard especially in survey research, because nonrespondents are likely to differ systematically from those who take the time to participate. If the response rate is low (say, below 65%), you should not assume that findings from even a random sample will be generalizable to the population.

Probability Sampling Methods

Introduced earlier, probability sampling methods are those in which the probability of selection is known and is not zero (so there is some chance of selecting each element). These methods randomly select elements and therefore have no systematic **bias**; nothing but chance determines which elements are included in the sample. This feature of probability samples makes them much more desirable than nonprobability samples when the goal is to generalize to a larger population.

However, even a randomly selected sample will have some degree of sampling error—some deviation from the characteristics of the population. In general, both the size of the sample and the homogeneity (sameness) of the population affect the degree of error due to chance. Interestingly, the *proportion* of the population represented by the sample (10%, 20%, etc.) does not affect its representativeness, unless that proportion is very large; it is the raw number of cases in the sample that is important. To elaborate,

> **Bias:** Sampling bias occurs when some population characteristics are over- or underrepresented in the sample because of particular features of the method of selecting the sample.

- *The larger the sample, the more confidence we can have in the sample's representativeness.* If we randomly pick 5 people to represent the entire population of our city, our sample is unlikely to be very representative of the entire population in terms of age, gender, race, attitudes, and so on. But if we randomly pick 100 people, the odds of having a representative sample are much better; with a random sample of 1,000, the odds become very good indeed.

- *The more homogeneous the population, the more confidence we can have in the representativeness of a sample of any particular size.* That's why blood testing works—blood is homogeneous in any specific individual's body. Or, let's say we plan to draw samples of 50 people from each of two communities to estimate mean family income. One community is very diverse, with family incomes varying from $12,000 to $85,000. In the other, more homogeneous community, family incomes are concentrated in a narrow range, from $41,000 to $64,000. The estimated mean family income based on the sample from the homogeneous community is more likely to be representative than is the estimate based on the sample from the more heterogeneous community. With less variation to represent, fewer cases are needed to represent the homogeneous community.

Encyclopedia Link 5.1
Read an overview
of probability sampling.

The fraction of the total population contained in a sample does not affect the sample's representativeness, unless that fraction is large. This isn't obvious, but it is mathematically true. The raw number of cases matters more than the proportion of the population. Other things being equal, a sample of 1,000 from a population of 1 million (with a sampling fraction of 0.001, or 0.1%) is much better than a sample of 100 from a population of 10,000 (although the sampling fraction in this case is 0.01, or 1%, which is 10 times higher). The larger size of the sample makes representativeness more likely, not the proportion of the whole that the sample represents. We can regard any sampling fraction under 2% with about the same degree of confidence (Sudman 1976:184). In fact, sample representativeness is not likely to increase much until the sampling fraction is quite a bit higher.

Journal Link 5.3
Read about survey data
collected through
probability sampling.

Research in the News

SAMPLE POLLS INDICATE LATINO TURNOUT LIKELY TO LAG

The debate over Arizona's tough law against illegal immigrants seems to have turned off many Latinos, rather than leaving them energized. According to a nationwide phone poll of 1,357 Latinos, including 618 registered voters, only 32% of the registered voters indicated they were likely to vote, compared to 50% of all registered voters. Conducted August 17 to September 19, 2010, the survey had a margin of sampling error of plus or minus 5 percentage points for registered voters.

Source: Lacey, Marc. 2010. Latino turnout likely to lag, new poll finds. *The New York Times,* October 6:A1.

Polls to predict presidential election outcomes illustrate both the value of random sampling and the problems that it cannot overcome. In most presidential elections, pollsters have predicted fairly accurately the outcomes of the actual votes by using random sampling and, these days, phone interviewing to learn for whom likely voters intend to vote. Exhibit 5.3 shows how accurate these sample-based predictions have been in the last 11 presidential elections. The exceptions were the 1980 and 1992 elections, when third-party candidates had a surprising effect. Otherwise, the small discrepancies between the votes predicted through random sampling and the actual votes can be attributed to random error.

The Gallup poll did quite well in predicting the result of the remarkable 2000 presidential election. The final Gallup prediction was that George W. Bush would win with 48% (Al Gore was predicted to receive only 46%, and Green Party candidate Ralph Nader was predicted to secure 4%). The race turned out to be much closer, with Gore actually winning the popular vote (before losing in the electoral college). In fairness, Gallup had spotted a late-breaking trend in favor of Gore (Newport 2000).

But even well-run election polls have produced some major errors in prediction. In 1948, pollsters mistakenly predicted that Thomas E. Dewey would beat Harry S. Truman, based on the random sampling method that George Gallup had used successfully since 1934. The problem? Pollsters stopped collecting data several weeks before the election, and in those weeks, many people changed their minds (Kenney 1987). So just as in the 2000 election, underrepresenting shifts in voter sentiment just before the election systematically biased the sample.

Now that we have sung the praises of probability-based samples in general, we need to introduce the different types of random samples. The four most common types of random sample are (1) simple random sampling, (2) systematic random sampling, (3) cluster sampling, and (4) stratified random sampling.

Simple Random Sampling

Simple random sampling identifies cases strictly on the basis of chance. Both flipping a coin and rolling a die can be used to identify cases strictly on the basis of chance, but these procedures are not very efficient tools for drawing a sample. A **random number table** simplifies the process considerably. The researcher numbers all the elements in the sampling frame and then uses a systematic procedure for picking corresponding numbers

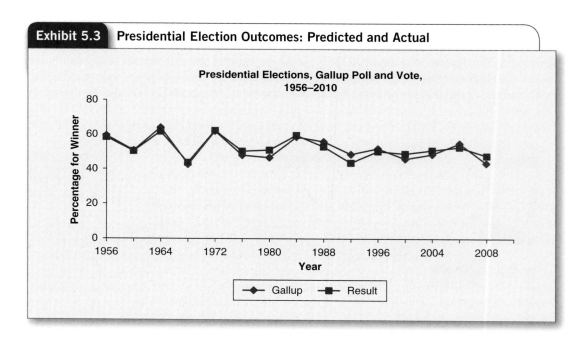

Exhibit 5.3 Presidential Election Outcomes: Predicted and Actual

from the random number table. (Exercise 1 under "Doing Research" at the end of this chapter explains the process step-by-step.) Alternatively, a researcher may use a lottery procedure. Each case number is written on a small card, and then the cards are mixed up and the sample selected from the cards. A computer program can also easily generate a random sample of any size.

Phone surveys often use a technique called **random digit dialing (RDD)** to draw a random sample. A machine dials random numbers within the phone prefixes corresponding to the area in which the survey is to be conducted. Random digit dialing is particularly useful when a sampling frame (list of elements) is unavailable, because the dialing machine can just skip ahead if a phone number is not in service.

In a true simple random sample, the probability of selection is equal for each element. If a sample of 500 is selected from a population of 17,000 (that is, a sampling frame of 17,000), then the probability of selection for each element is 500/17,000, or 0.03. Every element has an equal chance of being selected, just like the odds in a toss of a coin (1/2) or a roll of a die (1/6). Thus, simple random sampling is an *equal probability of selection method (EPSEM)*.

> **Simple random sampling:** A method of sampling in which every sample element is selected purely on the basis of chance, through a random process.
>
> **Random number table:** A table containing lists of numbers that are ordered solely on the basis of chance; it is used for drawing a random sample.
>
> **Random digit dialing (RDD):** The random dialing, by a machine, of numbers within designated phone prefixes, which creates a random sample for phone surveys.

Systematic Random Sampling

Systematic random sampling is an easy to use, efficient variant of simple random sampling. In this method, the first element is selected randomly from a list or from sequential files, and then every *n*th element is selected—for instance, every 7th name on an alphabetical list. This is a convenient method for drawing a random sample when the population elements are arranged sequentially. It is particularly efficient when the elements are not written down (that is, there is no written sampling frame) but instead are represented physically, say by folders in filing cabinets.

In almost all sampling situations, systematic random sampling yields what is essentially a simple random sample. The exception is a situation in which the sequence of elements is characterized by **periodicity**—that is, the sequence varies in some regular, periodic pattern. For example, in a new housing development with the same number of houses on each block (8, for example), houses may be listed by block, starting with the house in the northwest corner of each block and continuing clockwise. If the **sampling interval** is 8, the same as the periodic pattern, all the cases selected will be in the same position (Exhibit 5.4). Those houses may well be unusual—corner locations are typically more expensive, for instance. But usually, periodicity and the sampling interval are rarely the same, so this isn't a problem.

> **Systematic random sampling:** A method of sampling in which sample elements are selected from a list or from sequential files, with every *n*th element being selected after the first element is selected randomly.

> **Periodicity:** A sequence of elements (in a list to be sampled) that varies in some regular, periodic pattern.
>
> **Sampling interval:** The number of cases between one sampled case and another in a systematic random sample.

Cluster Sampling

Cluster sampling is useful when a sampling frame—a definite list—of elements is not available, as often is the case for large populations spread out across a wide geographic area or among many different organizations. We don't have a good list of all the Catholics in America, all the businesspeople in Arizona, or all the waiters in New York. A **cluster** is a naturally occurring, mixed aggregate of elements of the population, with each element (person, for instance) appearing in one and only one cluster. Schools could serve as clusters for sampling students, city blocks could serve as clusters for sampling residents, counties could serve as clusters for sampling the general population, and restaurants could serve as clusters for sampling waiters.

> **Cluster sampling:** Sampling in which elements are selected in two or more stages, with the first stage being the random selection of naturally occurring clusters and the last stage being the random selection of elements within clusters.
>
> **Cluster:** A naturally occurring, mixed aggregate of elements of the population.

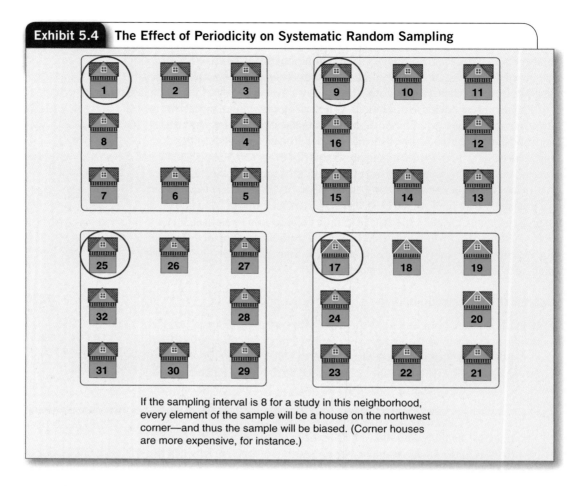

Exhibit 5.4   The Effect of Periodicity on Systematic Random Sampling

If the sampling interval is 8 for a study in this neighborhood, every element of the sample will be a house on the northwest corner—and thus the sample will be biased. (Corner houses are more expensive, for instance.)

Research|Social Impact
Link 5.1
Read more about cluster sampling methods.

Cluster sampling is at least a two-stage procedure. First, the researcher draws a random sample of clusters. (A list of clusters should be much easier to obtain than a list of all the individuals in each cluster in the population.) Next, the researcher draws a random sample of elements within each selected cluster. Because only a fraction of the total clusters is involved, obtaining the sampling frame at this stage should be much easier.

Cluster samples often involve multiple stages, with clusters within clusters, as when a national study of middle school students might involve first sampling states, then counties, then schools, and finally students within each selected school (Exhibit 5.5).

How many clusters and how many individuals within clusters should be selected? As a general rule, the more clusters you select, with the fewest individuals in each, the more representative your sampling will be. Unfortunately, this strategy also maximizes the cost of the sample. The more clusters selected, the higher the travel costs. Remember, too, that the more internally homogeneous the clusters, the fewer cases needed per cluster. Homogeneity within a cluster is good.

Cluster sampling is a very popular method among survey researchers, but it has one general drawback: Sampling error is greater in a cluster sample than in a simple random sample, because there are two steps involving random selection rather than just one. This sampling error increases as the number of clusters decreases, and it decreases as the homogeneity of cases per cluster increases. This is another way of restating the preceding points: It's better to include as many clusters as possible in a sample, and it's more likely that a cluster sample will be representative of the population if cases are relatively similar within clusters.

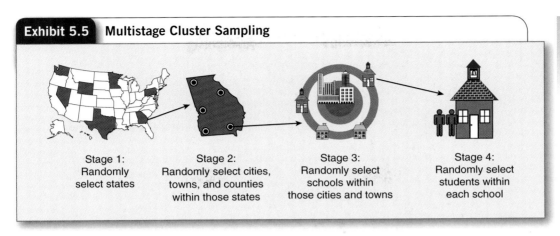

Exhibit 5.5 Multistage Cluster Sampling

Stage 1:
Randomly
select states

Stage 2:
Randomly select cities,
towns, and counties
within those states

Stage 3:
Randomly select
schools within
those cities and towns

Stage 4:
Randomly select
students within
each school

Audio Link 5.1
Listen to how sampling
methods impact research.

Stratified Random Sampling

Suppose you want to survey soldiers of an army to determine their morale. Simple random sampling would produce large numbers of enlisted personnel—that is, of lower ranks—but very few, if any, generals. But you want generals in your sample. **Stratified random sampling** ensures that various groups will be included.

First, all elements in the population (that is, in the sampling frame) are distinguished according to their value on some relevant characteristic (army rank, for instance: generals, captains, privates, etc.). That characteristic determines the sampling strata. Next, elements are sampled randomly from within these strata: so many generals, so many captains, and so on. Of course, to use this method, more information is required prior to sampling than is the case with simple random sampling. Each element must belong to one and only one stratum.

For *proportionate to size* sampling, the size of each stratum in the population must be known. This method efficiently draws an appropriate representation of elements across strata. Imagine that you plan to draw a sample of 500 from an ethnically diverse neighborhood. The neighborhood population is 15% black, 10% Hispanic, 5% Asian, and 70% white. If you drew a simple random sample, you might end up with somewhat disproportionate numbers of each group. But if you created sampling strata based on race and ethnicity, you could randomly select cases from each stratum in exactly the same proportions. This is termed **proportionate stratified sampling**, and it eliminates any possibility of sampling error in the sample's distribution of ethnicity. Each stratum would be represented exactly in proportion to its size in the population from which the sample was drawn (Exhibit 5.6).

In **disproportionate stratified sampling**, the proportion of each stratum that is included in the sample is intentionally varied from what it is in the population. In the case of the sample stratified by ethnicity, you might select equal numbers of cases from each racial or ethnic group: 125 blacks (25% of the sample), 125 Hispanics (25%), 125 Asians (25%), and 125 whites (25%). In this type of sample, the probability of selection of every case is known but unequal between strata. You know what the proportions are in the population, so you can easily adjust your combined sample statistics to reflect these true proportions. For instance, if you want to combine the ethnic groups and estimate the average income of the total population, you would have to weight each case in the sample to reflect its representation in the population.

> **Stratified random sampling:** A method of sampling in which sample elements are selected separately from population strata that the researcher identifies in advance.

> **Proportionate stratified sampling:** Sampling method in which elements are selected from strata in exact proportion to their representation in the population.

> **Disproportionate stratified sampling:** Sampling in which elements are selected from strata in proportions different from those that appear in the population.

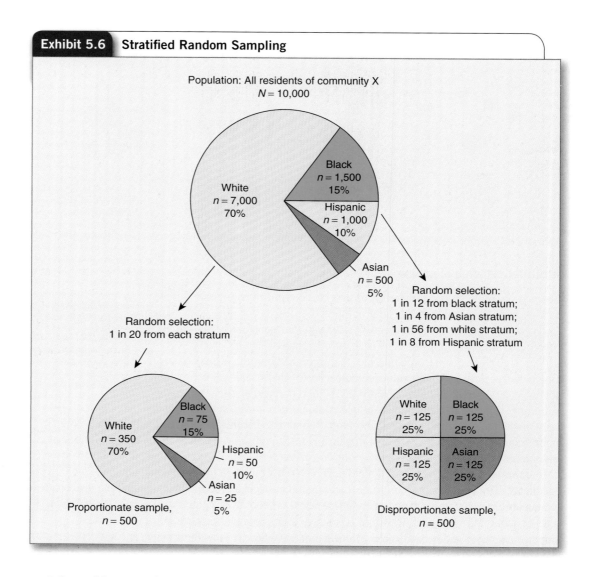

Exhibit 5.6 Stratified Random Sampling

Population: All residents of community X
N = 10,000

White
n = 7,000
70%

Black
n = 1,500
15%

Hispanic
n = 1,000
10%

Asian
n = 500
5%

Random selection:
1 in 20 from each stratum

Random selection:
1 in 12 from black stratum;
1 in 4 from Asian stratum;
1 in 56 from white stratum;
1 in 8 from Hispanic stratum

White
n = 350
70%

Black
n = 75
15%

Hispanic
n = 50
10%

Asian
n = 25
5%

Proportionate sample,
n = 500

White
n = 125
25%

Black
n = 125
25%

Hispanic
n = 125
25%

Asian
n = 125
25%

Disproportionate sample,
n = 500

   Why would anyone select a sample that is so unrepresentative in the first place? The most common reason is to ensure that cases from smaller strata are included in the sample in sufficient numbers to allow separate statistical estimates and to facilitate comparisons between strata. Remember that one of the determinants of sample quality is sample size. The same is true for subgroups within samples. If a key concern in a research project is to describe and compare the incomes of people from different racial and ethnic groups, then it is important that the researchers base the mean income of each group on enough cases to be a valid representation. If few members of a particular minority group are in the population, they need to be oversampled.

Nonprobability Sampling Methods

Nonprobability sampling methods are often used in qualitative research; they also are used in quantitative studies when researchers are unable to use probability selection methods. There are four common nonprobability sampling methods: (1) availability sampling, (2) quota sampling, (3) purposive sampling, and

(4) snowball sampling. Because they do not use a random selection procedure, we cannot expect a sample selected with any of these methods to yield a representative sample. Nonetheless, these methods are useful when random sampling is not possible; with a research question that calls for an intensive investigation of a small population; or for a preliminary, exploratory study.

Availability Sampling

Elements are selected for **availability sampling** (sometimes called *convenience sampling*) because they're available or easy to find. For example, sometimes people stand outside stores in a shopping mall asking passersby to answer a few questions about their shopping habits. That may make sense, but asking the same people for their views on the economy doesn't. In certain respects, regular mall shoppers are not representative people.

> **Availability sampling:** Sampling in which elements are selected on the basis of convenience.

An availability sample is often appropriate at key points in social research—for example, when a field researcher explores a new setting and tries to get some sense of prevailing attitudes or when a survey researcher conducts a preliminary test of a new set of questions. Intensive qualitative research efforts also often rely on availability samples. Howard Becker's (1963) classic work on jazz musicians, for instance, was based on groups in which Becker himself played.

Availability sampling often masquerades as a more rigorous form of research. Popular magazines periodically survey their readers by printing a questionnaire for readers to fill out and mail in. For many years, *Playboy* magazine has conducted a sex survey among its readers using this technique. But usually only a small fraction of readers return the questionnaire, and these respondents might—how to say it?—have more interesting sex lives than other readers of *Playboy*, not to mention the rest of us (or so they claim).

Quota Sampling

Quota sampling is intended to overcome the most obvious flaw of availability sampling—that the sample will just consist of whoever or whatever is available, whether or not it represents the population. In this approach, quotas are set to ensure that the sample represents certain characteristics in proportion to their prevalence in the population.

> **Quota sampling:** A nonprobability sampling method in which elements are selected to ensure that the sample represents certain characteristics in proportion to their prevalence in the population.

Suppose that you want to sample 500 adult residents of a town. You know from the town's annual report what the proportions of town residents are in terms of gender, employment status, and age. To draw a quota sample of a certain size, you then specify that interviews must be conducted with 500 residents who match the town population in terms of gender, employment status, and age.

The problem is that even when we know that a quota sample is representative of the particular characteristics for which quotas have been set, we have no way of knowing if the sample is representative in terms of any other characteristics. In Exhibit 5.7, for example, quotas have been set for gender only. Under the circumstances, it's no surprise that the sample is representative of the population only in terms of gender, not in terms of race.

Of course, you must know the relevant characteristics of the entire population to set the right quotas. In most cases, researchers know what the population looks like in terms of no more than a few of the characteristics relevant to their concerns. And in some cases, they have no such information on the entire population.

If you're now feeling skeptical of quota sampling, you've gotten the drift of our remarks. Nonetheless, in situations in which you can't draw a random sample, it may be better to establish quotas than to have no parameters at all.

> **Purposive sampling:** A nonprobability sampling method in which elements are selected for a purpose, usually because of their unique position.

Purposive Sampling

In **purposive sampling**, each sample element is selected for a purpose, usually because of the unique position of the sample elements. Purposive sampling may

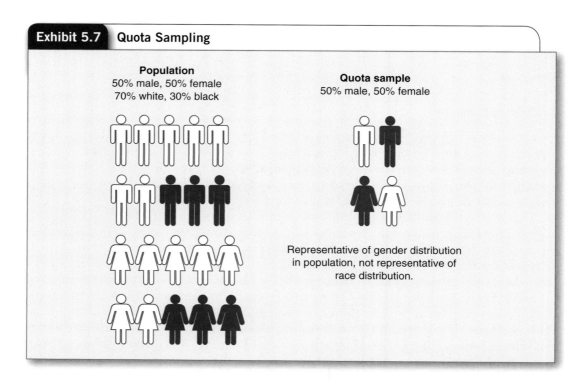

Exhibit 5.7 Quota Sampling

Population
50% male, 50% female
70% white, 30% black

Quota sample
50% male, 50% female

Representative of gender distribution in population, not representative of race distribution.

involve studying the entire population of some limited group (directors of shelters for homeless adults) or a subset of a population (midlevel managers with a reputation for efficiency). Or, a purposive sample may be a *key informant survey,* which targets individuals who are particularly knowledgeable about the issues under investigation.

Herbert Rubin and Irene Rubin (1995) suggest three guidelines for selecting informants when designing any purposive sampling strategy. Informants should be

1. knowledgeable about the cultural arena or situation or experience being studied

2. willing to talk

3. representative of the range of points of view (p. 66)

In addition, Rubin and Rubin (1995) suggest continuing to select interviewees until you can pass two tests:

1. Completeness—"What you hear provides an overall sense of the meaning of a concept, theme, or process."

2. Saturation—"You gain confidence that you are learning little that is new from subsequent interview[s]." (pp. 72–73)

Adhering to these guidelines will help to ensure that a purposive sample adequately represents the setting or issues studied.

Of course, purposive sampling does not produce a sample that represents some larger population, but it can be exactly what is needed in a case study of an organization, community, or some other clearly defined and relatively limited group.

Snowball Sampling

For **snowball sampling**, you identify and speak to one member of the population, and then ask that person to identify others in the population and speak to them, then ask them to identify others, and so on. The sample thus "snowballs" in size. This technique is useful for hard-to-reach or hard-to-identify, interconnected populations (at least some members of the population know each other). An example of a study using snowball sampling is Patricia Adler's (1993) study of Southern California drug dealers. Wealthy philanthropists, top business executives, or Olympic athletes, all of who may have reason to refuse a "cold call" from an unknown researcher, might be sampled effectively using the snowball technique. However, researchers using snowball sampling normally cannot be confident that their sample represents the total population of interest, so generalizations must be tentative.

> **Snowball sampling:** A method of sampling in which sample elements are selected as successive informants or interviewees identify them.

🔳 Conclusion

Sampling is a powerful tool for social science research. Probability sampling methods allow a researcher to use the laws of chance, or probability, to draw samples from which population parameters can be estimated with a high degree of confidence. A sample of just 1,000 or 1,500 individuals can be used to estimate reliably the characteristics of the population of a nation comprising millions of individuals.

But researchers do not come by representative samples easily. Well-designed samples require careful planning, some advance knowledge about the population to be sampled, and adherence to systematic selection procedures—all so that the selection procedures are not biased. And even after the sample data are collected, the researcher's ability to generalize from the sample findings to the population is not completely certain.

The alternatives to random, or probability-based, sampling methods are almost always much less palatable for quantitative studies, even though they are typically much cheaper. Without a method of selecting cases likely to represent the population in which the researcher is interested, research findings must be carefully qualified. Qualitative researchers whose goal is to understand a small group or setting in depth may necessarily have to use unrepresentative samples, but they must keep in mind that the generalizability of their findings will not be known. Additional procedures for sampling in qualitative studies will be introduced in Chapter 9.

Social scientists often seek to generalize their conclusions from the population that they studied to some larger target population. Careful design of appropriate sampling strategies makes such generalizations possible.

Key Terms

Availability sampling   97
Bias   91
Census   89
Cluster   93
Cluster sampling   93
Disproportionate stratified
    sampling   95
Elements   86
Nonprobability
    sampling method   90

Periodicity   93
Population   85
Probability of selection   90
Probability sampling method   90
Proportionate stratified sampling   95
Purposive sampling   97
Quota sampling   97
Random digit dialing (RDD)   93
Random number table   92
Random sampling   90

Representative sample   89
Sample   86
Sampling frame   86
Sampling interval   93
Sampling units   87
Simple random sampling   92
Snowball sampling   99
Stratified random sampling   95
Systematic random sampling   93
Target population   88

Highlights

- Sampling theory focuses on the generalizability of descriptive findings to the population from which the sample was drawn. It also considers whether statements can be generalized from one population to another.

- Sampling is unnecessary when the elements that would be sampled are identical, but the complexity of the social world often makes it difficult to argue that different elements are identical. Conducting a complete census of a population also eliminates the need for sampling, but the resources required for a complete census of a large population are usually prohibitive.

- Nonresponse undermines sample quality: The obtained sample, not the desired sample, determines sample quality.

- Probability sampling methods rely on a random selection procedure to ensure no systematic bias in the selection of elements. In a probability sample, the odds of selecting elements are known, and the method of selection is carefully controlled.

- A sampling frame (a list of elements in the population) is required in most probability sampling methods. The adequacy of the sampling frame is an important determinant of sample quality.

- Simple random sampling and systematic random sampling are equivalent probability sampling methods in most situations. However, systematic random sampling is inappropriate for sampling from lists of elements that have a regular, periodic structure.

- Stratified random sampling uses prior information about a population to make sampling more efficient. Stratified sampling may be either proportionate or disproportionate. Disproportionate stratified sampling is useful when a research question focuses on a stratum or on strata that make up a small proportion of the population.

- Cluster sampling is less efficient than simple random sampling but is useful when a sampling frame is unavailable. It is also useful for large populations spread out across a wide area or among many organizations.

- Nonprobability sampling methods can be useful when random sampling is not possible, when a research question does not concern a larger population, and when a preliminary exploratory study is appropriate. However, the representativeness of nonprobability samples cannot be determined.

STUDENT STUDY SITE

The Student Study Site, available at **www.sagepub.com/chambliss4e,** includes useful study materials including web exercises with accompanying links, eFlashcards, videos, audio resources, journal articles, and encyclopedia articles, many of which are represented by the media links throughout the text. The site also features Interactive Exercises—represented by the green icon here—to help you understand the concepts in this book.

Exercises

Discussing Research

1. When (if ever) is it reasonable to assume that a sample is not needed because "everyone is the same"—that is, the population is homogeneous? Does this apply to research such as Stanley Milgram's on obedience to authority? What about investigations of student substance abuse? How about investigations of how people (or their bodies) react to alcohol? What about research on likelihood of voting (the focus of Chapter 8)?

2. All adult U.S. citizens are required to participate in the decennial census, but some do not. Some social scientists have argued for putting more resources into getting a large representative sample so that census takers can secure higher rates of response from hard-to-include groups. Do you think that the U.S. Census should shift to a probability-based sampling design? Why or why not?

3. What increases sampling error in probability-based sampling designs? Stratified rather than simple random sampling? Disproportionate (rather than proportionate) stratified random sampling? Stratified rather than cluster random sampling?

Why do researchers select disproportionate (rather than proportionate) stratified samples? Why do they select cluster rather than simple random samples?

4. What are the advantages and disadvantages of probability-based sampling designs compared with nonprobability-based designs? Could any of the researches that are described in this chapter with a nonprobability-based design have been conducted instead with a probability-based design? What difficulties might have been encountered in an attempt to use random selection? How would you discuss the degree of confidence you can place in the results obtained from research using a nonprobability-based sampling design?

Finding Research

1. Locate one or more newspaper articles reporting the results of an opinion poll. What information does the article provide on the sample that was selected? What additional information do you need to determine whether the sample was a representative one?

2. From professional journals, select five articles that describe research using a sample drawn from some population. Identify the type of sample used in each study and note any strong and weak points in how the sample was actually drawn. Did the researchers have a problem due to nonresponse? Considering the sample, how confident are you in the validity of generalizations about the population based on the sample? Do you need any additional information to evaluate the sample? Do you think a different sampling strategy would have been preferable? To what larger population were the findings generalized? Do you think these generalizations were warranted? Why or why not?

3. Research on time use has been flourishing all over the world in recent years. Search the web for sites that include the words *time use* and see what you find. Choose one site and write a paragraph about what you learned from it.

4. Check out the "People and Households" section of the U.S. Census Bureau website (www.census.gov). Based on some of the data you find there, write a brief summary of some aspect of the current characteristics of the American population.

Critiquing Research

1. Shere Hite's popular book *Women and Love* (1987) is a good example of the claims that are often made based on an availability sample. In this case, however, the sample didn't necessarily appear to be an availability sample because it consisted of so many people. Hite distributed 100,000 questionnaires to church groups and many other organizations and received back 4.5%; 4,500 women took the time to answer some or all of her 127 essay questions regarding love and sex. Is Hite's sample likely to represent American women in general?

Why or why not? You might take a look at the book's empirical generalizations and consider whether they are justified.

2. In newspapers or magazines, find three examples of poor sampling, where someone's conclusions—either in formal research or in everyday reasoning—are weakened by the selection of cases the author has looked at. How is the author's sampling flawed, and how might that systematically distort the findings? Don't just say "the cases might not be typical"—try to guess, for instance, the direction of error. For example, did the person pick unusually friendly or accessible people? The most well-known examples? And how might their approach affect the findings?

Doing Research

1. Select a random sample using a table of random numbers (either one provided by your instructor or one from a website, such as www.bmra.com/extras/man-rand.htm). Compute a statistic based on your sample and compare it to the corresponding figure for the entire population. Here's how to proceed:

   a. First, select a very small population for which you have a reasonably complete sampling frame. One possibility would be the listing of some characteristic of states in a U.S. Census Bureau publication, such as average income or population size. Another possible population would be the list of asking prices for houses advertised in your local paper.

   b. Next, create a sampling frame, a numbered list of all the available elements in the population. If you are using a complete listing of all elements, as from a U.S. Census Bureau publication, the sampling frame is the same as the list. Just number the elements (states). If your population is composed of housing ads in the local paper, your sampling frame will be those ads that contain a housing price. Identify these ads, and then number them sequentially, starting with 1.

c. Decide on a method of picking numbers out of the random number table, such as taking every number in each row, row by row, or moving down or diagonally across the columns. Use only the first (or last) digit in each number if you need to select 1 to 9 cases or only the first (or last) two digits if you want 10 to 99 cases.

d. Pick a starting location in the random number table. It's important to pick a starting point in an unbiased way, perhaps by closing your eyes and then pointing to some part of the page.

e. Record the numbers you encounter as you move from the starting location in the direction you decided on in advance, until you have recorded as many random numbers as the number of cases you need in the sample. If you are selecting states, 10 might be a good number. Ignore numbers that are too large (or small) for the range of numbers used to identify the elements in the population. Discard duplicate numbers.

f. Calculate the average value in your sample for some variable that was measured (for example, population size in a sample of states or housing price for the housing ads). Calculate the average by adding up the values of all the elements in the sample and dividing by the number of elements in the sample.

g. Go back to the sampling frame and calculate this same average for all the elements in the list. How close is the sample average to the population average?

h. Estismate the range of sample averages that would be likely to include 90% of the possible samples.

Ethics Questions

1. How much pressure is too much pressure to participate in a probability-based sample survey? Is it okay for the U.S. government to mandate legally that all citizens participate in the decennial census? Should companies be able to require employees to participate in survey research about work-related issues? Should students be required to participate in surveys about teacher performance? Should parents be required to consent to the participation of their high school–age students in a survey about substance abuse and health issues? Is it okay to give monetary incentives for participation in a survey of homeless shelter clients? Can monetary incentives be coercive? Explain your decisions.

2. Federal regulations require special safeguards for research on persons with impaired cognitive capacity. Special safeguards are also required for research on prisoners and on children. Do you think special safeguards are necessary? Why or why not? Do you think it is possible for individuals in any of these groups to give "voluntary consent" to research participation? What procedures might help make consent to research truly voluntary in these situations? How could these procedures influence sampling plans and results?

Causation and Experimental Design

I dentifying causes—figuring out why things happen—is the goal of most social science research. Unfortunately, valid explanations of the causes of social phenomena do not come easily. Why did the homicide rate in the United States drop for 15 years and then start to rise in 1999 (Butterfield 2000:12)? Was it because of changes in the style of policing (Radin 1997) or because of changing attitudes among young people (Butterfield 1996a)? Was it due to variation in patterns of drug use (Krauss 1996), or to more stringent handgun regulations (Butterfield 1996b)? Did better emergency medical procedures result in higher survival rates for victims (Ramirez 2002)? If we are to evaluate these alternative explanations, we must design our research strategies carefully.

Video Link 6.1
Watch some ideas about planning social research.

This chapter considers the meaning of causation, the criteria for achieving causally valid explanations, the ways in which experimental and quasi-experimental research designs seek to meet these criteria, and the difficulties that can sometimes result in invalid conclusions. By the end of the chapter, you should have a good grasp of the meaning of causation and the logic of experimental design. Most social research, both academic and applied, uses data collection methods other than experiments. But because experimental designs are the best way to evaluate causal hypotheses, a better understanding of them will help you to be aware of the strengths and weaknesses of other research designs, which we will consider in subsequent chapters.

▣ Causal Explanation

A cause is an explanation of some characteristic, attitude, or behavior of groups, individuals, or other entities (such as families, organizations, or cities) or of events. For example, Sherman and Berk (1984) conducted a study to determine whether adults who were accused of a domestic violence offense would be less likely to repeat the offense if police arrested them rather than just warned them. Their conclusion that this hypothesis was correct meant that they believed police response had a causal effect on the likelihood of committing another domestic violence offense.

> **Causal effect:** The finding that change in one variable leads to change in another variable, *ceteris paribus* (other things being equal). *Example:* Individuals arrested for domestic assault tend to commit fewer subsequent assaults than similar individuals who are accused in the same circumstances but are not arrested.

More specifically, a **causal effect** is said to occur if variation in the independent variable is followed by variation in the dependent variable, when all other things are equal *(ceteris paribus)*. For instance, we know that for the most part, men earn more income than women do. But is this because they are men—or could it be due to higher levels of education or to longer tenure in their jobs (with no pregnancy breaks), or is it due to the kinds of jobs men go into as compared to those that women choose? We want to know if men earn more than women, *ceteris paribus*—other things (job, tenure, education, etc.) being equal.

> **Ceteris paribus:** Latin phrase meaning "other things being equal."

Of course, "all" other things can't literally be equal: We can't compare the same people at the same time in exactly the same circumstances except for the variation in the independent variable (King et al. 1994). However, we can design research to create conditions that are very comparable so that we can isolate the impact of the independent variable on the dependent variable.

▣ What Causes What?

Journal Link 6.1
Read an article that looks at religious participation and delinquency.

Five criteria should be considered in trying to establish a causal relationship. The first three criteria are generally considered as requirements for identifying a causal effect: (1) empirical association, (2) temporal priority of the independent variable, and (3) nonspuriousness. You must establish these three to claim a causal relationship. Evidence that meets the other two criteria—(4) identifying a causal mechanism and (5) specifying the context in which the effect occurs—can considerably strengthen causal explanations.

Research designs that allow us to establish these criteria require careful planning, implementation, and analysis. Many times, researchers have to leave one or more of the criteria unmet and are left with some important doubts about the validity of their causal conclusions, or they may even avoid making any causal assertions.

Research in the News

THE EFFECTS OF POLICE REFORM?

The Los Angeles Police Department (LAPD) announced in July 2011 that the violent crime rate had dropped by 9.6% in the preceding year, continuing a 9-year record of decline. Public approval of the LAPD has also risen to very high levels, including among Latinos and blacks. What accounts for these changes in a city where the police department was once intensely disliked by many? A "transformation" occurred under the leadership of a new police chief, William J. Bratton. Statistical models were used to track crime, police were encouraged to establish personal relationships with community residents and leaders, and gangs were targeted.

Source: Rice, Constance. 2011. A troubled police force has been transformed in Los Angeles. *The New York Times,* August 13:A9, A13.

Association

The first criterion for establishing a causal effect is an empirical (or observed) **association** (sometimes called a *correlation)* between the independent and dependent variables. They must vary together such that when one goes up (or down), the other goes up (or down) at the same time. Here are some examples: When cigarette smoking goes up, so does lung cancer. The longer you stay in school, the more money you will make later in life. Single women are more likely to live in poverty than married women. When income goes up, so does overall health. In all of these cases, a change in an independent variable correlates, or is associated with, a change in a dependent variable. If there is no association, there cannot be a causal relationship. For instance, empirically there seems to be no correlation between the use of the death penalty and a reduction in the rate of serious crime. That may seem unlikely to you, but empirically it is the case: There is no correlation. So there cannot be a causal relationship.

> **Association:** A criterion for establishing a causal relationship between two variables: Variation in one variable is empirically related to variation in another variable.

Time Order

Association is necessary for establishing a causal effect, but it is not sufficient. We must also ensure that the change in the independent variable came before change in the dependent variable—the cause must come before its presumed effect. This is the criterion of **time order,** or the temporal priority of the independent variable. Motivational speakers sometimes say that to achieve success (the dependent variable in our terms), you really need to believe in yourself (the independent variable). And it is true that many very successful politicians, actors, and businesspeople seem remarkably confident—there is an association. But it may well be that their confidence is the result of their success, not its cause. Until you know which came first, you can't establish a causal connection.

> **Time order:** A criterion for establishing a causal relationship between two variables: The variation in the presumed cause (the independent variable) must occur before the variation in the presumed effect (the dependent variable).

Nonspuriousness

The third criterion for establishing a causal effect is **nonspuriousness.** *Spurious* means false or not genuine. We say that a relationship between two variables is

Nonspuriousness: A criterion for establishing a causal relation between two variables; when a relationship between two variables is not due to variation in a third variable.

Spurious: Nature of a presumed relationship between two variables that is actually due to variation in a third variable.

Encyclopedia Link 6.1
Read more about spurious correlations.

spurious when it is actually due to changes in a third variable, so what appears to be a direct connection is in fact not one. Have you heard the old adage "Correlation does not prove causation"? It is meant to remind us that an association between two variables might be caused by something else. If we measure children's shoe sizes and their academic knowledge, for example, we will find a positive association. However, the association results from the fact that older children have larger feet as well as more academic knowledge. A third variable (age) is affecting both shoe size and knowledge so that they correlate, but one doesn't cause the other. Shoe size does not cause knowledge, or vice versa. The association between the two is, we say, spurious.

If this point seems obvious, consider a social science example. Do schools with better resources produce better students? There is certainly a correlation, but consider the fact that parents with more education and higher income tend to live in neighborhoods that spend more on their schools. These parents are also more likely to have books in the home and to provide other advantages for their children (Exhibit 6.1). Maybe parents' income causes variation in both school resources and student performance. If so, there would be an association between school resources and student performance, but it would be at least partially spurious. What we want, then, is nonspuriousness.

Mechanism

Mechanism: A discernible process that creates a causal connection between two variables.

A causal **mechanism** is the process that creates the connection between the variation in an independent variable and the variation in the dependent variable that it is hypothesized to cause (Cook & Campbell 1979:35; Marini & Singer 1988). Many social scientists (and scientists in other fields) argue that no causal explanation is adequate until a mechanism is identified.

For instance, there seems to be an empirical association at the individual level between poverty and delinquency: Children who live in impoverished homes seem more likely to be involved in petty crime. But why?

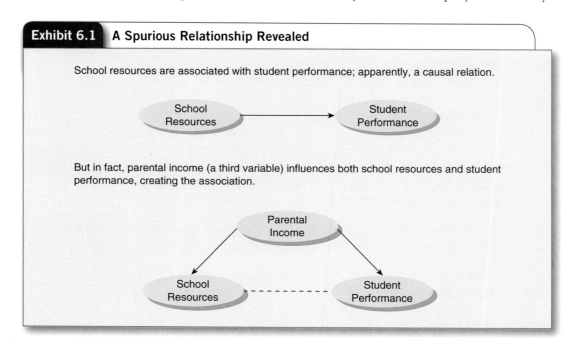

Exhibit 6.1   A Spurious Relationship Revealed

School resources are associated with student performance; apparently, a causal relation.

School Resources → Student Performance

But in fact, parental income (a third variable) influences both school resources and student performance, creating the association.

Parental Income

School Resources - - - - - Student Performance

Some researchers have argued for a *mechanism* of low parent–child attachment, inadequate supervision of children, and erratic discipline as the means by which poverty and delinquency are connected (Sampson & Laub 1994). In this way, figuring out some aspects of the process by which the independent variable influenced the variation in the dependent variable can increase confidence in our conclusion that a causal effect was at work (Costner 1989).

Research|Social Impact Link 6.1
Read a critique about a spurious relationship within scientific research.

Context

No cause has its effect apart from some larger **context** involving other variables. When, for whom, and in what conditions does this effect occur? A cause is really one among a set of interrelated factors required for the effect (Hage & Meeker 1988; Papineau 1978). Identification of the context in which a causal effect occurs is not itself a criterion for a valid causal conclusion, and it is not always attempted; but it does help us to understand the causal relationship.

> **Context:** The larger set of interrelated circumstances in which a particular outcome should be understood.

You may hypothesize, for example, that if you offer employees higher wages to work harder, they will indeed work harder. In the context of America, this seems indeed to be the case; incentive pay causes harder work. But in noncapitalist societies, workers often want only enough money to meet their basic needs and would rather work less than drive themselves hard just to have more money. In America, the correlation of incentive pay with greater effort seems to work; in medieval Europe, for instance, it did not (Weber 1930/1992).

As another example, in America in the 1960s, children of divorced parents ("from a broken home") were more likely to suffer from a variety of problems; they lived in a context of mostly intact families. In recent years, though, many parents are divorced, and the causal link between divorced parents and social pathology no longer seems to hold (Coontz 1997).

▣ Why Experiment?

Experimental research provides the most powerful design for testing causal hypotheses, because it allows us to establish confidently the first three criteria for causality—association, time order, and nonspuriousness. **True experiments** have at least three features that help us meet these criteria:

> **True experiment:** Experiment in which subjects are assigned randomly to an experimental group that receives a treatment or other manipulation of the independent variable and a comparison group that does not receive the treatment or receives some other manipulation. Outcomes are measured in a posttest.

1. Two comparison groups (in the simplest case, an experimental group and a control group), which establish association

2. Variation in the independent variable before assessment of change in the dependent variable, which establishes time order

3. Random assignment to the two (or more) comparison groups, which establishes nonspuriousness

We can determine whether an association exists between the independent and dependent variables in a true experiment, because two or more groups, the **comparison groups**, differ in terms of their value on the independent variable. One group receives some "treatment," which is a manipulation of the value of the

> **Comparison groups:** In an experiment, groups that have been exposed to different treatments, or values of the independent variable (e.g., a control group and an experimental group).

Experimental group: In an experiment, the group of subjects that receives the treatment or experimental manipulation.

Control group: A comparison group that receives no treatment.

Video Link 6.2
Watch a clip about control groups.

independent variable. This group is termed the **experimental group**. In a simple experiment, there may be one other group that does not receive the treatment; it is called a **control group**.

Consider an example in detail (Exhibit 6.2). Does drinking coffee improve one's writing of an essay? Imagine a simple experiment. Suppose you believe that drinking 2 cups of strong coffee before class will help you in writing an in-class essay. But other people think that coffee makes them too nervous and "wired" and so doesn't help in writing the essay. To test your hypothesis ("Coffee drinking causes improved performance."), you need to compare two groups of subjects, a control group and an experimental group. First, the two groups will sit and write an in-class essay. Then, the control group will drink no coffee, while the experimental group will drink 2 cups of strong coffee. Next, both groups will sit and write another in-class essay. At the end, all of the essays will be graded, and you will see which group improved more. Thus, you may establish *association*.

You may find an association outside such an experimental setting, of course, but it wouldn't establish time order. Perhaps good writers hang out in cafés and only then start drinking lots of coffee. So there would be an association, but not the causal relation we're looking for. By controlling who gets the coffee, and when, we establish *time order*.

Posttest: In experimental research, the measurement of an outcome (dependent) variable after an experimental intervention or after a presumed independent variable has changed for some other reason. The posttest is exactly the same "test" as the pretest, but it is administered at a different time.

All experiments have a **posttest**—that is, a measurement of the outcome in both groups after the experimental group has received the treatment. In our example, you grade the papers. Many true experiments also have **pretests**, which measure the dependent variable before the experimental intervention. A pretest is exactly the same as a posttest, just administered at a different time. Strictly speaking, though, a true experiment does not require a pretest. When researchers use random assignment, the groups' initial scores on the dependent variable and on all other variables are very likely to be similar. Any difference in outcome between the experimental and comparison groups is therefore likely to be due to the intervention (or to other processes occurring during the experiment), and the likelihood of a difference just on the basis of chance can be calculated.

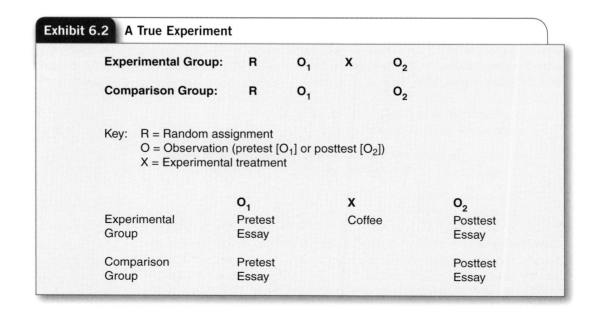

Exhibit 6.2 **A True Experiment**

		O₁	X	O₂
Experimental Group:	R	O_1	X	O_2
Comparison Group:	R	O_1		O_2

Key: R = Random assignment
      O = Observation (pretest [O_1] or posttest [O_2])
      X = Experimental treatment

	O_1	X	O_2
Experimental Group	Pretest Essay	Coffee	Posttest Essay
Comparison Group	Pretest Essay		Posttest Essay

Finally, remember that the two groups must be as equal as possible at the beginning of the study. If you let students choose which group to be in, ambitious students may pick the coffee group, hoping to stay awake and do better on the paper. Or, people who simply don't like the taste of coffee may choose the noncoffee group. Either way, your two groups won't be equivalent at the beginning of the study, and any difference in their writing may be the result of that initial difference (a source of spuriousness), not the drinking of coffee. Finally, as our colleague Stan Lieberson has pointed out to us, coffee affects coffee drinkers and nondrinkers quite differently. Ideally, we'd have similar proportions of each in our different groups.

So, you randomly sort the students into the two different groups. You can do this by flipping a coin for each student, by pulling names out of a hat, or by using a random number table as described in the previous chapter. In any case, the subjects themselves should not be free to choose, nor should you (the experimenter) be free to put them into whatever group you want. (If you did that, you might unconsciously put the better students into the coffee group, hoping to get the results you're looking for.) Thus, we hope to achieve nonspuriousness.

Note that the random assignment of subjects to experimental and comparison groups is not the same as random sampling of individuals from some larger population (Exhibit 6.3). In fact, **random assignment (randomization)** does not help at all to ensure that the research subjects are representative of some larger population; instead, representativeness is the goal of random sampling. What random assignment does—create two (or more) equivalent groups—is useful for ensuring internal validity, not generalizability.

Matching is another procedure sometimes used to equate experimental and comparison groups, but by itself, it is a poor substitute for randomization. Matching of individuals in a treatment group with those in a comparison group might involve pairing persons on the basis of similarity of gender, age, year in school, or some other characteristic. The basic problem is that, as a practical matter, individuals can be matched on only a few characteristics; unmatched differences between the experimental and comparison groups may still influence outcomes.

These defining features of true experimental designs give us a great deal of confidence that we can meet the three basic criteria for identifying causes: association, time order, and nonspuriousness. However, we can strengthen our understanding of causal connections, and increase the likelihood of drawing causally valid conclusions, by also investigating causal mechanism and causal context.

Pretest: In experimental research, the measurement of an outcome (dependent) variable prior to an experimental intervention or change in a presumed independent variable for some other reason. The pretest is exactly the same "test" as the posttest, but it is administered at a different time.

Journal Link 6.2
Read an article that uses matching as a way of comparing group treatment.

Random assignment (randomization): A procedure by which each experimental subject is placed in a group randomly.

Matching: A procedure for equating the characteristics of individuals in different comparison groups in an experiment. Matching can be done on either an individual or an aggregate basis. For individual matching, individuals who are similar in terms of key characteristics are paired prior to assignment, and then the two members of each pair are assigned to the two groups. For aggregate matching, groups are chosen for comparison that are similar in terms of the distribution of key characteristics.

▣ What If a True Experiment Isn't Possible?

Audio Link 6.1
Listen to this clip about the different elements of experimental design.

Often, testing a hypothesis with a true experimental design is not feasible. A true experiment may be too costly or take too long to carry out, it may not be ethical to randomly assign subjects to the different conditions, or the "treatment" events may already have occurred, so it may be too late to conduct a true experiment. Researchers may then instead use **quasi-experimental designs** that retain several components of experimental design but differ in important details.

Researcher Interview Link 6.1
Watch a researcher elaborate on experimental designs.

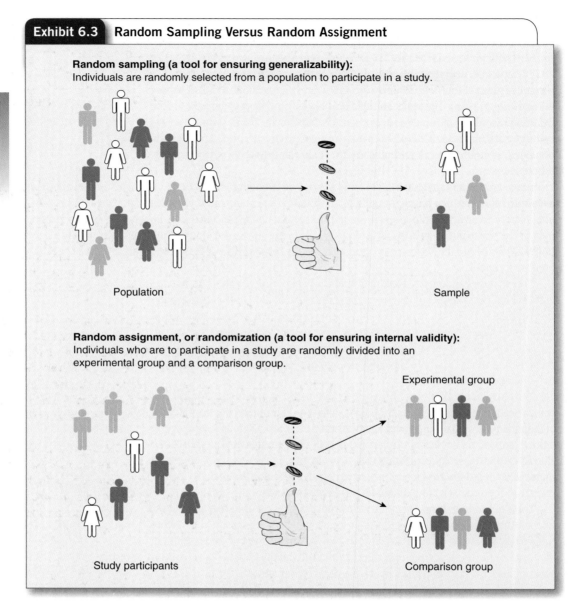

Exhibit 6.3 | **Random Sampling Versus Random Assignment**

Random sampling (a tool for ensuring generalizability):
Individuals are randomly selected from a population to participate in a study.

Population

Sample

Random assignment, or randomization (a tool for ensuring internal validity):
Individuals who are to participate in a study are randomly divided into an experimental group and a comparison group.

Experimental group

Study participants

Comparison group

In quasi-experimental design, a comparison group is predetermined to be comparable to the treatment group in critical ways, such as being eligible for the same services or being in the same school cohort (Rossi & Freeman 1989:313). Such research designs are only quasi-experimental, because subjects are not randomly assigned to the comparison and experimental groups. As a result, we cannot be as confident in the comparability of the groups as in true experimental designs. Nonetheless, to term a research design quasi-experimental, we have to be sure that the comparison groups meet specific criteria, to lessen the chance of preexisting differences between the groups.

We will discuss here the two major types of quasi-experimental designs, as well as one type—ex post facto (after the fact) control group design—that is often mistakenly termed quasi-experimental (other types can be found in Cook & Campbell 1979; Mohr 1992):

Quasi-experimental design:
A research design in which there is a comparison group that is comparable to the experimental group in critical ways but subjects are not randomly assigned to the comparison and experimental groups.

- *Nonequivalent control group designs*—**Nonequivalent control group designs** have experimental and comparison groups that are designated before the treatment occurs but are not created by random assignment.

- *Before-and-after designs*—**Before-and-after designs** have a pretest and posttest but no comparison group. In other words, the subjects exposed to the treatment serve, at an earlier time, as their own control group.

- *Ex post facto control group designs*—**Ex post facto control group designs** use nonrandomized control groups designated after the fact.

Exhibit 6.4 diagrams two studies, one using a nonequivalent control group design and another using the multiple group before-and-after design; the two studies are discussed subsequently. (An ex post facto control group design is the same as for a nonequivalent control group design, but the two types of experiment differ in how people are able to join the groups.)

Quasi-experiments can establish an association of variables: How well do they meet the other criteria for showing causal relationships? If quasi-experimental designs are longitudinal, they can establish time order. But these designs are weaker than true experiments in establishing nonspuriousness: They aren't good at ruling out the influence of some third, uncontrolled variable. Because quasi-experiments

> **Nonequivalent control group design:** A quasi-experimental design in which there are experimental and comparison groups that are designated before the treatment occurs but are not created by random assignment.
>
> **Before-and-after design:** A quasi-experimental design consisting of several before-and-after treatment comparisons involving the same variables but no comparison group.
>
> **Ex post facto control group design:** A nonexperimental design in which comparison groups are selected after the treatment, program, or other variation in the independent variable has occurred.

Exhibit 6.4 Quasi-Experimental Designs

Non-equivalent control group design: Interdependence and team performance (Wageman, 1995)				
Experimental group:		O_1	X_a	O_2
Comparison group 1:		O_1	X_b	O_2
Comparison group 2:		O_1	X_c	O_2
		Pretest	*Treatment*	*Posttest*
Team interdependence	Group	Team performance	Interdependent Tasks	Team performance
	Hybrid	Team performance	Mixed tasks	Team performance
	Individual	Team performance	Individual tasks	Team performance
Before-and-after design: Soap opera suicide and actual suicide (Phillips, 1982)				
Experimental group:		O_{11}	X_1	O_{21}
		O_{12}	X_2	O_{22}
		O_{13}	X_3	O_{23}
		O_{14}	X_4	O_{24}
		Pretest	*Treatment*	*Posttest*
		Suicide rate	Soap-opera suicides	Suicide rate
Key: O = Observation (pretest or posttest) X = Experimental treatment				

Journal Link 6.3
Read about a natural quasi-experimental design which examines rates of recidivism.

do not require random assignment, they can be conducted using more natural procedures in more natural settings, so we may gain a more complete understanding of causal context. However, quasi-experiments are neither better nor worse than experiments in identifying the mechanism of a causal effect.

Nonequivalent Control Group Designs

In this type of quasi-experimental design, a comparison group is selected so as to be as comparable as possible to the treatment group. Two selection methods can be used:

1. *Individual matching*—Individual cases in the treatment group are matched with similar individuals in the comparison group. This can sometimes create a comparison group that is very similar to the experimental group, such as when Head Start participants were matched with their siblings to estimate the effect of participation in Head Start. However, in many studies, it may not be possible to match on the most important variables.

2. *Aggregate matching*—In most situations when random assignment is not possible, the second method of matching makes more sense: identifying a comparison group that matches the treatment group in the aggregate rather than trying to match individual cases. This means finding a comparison group that has similar distributions on key variables: the same average age, the same percentage female, and so on. The upper part of Exhibit 6.4 diagrams a study done at Xerox Corporation by Ruth Wageman (1995), in which 152 technical service teams were divided into three experimental conditions. One emphasized a group orientation with interdependent tasks; another emphasized a "hybrid" style, with some interdependent and some individual tasks; the third group of teams worked as individual technicians. All were evaluated before and after on their performance. The groups were roughly—though not vigorously—equivalent before the study; their leaders chose which style they would pursue, so the procedure was not a true experiment. Interestingly, the hybrid condition proved less successful than either the group or individual approach.

Nonequivalent control group designs allow you to determine whether an association exists between the presumed cause and effect.

Before-and-After Designs

The common feature of before-and-after designs is the absence of a comparison group: All cases are exposed to the experimental treatment. The basis for comparison is instead provided by the pretreatment measures in the experimental group. These designs are thus useful for studies of interventions that are experienced by virtually every case in some population, such as total coverage programs like Social Security or single-organization studies of the effect of a new management strategy.

The simplest type of before-and-after design is the fixed-sample panel design. As you may recall from Chapter 2, in a panel design, the same individuals are studied over time; the research may entail one pretest and one posttest. However, this type of before-and-after design does not qualify as a quasi-experimental design, because comparing subjects to themselves at just one earlier point in time does not provide an adequate comparison group. Many influences other than the experimental treatment may affect a subject following the pretest—for instance, basic life experiences for a young subject.

A more powerful, **multiple group before-and-after design** is illustrated by David P. Phillips's (1982) study of the effect of TV soap opera suicides on the number of actual suicides in the United States. In this study, before-and-after comparisons were made of the same variables between different groups, as illustrated in the bottom half of Exhibit 6.4. Phillips identified 13 (fictional) soap opera suicides in 1977 and then recorded the actual U.S. suicide rate in the weeks prior to and following each TV story. In effect, the researcher had 13 different before-and-after studies, 1 for each suicide story. In 12 of these 13 comparisons, real deaths due to suicide increased from the week before each soap opera suicide to the week after (Exhibit 6.5). Phillips also found similar increases in motor vehicle deaths and crashes during the same period, some portion of which reflects covert suicide attempts.

Another type of before-and-after design involves multiple pretest and posttest observations of the same group. **Repeated measures panel designs** include several pretest and posttest observations, allowing the researcher to study the process by which an intervention or treatment has an impact over time; hence, they are better than simple before-and-after studies.

Time series designs include many (preferably 30 or more) such observations in both pretest and posttest periods. They are particularly useful for studying the impact of new laws or social programs that affect large numbers of people and that are readily assessed by some ongoing measurement. For example, we might use a time series design to study the impact of a new seat belt law on the severity of injuries in automobile accidents, using a monthly state government report on insurance claims. Special statistics are required to analyze time series data, but the basic idea is simple: Identify a trend in the dependent variable up to the date of the intervention, then project the trend into the postintervention period. This *projected* trend is then compared to the *actual* trend of the dependent variable after the intervention. A substantial disparity between the actual and projected trends is evidence that the intervention or event had an impact (Rossi & Freeman 1989:260–261, 358–363).

Multiple group before-and-after design: A type of quasi-experimental design in which several before-and-after comparisons are made involving the same independent and dependent variables but different groups.

Repeated measures panel design: A quasi-experimental design consisting of several pretest and posttest observations of the same group.

Time series design: A quasi-experimental design consisting of many pretest and posttest observations of the same group.

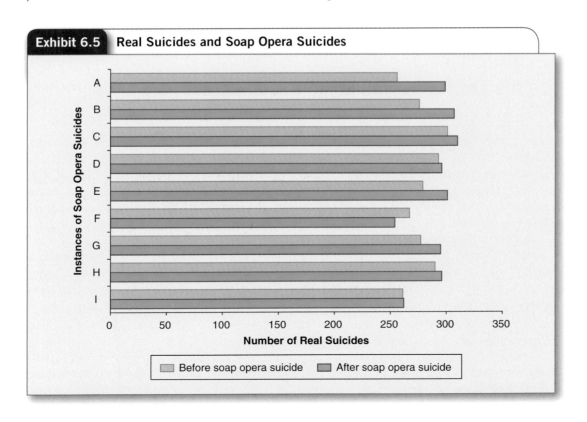

Exhibit 6.5   Real Suicides and Soap Opera Suicides

How well do these before-and-after designs meet the five criteria for establishing causality? The before-and-after comparison enables us to determine whether an *association* exists between the intervention and the dependent variable (because we can determine whether a change occurred after the intervention). They also clarify whether the change in the dependent variable occurred after the intervention, so *time order* is not a problem. However, there is no control group, so we cannot rule out the influence of extraneous factors as the actual cause of the change we observed; *spuriousness* may be a problem.

Overall, the longitudinal nature of before-and-after designs can help to identify causal *mechanisms*, while the loosening of randomization requirements makes it easier to conduct studies in natural settings, where we learn about the influence of *contextual* factors.

Ex Post Facto Control Group Designs

Groups in ex post facto designs are designated after the treatment has occurred; hence, ex post facto studies fail even to earn the quasi-experimental designation. The problem is that people were neither randomly assigned, nor were they picked for experimental treatments. They may well have selected themselves into (or out of) treatment groups. Of course, this makes it difficult to determine whether an association between group membership and outcome is spurious. However, the particulars will vary from study to study; in some circumstances, we may conclude that the treatment and control groups are so similar that causal effects can be tested (Rossi & Freeman 1989:343–344).

Susan Cohen and Gerald Ledford Jr.'s (1994) study of the effectiveness of self-managing teams used a well-constructed ex post facto design. They studied a telecommunications company with some work teams that were self-managing and some that were traditionally managed (meaning that a manager was responsible for the team's decisions). Cohen and Ledford found the self-reported quality of work life to be higher in the self-managing groups than in the traditionally managed groups.

▣ What Are the Threats to Validity in Experiments?

Experimental designs, like any research design, must be evaluated for their ability to yield valid conclusions. Remember, there are three kinds of validity: (1) internal (or causal), (2) external (or generalizability), and (3) measurement. True experiments are good at producing internal validity, but they fare less well in achieving external validity (generalizability). Quasi-experiments may provide more generalizable results than true experiments but are more prone to problems of internal invalidity. Measurement validity is a central concern for both kinds of research, but even true experimental design offers no special advantages or disadvantages in measurement.

In general, nonexperimental designs, such as those used in survey research and field research, offer less certainty of internal validity, a greater likelihood of generalizability, and no particular advantage or disadvantage in terms of measurement validity. We will introduce survey and field research designs in the following chapters; in this section, we focus on the ways in which experiments help (or don't help) to resolve potential problems of internal validity and generalizability.

Threats to Internal (Causal) Validity

The following sections discuss 10 *threats to validity* (also referred to as *sources of invalidity*) that occur frequently in social science research (Exhibit 6.6). These "threats" exemplify five major types of problems that arise in research design.

Exhibit 6.6 Threats to Internal Validity

Problem	Example	Type
Selection	Girls who choose to see a therapist are not representative of population.	Noncomparable Groups
Mortality	Students who most dislike college drop out, so aren't surveyed.	Noncomparable Groups
Instrument Decay	Interviewer tires, losing interest in later interviews, so poor answers result.	Noncomparable Groups
Testing	If someone has taken the SAT before, they are familiar with the format, so do better.	Endogenous Change
Maturation	Everyone gets older in high school; it's not the school's doing.	Endogenous Change
Regression	The lowest-ranking students on IQ must improve their rank; they can't do worse.	Endogenous Change
History	The Casey Anthony trial affects childcare workers.	History
Contamination	"John Henry" effect; people in study compete with one another.	Contamination
Experimenter Expectation	Researchers unconsciously help their subjects, distorting results.	Treatment Misidentification
Placebo Effect	Fake pills in medical studies produce improved health.	Treatment Misidentification
Hawthorne Effect	Workers enjoy being subjects and work harder.	Treatment Misidentification

Noncomparable Groups

The problem of noncomparable groups occurs when the experimental group and the control group are not really comparable—that is, when something interferes with the two groups being essentially the same at the beginning (or end) of a study.

- *Selection bias* —When the subjects in your groups are initially different, **selection bias** occurs. If the ambitious students decide to be in the "coffee" group, you'll think their performance was helped by coffee—but it could have been their ambition.

Examples of selection bias are everywhere. Harvard graduates are very successful people, but Harvard *admits* students who are likely to be successful anyway. Maybe Harvard itself had no effect on them. A few years ago, a psychotherapist named Mary Pipher wrote a best seller called *Reviving Ophelia* (1994) in which she described the difficult lives—as she saw it—of typical adolescent girls. Pipher painted a stark picture of depression, rampant eating disorders, low self-esteem, academic failure, suicidal thoughts, and even suicide itself. Where did she get this picture? From her patients—that is, from adolescent girls who were in deep despair,

Selection bias: A source of internal (causal) invalidity that occurs when characteristics of experimental and comparison group subjects differ in any way that influences the outcome.

or at least were unhappy enough to seek help. If Pipher had talked with a comparison sample of girls who hadn't sought help, perhaps the story would not have been so bleak.

In the Sherman and Berk (1984) domestic violence experiment in Minneapolis, some police officers sometimes violated the random assignment plan when they thought the circumstances warranted arresting a suspect who had been randomly assigned to receive just a warning; thus, they created a selection bias in the experimental group.

- *Mortality*—Even when random assignment works as planned, the groups can become different over time because of **differential attrition**, or **mortality**; this can also be called *deselection*. That is, the groups

> **Differential attrition (mortality):** A problem that occurs in experiments when comparison groups become different because subjects in one group are more likely to drop out for various reasons compared to subjects in the other group(s).

become different because subjects in one group are more likely to drop out for various reasons compared to subjects in the other group(s). At some colleges, satisfaction surveys show that seniors are more likely to rate their colleges positively than are freshmen. But remember that the freshmen who really hated the place may have transferred out, so their ratings aren't included with senior ratings. In effect, the lowest scores are removed; that's a mortality problem. This is not a likely problem in a laboratory experiment that occurs in one session, but some laboratory experiments occur over time, so differential attrition can become a problem. Subjects who experience the experimental condition may become more motivated to continue in the experiment than comparison subjects.

Note that whenever subjects are not assigned randomly to treatment and comparison groups, the threat of selection bias or mortality is very great. Even if the comparison group matches the treatment group on important variables, there is no guarantee that the groups were similar initially in terms of either the dependent variable or some other characteristic. However, a pretest helps the researchers to determine and control for selection bias.

- *Instrument decay*—Measurement instruments of all sorts wear out, in a process known as **instrument decay**, producing different results for cases studied later in the research. An ordinary spring-operated bathroom scale, for instance, becomes "soggy" after some years, showing slightly heavier weights than would be correct. Or a college teacher—a kind of instrument for measuring student performance—gets tired after reading too many papers one weekend and starts giving everyone a B. Research interviewers can get tired or bored, too, leading perhaps to shorter or less thoughtful answers from subjects. In all these cases, the measurement instrument has "decayed," or worn out.

Endogenous Change

The next three problems, subsumed under the label **endogenous change**, occur when natural developments in the subjects, independent of the experimental treatment itself, account for some or all of the observed change between pretest and posttest.

- *Testing*—Taking the pretest can itself influence posttest scores. As the Kaplan SAT prep courses attest, there is some benefit just to getting used to the test format. Having taken the test beforehand can be an advantage. Subjects may learn something or may be sensitized to an issue by the pretest and, as a result, respond differently the next time they are asked the same questions on the posttest.

- *Maturation*—Changes in outcome scores during experiments that involve a lengthy treatment period may be due to maturation. Subjects may age, gain experience, or grow in knowledge—all as part of a natural maturational experience—and thus respond differently on the posttest than on the pretest. In many high school yearbooks, seniors are quoted as saying, for instance, "I started at West Geneva High as a boy and leave as a man. WGHS made me grow up." Well, he probably would have grown up anyway, high school or not. WGHS wasn't the cause.

- *Regression*—Subjects who are chosen for a study because they received very low scores on a test may show improvement in the posttest, on average, simply because some of the low scorers were having a bad day. Whenever subjects are selected for study because of extreme scores (either very high or very low), the next time you take their scores, they will likely "regress," or move toward the average. After all, in a normal (bell curve) distribution, that's what the average *is:* the most likely score. For instance, suppose you give an IQ test to third graders and then pull out the bottom 20% of the class for special attention. The next time that group (the 20%) take the test, they'll almost certainly do better—and not just because of testing practice. In effect, they *can't* do worse—they were at the bottom already. On average, they must do better. A football team that goes 0–12 one season almost has to improve. A first-time novelist writes a wonderful book and gains worldwide acclaim and a host of prizes. The next book is not so good, and critics say, "The praise went to her head." But it didn't; she almost *couldn't* have done better. Whenever you pick people for being on an extreme end of a scale, odds are that next time, they'll be more average. This is called the **regression effect**.

> **Endogenous change:** A source of causal invalidity that occurs when natural developments or changes in the subjects (independent of the experimental treatment itself) account for some or all of the observed change from the pretest to the posttest.

> **Regression effect:** A source of causal invalidity that occurs when subjects chosen because of their extreme scores on a dependent variable become less extreme on a posttest due to mathematical necessity, not the treatment.

Testing, maturation, and regression effects are generally not a problem in experiments that have a control group, because they would affect the experimental group and the comparison group equally. However, these effects could explain any change over time in most before-and-after designs, because these designs do not have a comparison group. Repeated measures, panel studies, and time series designs are better in this regard, because they allow the researcher to trace the pattern of change or stability in the dependent variable up to and after the treatment. Ongoing effects of maturation and regression can thus be identified and taken into account.

History

History, or **external events** during the experiment (things that happen outside the experiment), could change subjects' outcome scores. Examples are newsworthy events that concern the focus of an experiment and major disasters to which subjects are exposed. If you were running liability-reduction workshops for child care workers in the spring of 2011, your success might have less to do with your workshop than with the hugely publicized Casey Anthony trial happening at the same time. This problem is often referred to as a **history effect**—history during the experiment, that is. Also called **effect of external events**, it is a particular concern in before-and-after designs.

> **History effect (effect of external events):** Events external to the study that influence posttest scores, resulting in causal invalidity.

Causal conclusions can be invalid in some true experiments because of the influence of external events. For example, in an experiment in which subjects go to a special location for the treatment, something at that location unrelated to the treatment could influence these subjects. External events are a major concern in studies that compare the effects of programs in different cities or states (Hunt 1985:276–277).

Contamination

Contamination occurs in an experiment when the comparison and treatment groups somehow affect each other. When comparison group members know they are being compared, they may increase their efforts just to be more competitive. This has been termed **compensatory rivalry**, or the **John Henry effect**, named after the "steel-driving man" of the folk song, who raced against a steam drill in driving railroad spikes and killed himself in the process. Knowing that they are being denied some advantage, comparison group subjects may as a result increase their efforts

> **Contamination:** A source of causal invalidity that occurs when either the experimental and/or the comparison group is aware of the other group and is influenced in the posttest as a result.

Compensatory rivalry (John Henry effect): A type of contamination in experimental and quasi-experimental designs that occurs when control group members are aware that they are being denied the treatment and modify their efforts by way of compensation.

Demoralization: A type of contamination in experimental and quasi-experimental designs that occurs when control group members feel that they have been left out of some valuable treatment, performing worse than expected as a result.

Treatment misidentification: A problem that occurs in an experiment when not the treatment itself, but rather some unknown or unidentified intervening process, is causing the outcome.

Expectancies of experiment staff (self-fulfilling prophecy): A source of treatment misidentification in experiments and quasi-experiments that occurs when change among experimental subjects is due to the positive expectancies of the staff who are delivering the treatment, rather than to the treatment itself.

Double-blind procedure: An experimental method in which neither the subjects nor the staff delivering experimental treatments know which subjects are getting the treatment.

Placebo effect: A source of treatment misidentification that can occur when subjects receive a treatment that they consider likely to be beneficial and improve as a result of that expectation rather than the treatment itself.

to compensate. On the other hand, comparison group members may experience **demoralization** if they feel that they have been left out of some valuable treatment, performing worse than expected as a result. Both compensatory rivalry and demoralization thus distort the impact of the experimental treatment.

The danger of contamination can be minimized if the experiment is conducted in a laboratory, if members of the experimental group and the comparison group have no contact while the study is in progress, and if the treatment is relatively brief. Whenever these conditions are not met, the likelihood of contamination increases.

Treatment Misidentification

Sometimes the subjects experience a "treatment" that wasn't intended by the researcher. The following are three possible sources of **treatment misidentification**:

1. *Expectancies of experiment staff*—Change among experimental subjects may be due to the positive expectancies of experiment staff who are delivering the treatment rather than to the treatment itself. Even well-trained staff may convey their enthusiasm for an experimental program to the subjects in subtle ways. This is a special concern in evaluation research, when program staff and researchers may be biased in favor of the program for which they work and are eager to believe that their work is helping clients. Such positive staff expectations, the **expectancies of experiment staff**, thus create a **self-fulfilling prophecy**. However, in experiments on the effects of treatments such as medical drugs, **double-blind procedures** can be used: Staff delivering the treatments do not know which subjects are getting the treatment and which are receiving a placebo—something that looks like the treatment but has no intrinsic effect.

2. *Placebo effect*—In medicine, a *placebo* is a chemically inert substance (a sugar pill, for instance) that looks like a drug but actually has no direct physical effect. Research shows that such a pill can actually produce positive health effects in two thirds of patients suffering from relatively mild medical problems (Goleman 1993:C3). In other words, if you wish that a pill will help, it often actually does. In social science research, such **placebo effects** occur when subjects think their behavior should improve through an experimental treatment and then it does—not from the treatment, but from their own belief. Researchers might then misidentify the treatment as having produced the effect.

3. *Hawthorne effect*—Members of the treatment group may change in terms of the dependent variable because their participation in the study makes them feel special. This problem could occur when treatment group members compare their situation to that of members of the control group who are not receiving the treatment, in which case it would be a type of contamination effect. But experimental group members could feel special simply because they are in the experiment. This is termed a **Hawthorne effect** after a classic worker productivity experiment conducted at the Hawthorne electric plant outside Chicago in the 1920s. No matter what conditions the researchers changed to improve or diminish productivity (for instance, increasing or

decreasing the lighting in the plant), the workers seemed to work harder simply because they were part of a special experiment. Oddly enough, some later scholars suggested that in the original Hawthorne studies, there was actually a selection bias, not a true Hawthorne effect—but the term has stuck (see Bramel & Friend 1981). Hawthorne effects are also a concern in evaluation research, particularly when program clients know that the research findings may affect the chances for further program funding.

> **Hawthorne effect:** A type of contamination in experimental and quasi-experimental designs that occurs when members of the treatment group change in terms of the dependent variable because their participation in the study makes them feel special.

Treatment misidentifications can sometimes be avoided through a technique called **process analysis** (Hunt 1985:272–274). Periodic measures are taken throughout an experiment to assess whether the treatment is being delivered as planned. For example, Drake and his colleagues (Drake, McHugo, Becker, Anthony, & Clark 1996) collected process data to monitor the implementation of two employment service models that they tested. One site did a poorer job of implementing the individual placement and support model than the other site, although the required differences between the experimental conditions were still achieved. Process analysis is often a special focus in evaluation research because of the possibility of improper implementation of the experimental program.

Research|Social Impact Link 6.2
Read more about the placebo effect.

> **Process analysis:** A research design in which periodic measures are taken to determine whether a treatment is being delivered as planned, usually in a field experiment.

Generalizability

Even true experimental designs do have one major weakness, an Achilles' heel: The design components essential for true experiments that minimize threats to causal validity simultaneously make it more difficult to achieve both sample generalizability—being able to apply the findings to some clearly defined larger population—and cross-population generalizability—generalizing across subgroups and to other populations and settings.

Sample Generalizability

Subjects who can be recruited for a laboratory experiment, randomly assigned to a group, and kept under carefully controlled conditions for the duration of the study are unlikely to be a representative sample of any large population of interest to social scientists. Can they be expected to react to the experimental treatment in the same way as members of the larger population? The generalizability of the treatment and of the setting for the experiment also must be considered (Cook & Campbell 1979:73–74): the more artificial the experimental arrangements, the greater the problem (Campbell & Stanley 1966:20–21).

Audio Link 6.2
Listen to how researchers use generalizable methods.

In some limited circumstances, a researcher may be able to sample subjects randomly for participation in an experiment and thus select a generalizable sample—one that is representative of the population from which it is selected. This approach is occasionally possible in **field experiments**. For example, some studies of the effects of income supports on the work behavior of poor persons have randomly sampled persons within particular states before randomly assigning them to experimental and comparison groups. Sherman and Berk's (1984) field experiment about the impact of arrest in actual domestic violence incidents (see Chapter 2) used a slightly different approach. In this study, all eligible cases were treated as subjects in the experiment during the data collection periods. As a result, we can place a good deal of confidence in the generalizability of the results to the population of domestic violence arrest cases in Minneapolis at the time.

> **Field experiment:** An experimental study conducted in a real-world setting.

One especially powerful type of field experiment is an audit (or paired testing) study, in which matched pairs of individuals (called *testers*) approach various organizations to discover how different people—for instance whites versus blacks, or men versus women—are treated. Audit studies were developed and widely used in the 1970s first to uncover housing discrimination. More recently, they have been used in research on employment (Cross, Kenney, Mell, & Zimmerman 1990), automobile purchases (Ayres & Siegelman 1995), restaurant hiring (women have more difficulty being hired in expensive restaurants) (Neumark 1996), and even taxicab rides (Ayres, Vars, & Zakariya 2005). Audit researchers try to make testers as similar as possible in every respect but the one trait they wish to test for (e.g., race or gender).

What effect, for example, might a criminal record noted on one's job application have on a man's chance of getting a job? A huge effect, as it happens—reducing the chance of getting a callback after submitting an application by at least 50%. In a study of 350 employers in Milwaukee, Wisconsin, Devah Pager (2003) used pairs of white and black testers, rotating which testers claimed a criminal record. Pager found that a (supposed) criminal record reduced white men's chances of a callback by one-half, and black men's chances by two-thirds. And black men—apart from the criminal record—were already seriously discriminated against. All told, in fact, a white man *with* a criminal record was more likely to be called than a black man *without* a criminal record. In a follow-up study, Pager and Lincoln Quillian (2005) found that the same employers who said they didn't discriminate against black men, or even against a criminal record, in fact—when faced with a live applicant—did discriminate against both, very significantly. The audit study showed that a survey was a poor indicator of what employers actually did.

Researchers using audit studies have to be careful to match the testers well, to make sure that no unintended differences (e.g., speech patterns, clothing styles) exist that might affect the results; and to train testers well so that they aren't inadvertently influencing people in the audited organizations, or seeing discrimination where there may be none. The generalizability of audit studies is also limited due to their focus on entry-level positions and, in employment studies, on "call back" outcomes (rather than, say, employment or salary offers) (Heckman & Siegelman 1993; Favreault 2008). And of course, the procedure used to select the employers or other organizations also determines the generalizability of an audit study's findings.

Cross-Population Generalizability

Researchers often are interested in determining whether treatment effects identified in an experiment hold true across different populations, times, or settings. When random selection is not feasible, the researchers may be able to increase the cross-population generalizability of their findings by selecting several different experimental sites that offer marked contrasts on key variables (Cook & Campbell 1979:76–77).

Within a single experiment, researchers also may be concerned with whether the relationship between the treatment and the outcome variable holds true for certain subgroups. This demonstration of *external validity* is important evidence about the conditions that are required for the independent variable(s) to have an effect. Price, Van Ryn, and Vinokur (1992) found that intensive job search assistance reduced depression among individuals who were at high risk for it because of other psychosocial characteristics; however, the intervention did not influence the rate of depression among individuals at low risk for depression. This is an important limitation on the generalizability of the findings, even if the sample Price and colleagues took was representative of the population of unemployed persons.

Finding that effects are consistent across subgroups does not establish that the relationship also holds true for these subgroups in the larger population, but it does provide supportive evidence. We have already seen examples of how the existence of treatment effects in particular subgroups of experimental subjects can help us predict the cross-population generalizability of the findings. For example, Sherman and Berk's research (1984; see Chapter 2) found that arrest did not deter subsequent domestic violence for unemployed individuals; arrest also failed to deter subsequent violence in communities with high levels of unemployment.

There is always an implicit tradeoff in experimental design between maximizing causal validity, on the one hand, and generalizability, on the other. Research subjects willing to be randomized into groups and experimented on are probably not representative of the larger population. College students, to take an important example, are easy to recruit and to assign to artificial but controlled manipulations, so they are frequently the subjects in experimental psychology research; but again, the generalizability to other groups may be uncertain. In a fascinating and clever series of experiments, Andrew Elliott and Daniela Nesta (2008) examined how the color red affected men's rating of a woman's attractiveness. They sorted male undergraduates randomly into two groups, then showed them head shots of a moderately attractive young woman, with the photograph bordered either by white (the control group) or by red (the treatment group). The woman in the red-framed picture was rated as significantly more attractive. The researchers then compared men with women raters, also looking at photos with differently colored frames; the female raters were unaffected by color. And, the ratings were found to be specifically on sexual attractiveness, not "likeability." In a series of studies, Elliott and Nesta tried different colors, controlled for sexual orientation, and ensured that subjects were not aware of the border color as a factor in their judgments. "Red," they found, "leads men to view women as more attractive and more sexually desirable" (p. 1150). The limitation may be that their research was on undergraduates; it may be that the "red" effect may not be generalizable or is less powerful, say, for older men—or, for that matter, older women who are being judged. From this research, we can't know.

Although we need to be skeptical about the generalizability of the results of a single experiment or setting, the body of findings accumulated from many experimental tests with different people in different settings can provide a very solid basis for generalization (Campbell & Russo 1999:143).

Interaction of Testing and Treatment

A variant on the problem of external validity occurs when the experimental treatment has an effect only when particular conditions created by the experiment occur. One such problem occurs when the treatment has an effect only if subjects have had the pretest. The pretest sensitizes the subjects to some issue so that when they are exposed to the treatment, they react in a way they would not have if they had not taken the pretest. In other words, testing and treatment interact to produce the outcome. For example, answering questions in a pretest about racial prejudice may sensitize subjects so that when they are exposed to the experimental treatment, seeing a film about prejudice, their attitudes are different from what they would have been. In this situation, the treatment truly had an effect, but it would not have had an effect if it were repeated without the sensitizing pretest. This possibility can be evaluated by using the Solomon four-group design to compare groups with and without a pretest (Exhibit 6.7). If testing and treatment do interact, the difference in outcome scores between the experimental and comparison groups will be different for subjects who took the pretest and those who did not.

As you can see, no single procedure establishes the external validity of experimental results. Ultimately, we must base our evaluation of external validity on the success of replications taking place at different times and places and using different forms of the treatment.

Exhibit 6.7	**Solomon Four-Group Design Testing the Interaction of Pretesting and Treatment**			
Experimental group:	R	O_1	X	O_2
Comparison group:	R	O_1		O_2
Experimental group:	R		X	O_2
Comparison group:	R			O_2

Key: R = Random assignment
O = Observation (pretest or posttest)
X = Experimental treatment

🔲 How Do Experimenters Protect Their Subjects?

Social science experiments often involve subject deception. Primarily because of this feature, some experiments have prompted contentious debates about research ethics. Experimental evaluations of social programs also pose ethical dilemmas, because they require researchers to withhold possibly beneficial treatment from

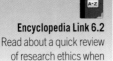

Encyclopedia Link 6.2
Read about a quick review
of research ethics when
working with human
subjects.

some of the subjects just on the basis of chance. Such research may also yield sensitive information about program compliance, personal habits, and even illegal activity—information that is protected from legal subpoenas only in some research concerning mental illness or criminal activity (Boruch 1997). In this section, we give special attention to the problems of deception and the distribution of benefits in experimental research.

Deception

Deception occurs when subjects are misled about research procedures to determine how they would react to the treatment if they were not research subjects. Deception is a critical component of many social experiments, in part because of the difficulty of simulating real-world stresses and dilemmas in a laboratory setting. Stanley Milgram's (1963) classic study of obedience to authority provides a good example. (If you have read Chapter 3 already, you'll be familiar with this example.) Volunteers were recruited for what they were told was a study of the learning process. The experimenter told the volunteers they were to play the role of "teacher" and to administer an electric shock to a "student" in the next room when the student failed a memory test. The shocks were phony (and the students were actors), but the real subjects, the volunteers, didn't know this. They were told to increase the intensity of the shocks, even beyond what they were told was a lethal level. Many subjects continued to obey the authority in the study (the experimenter), even when their obedience involved administering what they thought were potentially lethal shocks to another person.

> But did the experimental subjects actually believe that they were harming someone? Observational data suggest they did: "Persons were observed to sweat, tremble, stutter, bite their lips, and groan as they found themselves increasingly implicated in the experimental conflict" (Milgram 1965:66).

Verbatim transcripts of the sessions also indicated that participants were in much psychological agony about administering the "shocks." So it seems that Milgram's deception "worked." Moreover, it seemed "necessary," since Milgram could not have administered real electric shocks to the students, nor would it have made sense for him to order the students to do something that wasn't so troubling, nor could he have explained what he was really interested in before conducting the experiment. Here is the real question: Is this sufficient justification to allow the use of deception?

Aronson and Mills's study (1959) of severity of initiation (at an all-women's college in the 1950s), also mentioned in Chapter 3, provides a very different example of the use of deception in experimental research—one that does not pose greater-than-everyday risks to subjects. The students who were randomly assigned to the "severe initiation" experimental condition had to read a list of embarrassing words. Even in the 1950s, reading a list of potentially embarrassing words in a laboratory setting and listening to a taped discussion were unlikely to increase the risks to which students were exposed in their everyday lives. Moreover, the researchers informed subjects that they would be expected to talk about sex and could decline to participate in the experiment if this requirement would bother them. No one dropped out.

To further ensure that no psychological harm was caused, Aronson and Mills (1959) explained the true nature of the experiment to the subjects after the experiment, in what is called debriefing, also discussed in Chapter 3. The subjects' reactions were typical: "None of the Ss expressed any resentment or annoyance at having been misled. In fact, the majority were intrigued by the experiment, and several returned at the end of the academic quarter to ascertain the result" (p. 179). Although the American Sociological Association's (1997) *Code of Ethics* does not discuss experimentation explicitly, one of its principles highlights the ethical dilemma deceptive research poses:

> (a) Sociologists do not use deceptive techniques (1) unless they have determined that their use will not be harmful to research participants; is justified by the study's prospective scientific, educational, or applied value; and that equally effective alternative procedures that do not use deception are not

feasible, and (2) unless they have obtained the approval of institutional review boards or, in the absence of such boards, with another authoritative body with expertise on the ethics of research.

(b) Sociologists never deceive research participants about significant aspects of the research that would affect their willingness to participate, such as physical risks, discomfort, or unpleasant emotional experiences. (p. 3)

Selective Distribution of Benefits

Field experiments conducted to evaluate social programs also can involve issues of informed consent (Hunt 1985:275–276). One ethical issue that is somewhat unique to field experiments is the **distribution of benefits**: How much are subjects harmed by the way treatments are distributed in the experiment? For example, Sherman and Berk's (1984) experiment, and its successors, required police to make arrests in domestic violence cases largely on the basis of a random process. When arrests were not made, did the subjects' abused spouses suffer? Price and colleagues (1992) randomly assigned unemployed individuals who had volunteered for job-search help to an intensive program. Were the unemployed volunteers who were assigned to the comparison group at a big disadvantage?

> **Distribution of benefits:** An ethical issue about how much researchers can influence the benefits subjects receive as part of the treatment being studied in a field experiment.

Is it ethical to give some potentially advantageous or disadvantageous treatment to people on a random basis? Random distribution of benefits is justified when the researchers do not know whether some treatment actually is beneficial or not—and, of course, it is the goal of the experiment to find out. Chance is as reasonable a basis for distributing the treatment as any other. Also, if insufficient resources are available to fund fully a benefit for every eligible person, distribution of the benefit on the basis of chance to equally needy persons is ethically defensible (Boruch 1997:66–67).

🔲 Conclusion

Causal (internal) validity is the last of the three legs on which the validity of research rests (the first two being valid measurement and generalizability). In this chapter, you have learned about the five criteria used to evaluate the causal validity of particular research designs. You have seen the problem of spuriousness and the way that randomization deals with it.

True experiments help greatly to achieve more valid causal conclusions—they are the "gold standard" for testing causal hypotheses. But even when conditions preclude a true experiment, adding experimental components can improve many research designs. However, although it may be possible to test a hypothesis with an experiment, it is not always desirable to do so. Laboratory experiments may be inadvisable when they do not test the real hypothesis of interest but test instead a limited version that is amenable to laboratory manipulation. It also may not make sense to test the impact of social programs that cannot actually be implemented because of financial or political problems (Rossi & Freeman 1989:304–307). Yet the virtues of experimental designs mean that they should always be considered when explanatory research is planned.

Understandings of causal relationships are always partial. Researchers must always wonder whether they have omitted some relevant variables from their controls or whether their experimental results would differ if the experiment were conducted in another setting or at another time in history. But the tentative nature of causal conclusions means that we must give more—not less—attention to evaluating the causal validity of social science research whenever we need to ask the simple question, what caused variation in this social phenomenon?

Key Terms

Highlights

- Three criteria generally are viewed as necessary for identifying a causal relationship: association between the variables, proper time order, and nonspuriousness of the association. In addition, identification of a causal mechanism and the context strengthens the basis for concluding that a causal relationship exists.

- Association between two variables by itself is insufficient evidence of a causal relationship. This point is commonly made by the expression, "Correlation does not prove causation."

- The independent variable in an experiment is represented by a treatment or other intervention. Some subjects receive one type of treatment; others may receive a different treatment or no treatment. In true experiments, subjects are assigned randomly to comparison groups.

- Experimental research designs have three essential components: use of at least two groups of subjects for comparison, measurement of the change that occurs as a result of the experimental treatment, and use of random assignment. In addition, experiments may include identification of a causal mechanism and control over experimental conditions.

- Random assignment of subjects to experimental and comparison groups eliminates systematic bias in group assignment. The odds of there being a difference between the experimental and

comparison groups on the basis of chance can be calculated. They become very small for experiments with at least 30 subjects per group.

- Random assignment and random sampling both rely on a chance selection procedure, but their purposes differ. Random assignment involves placing predesignated subjects into two or more groups on the basis of chance; random sampling involves selecting subjects out of a larger population on the basis of chance. Matching of cases in the experimental and comparison groups is a poor substitute for randomization, because identifying in advance all important variables on which to make the match is not possible. However, matching can improve the comparability of groups when it is used to supplement randomization.

- Ethical and practical constraints often preclude the use of experimental designs.

- A quasi-experimental design can be either a nonequivalent control group design or a before-and-after design. Nonequivalent control groups can be created through either individual matching of subjects or matching of group characteristics. In either case, these designs can allow us to establish the existence of an association and the time order of effects, but they do not ensure that some unidentified extraneous variable did not cause what we think of as the effect of the independent variable.

Before-and-after designs can involve one or more pretests and posttests. Although multiple pretests and posttests make it unlikely that another, extraneous influence caused the experimental effect, they do not guarantee it.

- Ex post facto control group designs include a comparison group that individuals could have decided to join precisely because they prefer this experience rather than what the experimental group offers. This creates differences in subject characteristics between the experimental and control groups, which might very well result in a difference in the dependent variable. Because of this possibility, this type of design is not considered a quasi-experimental design.

- Causal conclusions derived from experiments can be invalid because of selection bias, endogenous change, the effects of external events, cross-group contamination, or treatment misidentification. In true experiments, randomization should eliminate selection bias and bias due to endogenous change. External events, cross-group contamination, and treatment misidentification can threaten the validity of causal conclusions in both true experiments and quasi-experiments.

- Process analysis can be used in experiments to identify how the treatment had (or didn't have) an effect—a matter of particular

concern in field experiments. Treatment misidentification is less likely when process analysis is used.

- The generalizability of experimental results declines if the study conditions are artificial and the experimental subjects are unique. Field experiments are likely to produce more generalizable results than experiments conducted in the laboratory.

- The external validity of causal conclusions is determined by the extent to which they apply to different types of individuals and settings. When causal conclusions do not apply to all the subgroups in a study, they are not generalizable to corresponding subgroups in the population; consequently, they are not externally valid with respect to those subgroups. Causal conclusions can also be considered externally invalid when they occur only under the experimental conditions.

- Subject deception is common in laboratory experiments and poses unique ethical issues. Researchers must weigh the potential harm to subjects and debrief subjects who have been deceived. In field experiments, a common ethical problem is selective distribution of benefits. Random assignment may be the fairest way of allocating treatment when treatment openings are insufficient for all eligible individuals and when the efficacy of the treatment is unknown.

STUDENT STUDY SITE

The Student Study Site, available at **www.sagepub.com/chambliss4e,** includes useful study materials including web exercises with accompanying links, eFlashcards, videos, audio resources, journal articles, and encyclopedia articles, many of which are represented by the media links throughout the text. The site also features Interactive Exercises—represented by the green icon here—to help you understand the concepts in this book.

Exercises

Discussing Research

1. There's a lot of "sound and fury" in the social science literature about units of analysis and levels of explanation. Some social researchers may call another a "reductionist" if the researcher explains a problem, such as substance abuse, as due to "lack of self-control." The idea is that the behavior requires consideration of social structure—a group level of analysis rather than an individual level of analysis. Another researcher may be said to commit an "ecological fallacy" if she assumes that group-level characteristics explain behavior at the individual level (such as saying that "immigrants are more likely to commit crime" because the neighborhoods with higher proportions of immigrants have higher crime rates). Do you favor causal explanations at the individual or the group (or social structural) level? If you were forced to mark on a scale from 0 to 100 the percentage of crime that is due to problems with individuals rather than to problems with the settings in which they live and other aspects of social structure, where would you make your mark? Explain your decision.

2. Researchers often try to figure out how people have changed over time by conducting a cross-sectional survey of people of different ages. The idea is that if people who are in their 60s tend to be happier than people who are in their 20s, it is because people tend to "become happier" as they age. But maybe people who are in their 60s now were just as happy when they were in their 20s and people in their 20s now will be just as unhappy when they are in their 60s. (That's called a *cohort effect*.) We can't be sure unless we conduct a panel study (survey the same people at different ages). What, in your experience, are the major differences between the generations today in social attitudes and behaviors? Which would you attribute to changes as people age and which to differences between cohorts in what they have experienced (such as common orientations among baby boomers)? Explain your reasoning.

3. The chapter begins with some alternative explanations for recent changes in the homicide rate. Which of the explanations make the most sense to you? Why? How could you learn more about the effect on crime of one of the "causes" you have identified in a laboratory experiment? What type of study could you conduct in the community to assess its causal impact?

4. This chapter discusses both experimental and quasi-experimental approaches to identifying causes. What are the advantages and disadvantages of both approaches for achieving each of the five criteria identified for causal explanations?

Finding Research

1. Read an original article describing a social experiment. (Social psychology *readers,* collections of such articles for undergraduates, are a good place to find interesting studies.) Critique the article, using as your guide the article review questions presented in Exhibit 12.2 on page 261. Focus on the extent to which experimental conditions were controlled and the causal mechanism was identified. Did inadequate control over conditions or inadequate identification of the causal mechanism make you feel uncertain about the causal conclusions?

2. Go to the website of the Community Policing Consortium (www.policing.com/links/index.html). What causal assertions are made? Pick one of these assertions and propose a research design with which to test this assertion. Be specific.

3. Go to Sociosite (www.sociosite.net/). Choose "Subject Areas," and pick a sociological subject area you are interested in. Find an example of research that has been done using experimental methods in this subject. Explain the experiment. Choose at least five of the Key Terms listed at the end of this chapter that are relevant to and incorporated in the research experiment you have located on the Internet. Explain how each of the five Key Terms you have chosen plays a role in the research example you found on the web.

Critiquing Research

1. From newspapers or magazines, find two recent studies of education (reading, testing, etc.). For each study, list in order what you see as the most likely sources of internal invalidity (selection, mortality, etc.).

2. Select a true experiment, perhaps from the *Journal of Experimental and Social Psychology,* the *Journal of Personality and Social Psychology,* or sources suggested in class. Diagram the experiment using the exhibits in this chapter as a model. Discuss the extent to which experimental conditions were controlled and the causal mechanism was identified. How confident can you be in the causal conclusions from the study, based on review of the threats to internal validity discussed in this chapter: selection bias, endogenous change, external events, contamination, and treatment misidentification? How generalizable do you think the study's results are to the population from which the cases were selected? To specific subgroups in the study? How thoroughly do the researchers discuss these issues?

3. Repeat the previous exercise with a quasi-experiment.

4. Critique the ethics of one of the experiments presented in this chapter or some other experiment you have read about. What specific rules do you think should guide researchers' decisions about subject deception and the selective distribution of benefits?

Doing Research

1. Try out the process of randomization. Go to the Researcher Randomizer website (www.randomizer.org). Now just type numbers into the randomizer for an experiment with two groups and 20 individuals per group. Repeat the process for an

experiment with four groups and 10 individuals per group. Plot the numbers corresponding to each individual in each group. Does the distribution of numbers within each group truly seem to be random?

2. Participate in a social psychology experiment on the Internet at the Social Psychology Network website (www.socialpsychology .org/expts.htm). Pick an experiment in which to participate and follow the instructions. After you finish, write a description of the experiment and evaluate it using the criteria discussed in the chapter.

3. Volunteer for an experiment. Contact the psychology department at your school and ask about opportunities for participating in laboratory experiments. Discuss the experience with your classmates.

Ethics Questions

1. Randomization is a key feature of experimental designs that are often used to investigate the efficacy of new treatments for serious and often incurable terminal diseases. What ethical issues do these techniques raise in studies of experimental treatments for incurable, terminal diseases? Would you make an ethical argument that in some situations, it is more ethical to use random assignment than the usual procedures for deciding whether patients receive a new treatment?

2. In their study of "neighborhood effects" on crime, sociologists Sampson and Raudenbush (1999) had observers drive down neighborhood streets in Chicago and record the level of disorder they observed. What should have been the observers' response if they observed a crime in progress? What if they just suspected that a crime was going to occur? What if the crime was a drug dealer interacting with a driver at the curb? What if it was a prostitute soliciting a customer? What, if any, ethical obligation does a researcher studying a neighborhood have to residents in that neighborhood? Should research results be shared at a neighborhood forum?

Survey Research

Some 6 months after the September 11, 2001, attacks on the World Trade Center and the Pentagon, a small group of students at Hamilton College and their professor, Dennis Gilbert (2002), conducted a nationwide survey of American Muslims. The survey found that nearly 75% of the respondents either knew someone who had, or had themselves, experienced anti-Muslim discrimination since the attacks. "You are demons," "Pig religion," "You guys did it," some were told. Respondents described actions such as "He spit in my face," "He pulled off my daughter's hajib [her head covering]"—the list of abuses went on. In all, 517 American Muslims were contacted, through a careful sampling procedure, and were interviewed via telephone by Gilbert's students and by employees of the Zogby International polling firm. This survey provided a snapshot of the views of an important segment of American society.

In this chapter, we will use the Muslim America project, a "youth and guns" survey also done by Gilbert, and other surveys to illustrate some key features of survey research. We explain the major steps in questionnaire design and then consider the features of four types of surveys, highlighting the unique problems attending each one and suggesting some possible solutions. (For instance, how do we develop an initial list—a sampling frame—of American Muslims?) We discuss ethics issues in the final section. By the chapter's end, you should be well on your way to becoming an informed consumer of survey reports and a knowledgeable developer of survey designs.

▣ Why Is Survey Research So Popular?

Survey research collects information from a *sample of individuals* through their responses to *standardized questions*. As you probably have observed, a great many social scientists rely on surveys as their primary method of data collection. In fact, surveys have become so common that we cannot evaluate much of what we read in the newspaper or see on TV without having some understanding of this method of data collection (Converse 1984).

> **Survey research:** Research in which information is collected from a sample of individuals through their responses to a set of standardized questions.

Survey research owes its popularity to three advantages: (1) versatility, (2) efficiency, and (3) generalizability. The *versatility* of surveys is apparent in the wide range of uses to which they are put, including opinion polls, election campaigns, marketing surveys, community needs assessments, and program evaluations. Surveys are *efficient* because they are a relatively fast means of collecting data on a wide range of issues at relatively little cost—ranging from about $10 to $15 per respondent in mailed surveys of the general population to $30 for a telephone survey and then as much as $300 for in-person interview surveys (F. J. Fowler, personal communication, January 7, 1998; see also Dillman 1982/1991; Groves & Kahn 1979/1991). Because they can be widely distributed to representative samples (see Chapter 5), surveys also help in achieving *generalizable* results.

Perhaps the most efficient type of survey is an **omnibus survey**, which includes a range of topics of interest to different social scientists or to other sponsors. The General Social Survey (GSS) of the National Opinion Research Center at the University of Chicago is a prime example of an omnibus survey. It is a 90-minute interview administered biennially to a probability sample of almost 3,000 Americans, with a wide range of questions and topic areas chosen by a board of overseers. The resulting datasets are made available to many universities, instructors, and students (Davis & Smith 1992; National Opinion Research Center 1992).

> **Omnibus survey:** A survey that covers a range of topics of interest to different social scientists.

Research in the News

POLL-DRIVEN POLITICIANS

Kremlin insiders are relying heavily on survey data to provide feedback on the popularity of their policies. Russia's Public Opinion Foundation delivers briefings weekly at the Kremlin, based on findings from a regular survey of 60,000 Russians and a weekly poll of about 3,000. The Kremlin's own sociology department helped to identify popular support for strong leaders and shows of force. Yet another more liberal group of social scientists argue that survey data indicates popular sentiment in favor of a more open political system.

Source: Barry, Ellen. 2011. Before voting, Russian leaders go to the polls. *The New York Times,* August 17:A1, A8.

How Should We Write Survey Questions?

Questions are the centerpiece of survey research, so selecting good questions is the single most important concern for survey researchers. All hope for achieving measurement validity is lost unless the questions in a survey are clear and convey the intended meaning to respondents.

Video Link 7.1
Watch to learn about creating surveys.

Question writing for a particular survey might begin with a brainstorming session or a review of previous surveys. The Muslim America survey began with students formulating questions with help from Muslim students and professors. Most professionally prepared surveys contain previously used questions as well as some new ones, but every question that is considered for inclusion must be reviewed carefully for clarity and for its ability to convey the intended meaning to the respondents.

Adherence to the following basic principles will go a long way toward ensuring clear and meaningful questions.

Be Clear; Avoid Confusing Phrasing

In most cases, a *simple, direct approach* to asking a question minimizes confusion ("Overall, do you enjoy living in Ohio?"). Use shorter rather than longer words and sentences: *brave* rather than *courageous; job concerns* rather than *work-related employment issues* (Dillman 2000:52). On the other hand, questions shouldn't be abbreviated so much that the results are ambiguous. The simple statement,

Residential location: _____

is *too* simple. Does it ask for town? Country? Street address? In contrast, asking, "In what city or town do you live?" focuses attention clearly on a specific geographic unit, a specific time, and a specific person.

Avoid *negative phrases or words,* especially **double negatives**: "Do you disagree that there should not be a tax increase?" Respondents have a hard time figuring out which response matches their sentiments. Such errors can easily be avoided with minor wording changes, but even experienced survey researchers can make this mistake.

> **Double negative:** A question or statement that contains two negatives, which can muddy the meaning of the question.

Avoid **double-barreled questions**; these actually ask two questions but allow only one answer. For instance, "Our business uses reviews and incentive plans to drive employee behavior. Do you agree or disagree?" What if the business uses only reviews? How should respondents answer? Double-barreled questions can lead to dramatically misleading results. For example, during the Watergate scandal in the 1970s, the Gallup poll asked, "Do you think President Nixon should be impeached and compelled to leave the presidency, or not?" Only about a third of Americans said yes. But when the wording was changed to ask whether President Nixon should be brought to trial before the Senate, more than half answered yes. The first version combined impeachment—trial—with conviction and may have confused people (Kagay 1992:E5).

> **Double-barreled question:** A single survey question that actually asks two questions but allows only one answer.

It is also important to identify clearly what kind of information each question is to obtain. Some questions focus on attitudes, or on what people say they want or how they feel. Some questions focus on beliefs, or what people think is true. Some questions focus on behavior, or on what people do. And some questions focus on attributes, or on what people are like or have experienced (Dillman 1978:79–118; Gordon 1992). Rarely can a single question effectively address more than one of these dimensions at a time.

Minimize Bias

The words used in survey questions should not trigger biases, unless doing so is the researcher's conscious intent. Biased words and phrases tend to produce misleading answers. Some polls ask obviously loaded questions, such as "Isn't it time for Americans to stand up for morality and stop the shameless degradation of the airwaves?" Especially when describing abstract ideas (e.g., *freedom, justice, fairness*), your choice of words can dramatically affect how respondents answer. Take the difference between *welfare* and *assistance for the poor*. On average, surveys have found that public support for more assistance for the poor is about *39 percentage points higher* than for welfare (Smith 1987). Most people favor helping the poor; most people oppose welfare. The "truly needy" gain our sympathy, but "loafers and bums" do not.

Sometimes responses can be distorted through the lack of good alternative answers. For example, the Detroit Area Study (Turner & Martin 1984:252) asked the following question: "People feel differently about making changes in the way our country is run. In order to keep America great, which of these statements do you think is best?" When the only two response choices were "We should be very cautious of making changes," or "We should be free to make changes," only 37% said that we should be free to make changes. However, when a stronger response choice was added suggesting that we should "constantly" make changes, 24% chose that response, and another 32% still chose the "free to make changes" response. So instead of 37%, we now had a total of 56% who seemed open to making changes in the way our country is run (Turner & Martin:252). Including the more extreme positive alternative (constantly make changes) made the less extreme positive alternative more attractive.

To minimize biased responses, researchers have to test reactions to the phrasing of a question.

Allow for Disagreement

Some respondents tend to "agree" with a statement just to avoid disagreeing. In a sense, they want to be helpful. You can see the impact of this human tendency in a 1974 Michigan Survey Research Center survey about crime and lawlessness in the United States (Schuman & Presser 1981). When one question stated that individuals were more to blame for crime than were social conditions, 60% of the respondents agreed. But when the question was rephrased so respondents were asked, "In general, do you believe that individuals or social conditions are more to blame for crime and lawlessness in the United States?" only 46% chose individuals.

As a rule, you should present both sides of attitude scales in the question itself (Dillman 2000:61–62). The response choices themselves should be phrased to make each one seem as socially approved, as "agreeable," as the others.

Most people, for instance, won't openly admit to having committed a crime or other disreputable activities. In this situation, you should write questions that make agreement seem more acceptable. Rather than ask, "Have you ever shoplifted something from a store?" Dillman (2000) suggests "Have you ever taken anything from a store without paying for it?" (p. 25). Asking about a range of behaviors or attitudes can also facilitate agreeing with those that are socially unacceptable.

Don't Ask Questions They Can't Answer

Respondents should be *competent* to answer questions. Too many surveys expect accurate answers from people who couldn't reasonably know the answers. One campus survey we've seen asked professors to agree or disagree with statements such as the following:

"Minority students are made to feel they are second-class citizens."

"The Campus Center does a good job of meeting the informal needs of students."

"The Campus Center is where students go to meet one another and socialize informally."

"Alcohol contributes to casual sex among students."

But of course, most professors are in no position to know the answers to these questions about students' lives. To know what students do or feel, one should ask students, not professors. You should also realize that memory isn't a perfect tool—most of us, for instance, cannot accurately report what we ate for lunch on a Tuesday 2 weeks ago. To get accurate lunch information, ask about today's meal.

Sometimes your survey itself can sort people by competence so that they answer the appropriate questions. For instance, if you include a question about job satisfaction in a survey of the general population, first ask respondents whether they have a job. These **filter questions** create **skip patterns**. For example, respondents who answer no to one question are directed to skip ahead to another question, but respondents who answer yes go on to the **contingent question**. Skip patterns should be indicated clearly, as demonstrated in Exhibit 7.1.

Allow for Uncertainty

Some respondents just don't know—about your topic, about their own feelings, about what they think. Or they like to be neutral and won't take a stand on anything. Or they don't have any information. All of these choices are okay, but you should recognize and allow for them.

Many people, for instance, are **floaters**: respondents who choose a substantive answer even when they really don't know. Asked for their opinion on a law of which they're completely ignorant, a third of the public will give an opinion anyway, if "Don't know" isn't an option. But if it *is* an option, 90% of that group will pick that answer. You should give them the chance to say that they don't know (Schuman & Presser 1981:113–160).

Because there are so many floaters in the typical survey sample, the decision to include an explicit "Don't know" option for a question is important, especially with surveys of less educated populations. "Don't know" responses are chosen more often by those with less education (Schuman & Presser 1981:113–146). Unfortunately, the inclusion of an explicit "Don't know" response choice also allows some people who *do* have a preference to take the easy way out and choose "Don't know."

Fence-sitters, people who see themselves as being neutral, may skew the results if you force them to choose between opposites. In most cases, about 10% to 20% of respondents—those who do not have strong feelings on an issue—will choose an explicit middle, neutral alternative (Schuman & Presser 1981:161–178). Adding an explicit neutral response option is appropriate when you want to find out who is a fence-sitter.

Filter question: A survey question used to identify a subset of respondents who then are asked other questions.

Skip pattern: The unique combination of questions created in a survey by filter questions and contingent questions.

Contingent question: A question that is asked of only a subset of survey respondents.

Floaters: Survey respondents who provide an opinion on a topic in response to a closed-ended question that does not include a "Don't know" option but who will choose "Don't know" if it is available.

Fence-sitters: Survey respondents who see themselves as being neutral on an issue and choose a middle (neutral) response that is offered.

Exhibit 7.1 Filter Questions and Skip Patterns

9. (GUNSHOT) Not including military combat, have you or anyone close to you ever been shot by a gun?
   1. Yes   2. No (**skip to 11**)   3. Not sure (**do not read**)
10. (OPENSHOT) Could you explain the circumstances? _____
11. (GUNLAWS) In general, do you feel that laws covering the sale of firearms should be made more strict, less strict, or kept as they are now?
    1. More strict
    2. Less strict
    3. Kept as are
    4. Not sure (**do not read**)

Fence-sitting and floating can be managed by including an explicit "no opinion" category after all the substantive responses. If neutral sentiment is a possibility, also include a neutral category in the middle of the substantive responses (such as "neither agree nor disagree") (Dillman 2000:58–60). Finally, adding an open-ended question in which respondents are asked to discuss their opinions (or reasons for having no opinion) can help by shedding some light on why some persons choose "Don't know" in response to a particular question (Smith 1984).

Make Response Categories Exhaustive and Mutually Exclusive

Questions with fixed response choices must provide one and only one possible response for everyone who is asked the question. First, all of the possibilities should be offered (choices should be *exhaustive*). In one survey of employees who were quitting their jobs at a telecommunications company, respondents were given these choices for "Why are you leaving [the company]?": (a) poor pay, (b) poor working environment, (c) poor benefits, or (d) poor relations with my boss. Clearly, there may be other reasons (e.g., family or health reasons, geographical preferences) to leave an employer. The response categories were not exhaustive. Or when asking college students their class (senior, junior, etc.), you should probably consider having an "other" category for nontraditional matriculants who may be on an unusual track.

Second, response choices shouldn't overlap—they should be mutually exclusive so that picking one rules out picking another. If I say, for instance, that I'm 25-years-old, I cannot also be 50-years-old; but I may claim to be both "young" and "mature." Those two choices aren't mutually exclusive, so they shouldn't be used as response categories for a question about age.

There are two exceptions to these principles: Filter questions may tell some respondents to skip over a question (the response choices do not have to be exhaustive), and respondents may be asked to "check all that apply" (the response choices are not mutually exclusive). Even these exceptions should be kept to a minimum. Respondents to a self-administered questionnaire should not have to do a lot of skipping around, or else they may lose interest in completing carefully all the applicable questions. And, some survey respondents react to a "check all that apply" request by just checking enough responses so that they feel they have "done enough" for that question and then ignoring the rest of the choices (Dillman 2000:63).

🔲 How Should Questionnaires Be Designed?

Survey questions are asked as part of a **questionnaire**—or **interview schedule**, in interview-based studies; they are not isolated from other questions. The context the questionnaire creates as a whole has a major impact on how individual questions are interpreted and answered. Therefore, survey researchers must carefully design the questionnaire itself, not just each individual question. Several steps, explained in the following sections, will help you design a good questionnaire.

> **Questionnaire:** A survey instrument containing the questions in a self-administered survey.
>
> **Interview schedule:** A survey instrument containing the questions asked by the interviewer in an in-person or phone survey.

Build on Existing Instruments

If another researcher has already designed a set of questions to measure a key concept and previous surveys indicate that this measure is reliable and valid, then by all means use that instrument. Resources such as the *Handbook of Research Design and Social Measurement, 6th edition* (Miller & Salkind 2002) can give you many ideas about existing question-naires; your literature review at the start of a research project should be an even better source.

But there is a tradeoff here. Questions used previously may not concern quite the right concept or may not be appropriate in some ways for your population. A good rule of thumb is to use a previously designed instrument if it measures the concept of concern to you and it seems appropriate for your survey population.

Refine and Test Questions

Audio Link 7.1
Listen to learn more
about questionnaires.

The only good question is a pretested question. Before you rely on a question in your research, you need evidence that your respondents will understand what it means. So try it out on a few people (Dillman 2000:140–147).

One important form of pretesting is discussing the questionnaire with colleagues. You can also review prior research in which your key questions or indexes have been used. Another increasingly popular form of pretesting comes from guided discussions among potential respondents. Such *focus groups* let you check for consistent understanding of terms and identify the range of events or experiences about which people will be asked to report (Fowler 1995). (See Chapter 9 for more about this technique.)

Cognitive interview: A technique for evaluating questions in which researchers ask people test questions, and then probe with follow-up questions to learn how they understood the question and what their answers mean.

Professional survey researchers have also developed a technique for evaluating questions called the **cognitive interview** (Fowler 1995). Although the specifics vary, the basic approach is to ask people to "think aloud" as they answer questions. The researcher asks a test question and then probes with follow-up questions to learn how the question was understood and whether its meaning varied for different respondents. This method can identify many potential problems.

Conducting a pilot study is the final stage of questionnaire preparation. For the Muslim America study, students placed 550 telephone calls and in the process learned (1) the extent of fear that many respondents felt about such a poll; (2) that females were, for cultural reasons, less likely to respond in surveys of the Muslim population; and (3) that some of their questions were worded ambiguously.

To do a pilot study, draw a small sample of individuals from the population you are studying or one very similar to it (it is best to draw a sample of at least 100 respondents) and carry out the survey procedures with them. You may include in the pretest version of a written questionnaire some space for individuals to add comments on each key question or, with in-person interviews, audiotape the test interviews for later review.

Interpretive questions: Questions included in a questionnaire or interview schedule to help explain answers to other important questions.

Review the distribution of responses to each question and revise any that respondents do not seem to understand.

A survey researcher also can try to understand what respondents mean by their responses after the fact—that is, by including additional questions in the survey itself. Adding such **interpretive questions** after key survey questions is always a good idea, but it is of utmost importance when the questions in a survey have not been thoroughly pretested (Labaw 1980).

Maintain Consistent Focus

A survey (with the exception of an omnibus survey) should be guided by a clear conception of the research problem under investigation and the population to be sampled. Remember to have measures of all of the independent and dependent variables you plan to use. Of course, not even the best researcher can anticipate the relevance of every question. Researchers tend to try to avoid "missing something" by erring on the side of extraneous questions (Labaw 1980:40).

At the same time, long lists of redundant or unimportant questions dismay respondents, so respect their time and make sure that each question counts. Surveys too often include too many irrelevant questions.

Order the Questions

The sequence of questions on a survey matters. As a first step, the individual questions should be sorted into broad thematic categories, which then become separate sections in the questionnaire. Both the sections and the questions within the sections must then be organized in a logical order that would make sense in a conversation.

The first question deserves special attention, particularly if the questionnaire is to be self-administered. This question signals to the respondent what the survey is about, whether it will be interesting, and how easy it will be to complete ("Overall, would you say your physical health right now is excellent, good, fair, or poor?"). The first question should be connected to the primary purpose of the survey, it should be interesting, it should be easy, and it should apply to everyone in the sample (Dillman 2000:92–94). Don't try to jump right into sensitive issues ("In general, how well do you think your marriage is working?"); respondents have to "warm up" before they will be ready for such questions. As a standard practice, for instance, most researchers ask any questions about income or finances near the end of a survey, because many people are cautious about discussing such matters.

Question order can lead to **context effects** when one or more questions influence how subsequent questions are interpreted (Schober 1999:89–98). The potential for context effects is greatest when two or more questions concern the same issue or closely related issues. For example, if an early question asks respondents to state for whom they plan to vote in an election, they may hesitate in later questions to support views that are clearly not those of that candidate. In general, people try to appear consistent (even if they are not); be sensitive to this and realize that earlier questions may "commit" respondents to answers on later questions.

> **Context effects:** In survey research, refers to the influence that earlier questions may have on how subsequent questions are answered.

Make the Questionnaire Attractive

An attractive questionnaire—neat, clear, clean, and spacious—is more likely to be completed and less likely to confuse either the respondent or, in an interview, the interviewer.

An attractive questionnaire does not look cramped; plenty of white space—more between questions than within question components—makes the questionnaire appear easy to complete. Response choices are listed vertically and are distinguished clearly and consistently, perhaps by formatting them in all capital letters and keeping them in the middle of the page. Skip patterns are indicated with arrows or other graphics. Some distinctive type of formatting should be used to identify instructions. Printing a multipage questionnaire in booklet form usually results in the most attractive and simple-to-use questionnaire (Dillman 2000:80–86).

Exhibit 7.2 contains portions of a telephone interview questionnaire that illustrates these features, making it easy for the interviewer to use.

Audio Link 7.2
Listen to information about crafting an effective survey.

What Are the Alternatives for Administering Surveys?

Surveys can be administered in at least five different ways. They can be *mailed* or *group-administered* or conducted *by telephone, in person,* or *electronically.* (Exhibit 7.3 summarizes the typical features of each.) Each approach differs from the others in one or more important features:

- *Manner of administration*—The respondents themselves complete mailed, group, and electronic surveys. During phone and in-person interviews, however, the researcher or a staff person asks the questions and records the respondent's answers.

- *Questionnaire structure*—Most mailed, group, phone, and electronic surveys are highly structured, fixing in advance the content and order of questions and response choices. In-person interviews may be highly structured, but they also may include many questions without fixed response choices.

- *Setting*—Mailed, electronic, and phone interviews are usually intended for only one respondent. The same is usually true of in-person interviews, although sometimes researchers interview several family members at once. On the other hand, some surveys are distributed simultaneously to a group of respondents, who complete the survey while the researcher (or assistant) waits.

- *Cost*—As mentioned earlier, in-person interviews are clearly the most expensive type of survey. Phone interviews are much less expensive, and surveying by mail is cheaper yet. Electronic surveys are now the least expensive method, because there are no interviewer costs; no mailing costs; and, for many designs, almost no costs for data entry. (Of course, extra staff time and expertise are required to prepare an electronic questionnaire.)

Exhibit 7.2 Sample Interview Guide

Hi, my name is _____. I am calling on behalf of (I am a student at) Hamilton College in New York. We are conducting a national opinion poll of high school students.

SCREENER: Is there a sophomore, junior, or senior in high school in your household with whom I may speak?

    1. Yes   2. No/not sure/refuse **(End)**

(If student not on phone, ask:) Could he or she come to the phone?

(When student is on the phone) Hi, my name is _____. I am calling on behalf of (I am a student at) Hamilton College in New York. We are conducting a national opinion poll of high school students about gun control. Your answers will be completely anonymous. Would you be willing to participate in the poll?

    1. Yes   2. No/not sure/refuse **(End)**

1. (SKOLYR) What year are you in school?

    1. Sophomore
    2. Junior
    3. Senior
    4. Not sure/refuse **(do not read) (End)**

Now some questions about your school:

2. (SKOL) Is it a public, Catholic, or private school?

    1. Public   2. Catholic   3. Private   4. Not sure **(do not read)**

Exhibit 7.3 Typical Features of the Five Survey Designs

Design	Manner of Administration	Setting	Questionnaire Structure	Cost
Mailed survey	Self	Individual	Structured	Low
Group survey	Self	Group	Structured	Very low
Phone survey	Professional	Individual	Structured	Moderate
In-person interview	Professional	Individual or unstructured	Mostly Structured	High
Electronic survey	Self	Individual	Structured	Very low

Because of their different features, the five administrative options vary in the types of error to which they are most prone and the situations in which they are most appropriate. The rest of this section focuses on each format's unique advantages and disadvantages.

Mailed, Self-Administered Surveys

Video Link 7.2
Watch to learn more
about survey design.

A **mailed (self-administered) survey** is conducted by mailing a questionnaire to respondents, who then take the survey by themselves. The central problem for a mailed survey is maximizing the response rate. Even an attractive questionnaire with clear questions will probably be returned by no more than 30% of a sample unless extra steps are taken. A response rate of 30%, of course, is a disaster, destroying any hope of a representative sample. That's because people who *do* respond are often systematically different from people who *don't* respond—women respond more often, for instance, to most surveys; people with very strong opinions respond more than those who are indifferent; very wealthy and very poor people, for different reasons, are less likely to respond.

> **Mailed (self-administered) survey:** A survey involving a mailed questionnaire to be completed by the respondent.

Fortunately, the conscientious use of systematic techniques can push the response rate to 70% or higher for most mailed surveys (Dillman 2000:27), which is acceptable. Sending follow-up mailings to nonrespondents is the single most important technique for obtaining an adequate response rate. The follow-up mailings explicitly encourage initial nonrespondents to return a completed questionnaire; implicitly, they convey the importance of the effort. Dillman (155–158, 177–188) has demonstrated the effectiveness of a standard procedure for the mailing process: a preliminary introductory letter, a well-packaged survey mailing with a personalized **cover letter**, a reminder postcard 2 weeks after the initial mailing, and then new cover letters and replacement questionnaires 2 to 4 weeks and 6 to 8 weeks after that mailing.

The cover letter, actually, is critical to the success of a mailed survey. This statement to respondents sets the tone for the entire questionnaire. The cover letter or introductory statement must establish the credibility of the research and the researcher, it must be personalized (including a personal salutation and an original signature), it should be interesting to read, and it must explain issues about voluntary participation and maintaining subject confidentiality (Dillman 1978:165–172). A carefully prepared cover letter should increase the response rate and result in more honest and complete answers to the survey questions; a poorly prepared cover letter can have the reverse effects. Exhibit 7.4 is an example of a cover letter for a questionnaire.

> **Cover letter:** The letter sent with a mailed questionnaire. It explains the survey's purpose and auspices and encourages the respondent to participate.

Other steps that help to maximize the response rate include clear and understandable questions, not many open-ended questions, a credible research sponsor, a token incentive (such as a $1 coupon), and presurvey advertising (Fowler 1988:99–106; Mangione 1995:79–82).

Group-Administered Surveys

A **group-administered survey** is completed by individual respondents assembled in a group. The response rate is usually high because most group members will participate. Unfortunately, this method is seldom feasible because it requires a captive audience. With the exception of students, employees, members of the armed forces, and some institutionalized populations, most people cannot be sampled in such a setting.

Whoever is responsible for administering the survey to the group must be careful to minimize comments that might bias answers or that could vary between different groups in the same survey (Dillman 2000:253–256). A standard introductory statement should be read to the group that expresses appreciation for their participation, describes the steps of the survey, and emphasizes (in classroom surveys) that the survey is not the same as a test. A cover letter like that used in mailed surveys

> **Group-administered survey:** A survey that is completed by individual respondents who are assembled in a group.

Exhibit 7.4 **Sample Questionnaire Cover Letter**

University of Massachusetts at Boston
Department of Sociology
May 24, 2003

Jane Doe
AIDS Coordinator
Shattuck Shelter

Dear Jane:

   AIDS is an increasing concern for homeless people and for homeless shelters. The enclosed survey is about the AIDS problem and related issues confronting shelters. It is sponsored by the Life Lines AIDS Prevention Project for the Homeless—a program of the Massachusetts Department of Public Health.
   As an AIDS coordinator/shelter director, you have learned about homeless persons' problems and about implementing programs in response to those problems. The Life Lines Project needs to learn from your experience. Your answers to the questions in the enclosed survey will improve substantially the base of information for improving AIDS prevention programs.
   Questions in the survey focus on AIDS prevention activities and on related aspects of shelter operations. It should take about 30 minutes to answer all the questions.
   Every shelter AIDS coordinator (or shelter director) in Massachusetts is being asked to complete the survey. And every response is vital to the success of the survey: The survey report must represent the full range of experiences.
   You may be assured of complete confidentiality. No one outside of the university will have access to the questionnaire you return. (The ID number on the survey will permit us to check with nonrespondents to see if they need a replacement survey or other information.) All information presented in the report to Life Lines will be in aggregate form, with the exception of a list of the number, gender, and family status of each shelter's guests.
   Please mail the survey back to us by Monday, June 4, and feel free to call if you have any questions.

Thank you for your assistance.

Yours sincerely,

Russell K. Schutt

Russell K. Schutt, PhD

Project Director

Stephanie Howard

Stephanie Howard

Project Assistant

---

also should be distributed with the questionnaires. To emphasize confidentiality, respondents should be given envelopes in which to seal their questionnaires after they are completed.

Another issue of special concern with group-administered surveys is the possibility that respondents will feel coerced to participate and, therefore, will be less likely to answer questions honestly. Also, because administering group surveys requires approval of the authorities—and this sponsorship is made quite obvious, because the survey is conducted on the organization's premises—respondents may infer that the researcher is in league with the sponsor. No complete solution to this problem exists, but it helps to make an introductory statement emphasizing the researcher's independence and giving participants a chance to ask questions about the survey. The sponsor should keep a low profile and allow the researcher both control over the data and autonomy in report writing.

Journal Link 7.1
Read an article about telephone surveys.

Telephone Surveys

In a **phone survey**, interviewers question respondents over the phone and then record respondents' answers. Phone interviewing is traditionally a very popular method of conducting surveys in the United States because almost all families have phones. But two problems often threaten the validity of a phone survey: not reaching the proper sampling units (or *coverage error*) and not getting enough successfully completed responses to make the results generalizable.

> **Phone survey:** A survey in which interviewers question respondents over the phone and record their answers.

Reaching Sampling Units

The first big problem lies in the difficulty of actually contacting the sample units (typically households). Most telephone surveys use random digit dialing (RDD) at some point in the sampling process (Lavrakas 1987) to contact a random sample of households. A machine calls random phone numbers within the designated exchanges, whether or not the numbers are published. RDD is a good way to "capture" unlisted numbers, whose owners are systematically different (often they are wealthier than the general population). When the machine reaches an inappropriate household (such as a business, in a survey of individuals), the phone number is simply replaced with another.

Researcher Interview Link 7.1
Watch a researcher describe using telephone surveys in order to receive valid estimates of sexually transmitted infection prevalence rates.

But the tremendous recent (since 2000) popularity of cellular, or mobile, telephones has made accurate coverage of random samples almost impossible, for several reasons (Tavernise 2011:A13; Tourangeau 2004:781–792): (1) Cell phones are typically not listed in telephone directories, so they can't be included in prepared calling lists; (2) laws generally forbid the use of automated (RDD) dialers to contact cell phones; (3) close to 27% of the U.S. population now has only a cell phone (no landline) and therefore can't be reached by either RDD or many directories; and (4) for 18- to 30-year-olds, some 44% have cell phones only. Cell phone-only households are also more common among non-English speakers and among poor people.

The net effect, then, of widespread cell phone usage is to underrepresent young, poor, and non-English speaking people in particular from inclusion in most large telephone surveys, obviously damaging the results.

Even if an appropriate (for sampling) number is dialed, responses may not be completed. First, because people often are not home, multiple callbacks will be needed for many sample members. With large numbers of single-person households, dual-earner families, and out-of-home activities, survey research organizations have had to increase the usual number of phone contact attempts from just 4 to 8 tries to 20—a lot of attempts just to reach one person. Response rates can also be much lower in harder-to-reach populations (Exhibit 7.5). In a recent phone survey of low-income women in a public health program (Schutt & Fawcett 2005), the University of Massachusetts Center for Survey Research (CSR) achieved a 55.1% response rate from all eligible sampled clients after a protocol that included up to 30 contact attempts, although the response rate rose to 72.9% when it was calculated as a percentage of clients who CSR was able to locate (Roman 2005:7). Caller ID and call waiting allow potential respondents to avoid answering calls from strangers, including researchers. The growth of telemarketing has accustomed individuals nowadays to refuse calls from unknown individuals and organizations or to use their answering machines to screen calls (Dillman 2000:8, 28).

In the Muslim America study, many people were afraid to talk with the researchers or were actively hostile; after all, respondents don't really know who is calling and may have good reason to be suspicious. And, since a huge number of cell phone users are children, and therefore legally unavailable for surveys, calls made to them are all wasted efforts for researchers.

Such problems mean that careful training and direction of interviewers is essential in phone surveys. The instructions shown in Exhibit 7.6 were developed to clarify procedures for asking and coding a series of questions in the phone interviews conducted for the youth and guns survey.

Research|Social Impact Link 7.1
Read more about reaching sample units.

Phone surveying is the method of choice for relatively short surveys of the general population. Response rates in phone surveys traditionally have tended to be very high—often above 80%—because few individuals would hang up on a polite caller or refuse to stop answering questions (at least within the first 30 minutes or so). But the problems we have noted, especially those connected with cell phone usage, makes this method of surveying populations increasingly difficult. The long-term decline in response rates to household surveys is such a problem for survey researchers that they have devoted entire issues of major journals to it (Singer 2006:637–645). Traditionally, a high response rate, since it preserves the sample selected, has been considered

Exhibit 7.5 **Phone Survey Response Rates by Year, 1979–2003**

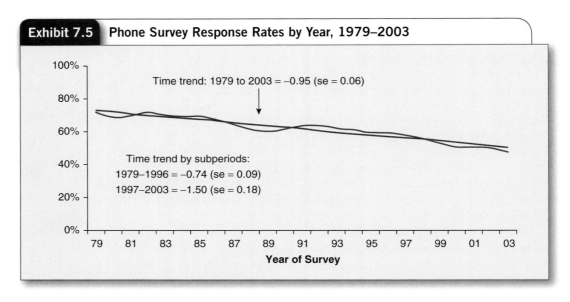

Time trend: 1979 to 2003 = −0.95 (se = 0.06)

Time trend by subperiods:
1979–1996 = −0.74 (se = 0.09)
1997–2003 = −1.50 (se = 0.18)

Year of Survey

Exhibit 7.6 **Sample Interviewer Instructions**

Sample Interviewer Instructions, Youth and Guns Survey, 2000

22. (CONSTIT) To your knowledge, does the U.S. Constitution guarantee citizens the right to own firearms?

   1. Yes   2. No **(skip to 24)**   3. Not sure **(do not read)**

23. (CONLAW) Do you believe that laws regulating the sale and use of handguns violate the constitutional rights of gun owners?

   1. Yes   2. No   3. Not sure **(do not read)**

24. (PETITION) In some localities, high school students have joined campaigns to change the gun laws, and sometimes they have been successful. Earlier you said that you thought that the current gun control laws were (**if Q11 = 1, insert "not strict enough"; if Q11 = 2, insert "too strict"**). Suppose a friend who thinks like you do about this asked you to sign a petition calling for (**if Q11 = 1, insert "stronger gun control laws"; if Q11 = 2, insert "less restrictive gun control laws"**). On a scale from 1 to 5, with 1 being very unlikely and 5 being very likely, how likely is it that you would sign the petition?

   1. (Very unlikely)
   2.
   3.
   4.
   5. (Very likely)
   6. Not sure (do not read)

---

preferable. But given the difficulty nowadays of getting responses for some people, it may be that high response rates may themselves—oddly enough—introduce bias: If someone is so difficult to persuade, they may not be a typical person. And in certain cases, it's not clear that low response rates actually bias the sample. These are issues over which sophisticated professionals differ. But usually, a high response is better overall.

An interesting variant of telephone surveys that you may have encountered is the IVR survey. Computerized **interactive voice response (IVR)** survey technology allows great control over interviewer-respondent interaction. In an IVR survey, respondents receive automated calls and answer questions by pressing numbers on their touch-tone phones or speaking numbers that are interpreted by computerized voice recognition software. These surveys can also record verbal responses to open-ended questions for later transcription. Although they present some difficulties when many answer choices must be used or skip patterns must be followed, IVR surveys have been used successfully with short questionnaires and when respondents are highly motivated to participate (Dillman 2000:402–411). When these conditions are not met, potential respondents may be put off by the impersonality of this computer-driven approach.

> **Interactive voice response (IVR):** A survey in which respondents receive automated calls and answer questions by pressing numbers on their touch-tone phones or speaking numbers that are interpreted by computerized voice recognition software.

In-Person Interviews

What is unique to the **in-person interview**, compared to the other survey designs, is the face-to-face social interaction between interviewer and respondent. If money is no object, in-person interviewing is often the best survey design.

> **In-person interview:** A survey in which an interviewer questions respondents face-to-face and records their answers.

In-person interviewing has several advantages: Response rates are higher than with any other survey design; questionnaires can be much longer than with mailed or phone surveys; the questionnaire can be complex, with both open-ended and closed-ended questions and frequent branching patterns; the interviewer can control the order in which questions are read and answered; the physical and social circumstances of the interview can be monitored; and respondents' interpretations of questions can be probed and clarified. The interviewer, therefore, is well placed to gain a full understanding of what the respondent really wants to say.

Researcher Interview Link 7.2
Watch a researcher describe using interviews to understand how women experience reintegration into a community after incarceration.

However, researchers must be alert to some special hazards due to the presence of an interviewer. Ideally, every respondent should have the same interview experience—that is, each respondent should be asked the same questions in the same way by the same type of person, who reacts similarly to the answers. Suppose one interviewer is smiling and pleasant while another is gruff and rude; the two interviewers will likely elicit very different results in their surveys, if only in the length of responses. Careful training and supervision are essential (Groves 1989:404–406).

Computers can be used to increase control of the in-person interview. In a **computer-assisted personal interview (CAPI)** project, interviewers carry a laptop computer that is programmed to display the interview questions and to process the responses that the interviewer types in, as well as to check that these responses fall within allowed ranges (Tourangeau 2004:790–791). Interviewers seem to like CAPI, and the data obtained are comparable in quality to data obtained in a noncomputerized interview (Shepherd, Hill, Bristor, & Montalvan 1996). A CAPI approach also makes it easier for the researcher to develop skip patterns and experiment with different types of questions for different respondents without increasing the risk of interviewer mistakes (Couper et al. 1998).

> **Computer-assisted personal interview (CAPI):** A personal interview in which the laptop computer is used to display interview questions and to process responses that the interviewer types in, as well as to check that these responses fall within allowed ranges.

The presence of an interviewer may make it more difficult for respondents to give honest answers to questions about socially undesirable behaviors such as drug use, sexual activity, and not voting (Schaeffer & Presser 2003:75). CAPI is valued for this reason, since respondents can enter their answers directly in the laptop without the interviewer knowing what their response is. Alternatively, interviewers can simply hand respondents a separate self-administered questionnaire containing the more sensitive questions. After answering those questions, the respondent seals the separate questionnaire in an envelope so that the interviewer does not know the answers. When this approach was used for the GSS questions about sexual activity, about 21% of men and 13% of women who were married or had been married admitted to having cheated on a spouse ("Survey on Adultery" 1993:A20).

Interviewing Online

Journal Link 7.2
Read about the importance of how a survey is administered and what effects may come from the mode chosen.

Our social world now includes many connections initiated and maintained through e-mail and other forms of web-based communication, so it is only natural that interviewing has also moved online. Interviewing online can facilitate interviews with others who are separated by physical distance; it also is a means to conduct research with those who are only known through such online connections as a discussion group or an e-mail distribution list (James & Busher 2009:14).

Online interviews can be either synchronous—in which the interviewer and interviewee exchange messages as in online chatting—or asynchronous—in which the interviewee can respond to the interviewer's questions whenever it is convenient, usually through e-mail. Both styles of online interviewing have advantages and disadvantages (James & Busher 2009:13–16). Synchronous interviewing provides an experience more similar to an in-person interview, thus giving more of a sense of obtaining spontaneous reactions, but it requires careful attention to arrangements and is prone to interruptions. Asynchronous interviewing allows interviewees to prove more thoughtful and developed answers, but it may be difficult to maintain interest and engagement if the exchanges continue over many days. The online asynchronous interviewer should plan carefully how to build rapport as well as how to terminate the online relationship after the interview is concluded (King & Horrocks 2010:86–93).

Whether a synchronous or asynchronous approach is used, online interviewing can facilitate the research process by creating a written record of the entire interaction without the need for typed transcripts. The relative anonymity of online communications can also encourage interviewees to be more open and honest about their feelings than they would be if interviewed in person (James & Busher 2009:24–25). However, online interviewing lacks some of the most appealing elements of qualitative methods: The revealing subtleties of facial expression, intonation, and body language are lost, and the intimate rapport that a good intensive interviewer can develop in a face-to-face interview cannot be achieved. In addition, those who are being interviewed have much greater ability to present an identity that is completely removed from their in-person persona; for instance, basic characteristics such as age, gender, and physical location can be completely misrepresented.

Maximizing Response to Interviews

Several factors affect the response rate in interview studies. Contact rates tend to be lower in central cities, in part because of difficulties in finding people at home and gaining access to high-rise apartments, and, in part, because of interviewer reluctance to visit some areas at night, when people are more likely to be home (Fowler 1988:45–60). Households with young children or elderly adults tend to be easier to contact, whereas single-person households are more difficult to reach (Groves & Couper 1998:119–154).

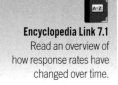

Encyclopedia Link 7.1
Read an overview of how response rates have changed over time.

Refusal rates vary with some respondent characteristics. People with less education participate somewhat less in surveys of political issues (perhaps because they are less aware of current political issues). Less education is also associated with higher rates of "Don't know" responses (Groves 1989). On the other hand, wealthy people often refuse to be surveyed about their income or buying habits, perhaps to avoid being plagued by sales calls. Such problems can be lessened with an advance letter introducing the survey project and by multiple contact attempts throughout the day and evening, but they cannot be entirely avoided (Fowler 1988:52–53).

Electronic Surveys

The widespread use of personal computers and the growth of the Internet have created new possibilities for survey research. **Electronic surveys** can be prepared in two ways (Dillman 2000:352–354). **E-mail surveys** can be sent as messages to respondent e-mail addresses. Respondents then mark their answers in the message and send them back to the researcher. This approach is easy for researchers to develop and for respondents to use. However, this approach is cumbersome for surveys that are more than four or five pages long. **Web surveys** are stored on a server that the researcher controls; respondents are then asked to visit the website (often by just clicking an e-mailed link) and respond to the questionnaire by checking answers. Web surveys require more programming by the researcher, but a well-designed web survey can tailor its questions to a given respondent and thus seem shorter, more interesting, and more attractive.

Electronic survey: A survey that is sent and answered by computer, either through e-mail or on the web.

E-mail survey: A survey that is sent and answered through e-mail.

Web survey: A survey that is accessed and responded to on the World Wide Web.

Web surveys are becoming a very popular form of electronic survey, so we devote a longer discussion to them. They are flexible and inexpensive (Exhibit 7.7). The questionnaire design can feature appealing graphic and typographic elements. By clicking on linked terms, respondents can view definitions of words or instructions for answering questions. Lengthy sets of response choices can be presented with pull-down menus. Pictures and audio segments can be added. Because answers are recorded directly in the researcher's database, data entry errors are virtually eliminated, and results can be reported quickly. Many specific populations have very high rates of Internet use, so a web-based survey can be a good option for groups such as professionals, middle-class communities, members of organizations and, of course, college students. Due to the Internet's global reach, web-based surveys also make it possible to conduct large, international surveys. However, coverage remains a problem with many populations. About 30% of American households are not connected to the Internet (U.S. Bureau of the Census 2010e), so it is not yet possible to survey directly a representative sample of the U.S. population on the web—and given a plateau in the rate of Internet connections, this coverage problem may persist for the near future (Couper & Miller 2008:832). Rates of Internet usage are much lower in other parts of the world, with a worldwide average of 28.7% and rates as low as 10.9% in Africa and 21.5% in all of Asia (Internetworldstats.com 2011). Since households without Internet access also tend to be older, poorer and less educated than those who are connected, web-based surveys of the general population can result in seriously biased estimates (Tourangeau 2004:792–793).

There are several different approaches to conducting web-based surveys, each with unique advantages and disadvantages and somewhat different effects on the coverage problem. Many web-based surveys begin with an e-mail message to potential respondents that contains a direct "hotlink" to the survey website (Gaiser & Schreiner 2009:70). This approach is particularly useful when a defined population with known e-mail addresses is to be surveyed. The researcher can then send e-mail invitations to a representative sample without difficulty. To ensure that the appropriate people respond to a web-based survey, researchers may require that respondents enter a PIN (personal identification number) to gain access to the web survey (Dillman 2000:378). However, many people have more than one e-mail address and often there is no apparent link between an e-mail address and the name or location of the person to whom it is assigned. As a result, there is no available method for drawing a random sample of e-mail addresses for people from any general population, even if the focus is only on those with Internet access (Dillman 2007:449).

Web-based surveys that use volunteer samples may instead be linked to a website that the intended population uses, and everyone who visits that site is invited to complete the survey. This was the approach used in the international web survey sponsored by the National Geographic Society in 2000 (Witte, Amoroso, & Howard 2000). However, although this approach can generate a very large number of respondents (50,000 persons completed National Geographic's Survey 2000), the resulting sample will necessarily reflect the type of people who visit that website (more middle class, young North Americans, in Survey 2000) and thus be a biased representation of the larger population (Couper 2000:486–487; Dillman 2000:355). Some control over the resulting sample can be maintained by requiring participants to meet certain inclusion criteria (Selm & Jankowski 2006:440).

Some web-based surveys are designed to reduce coverage bias by providing computers and Internet connections to those who do not have them. This design-based recruitment method begins by contacting people

Journal Link 7.3
Read an article about a web survey.

Research|Social Impact Link 7.2
Read about how different survey methods can achieve different results.

Exhibit 7.7   Survey.Net—Year 2000 Presidential Election Survey

Your source for information, opinions & demographics from the Net Community!

Year 2000 Presidential Election Survey

Take the year 2000 presidential election survey!

1. **What is your age?**
   [No Answer ⬍]

2. **Your Sex:**
   [No Answer ⬍]

3. **Your highest level of education completed:**
   [No Answer ⬍]

4. **Your political affiliation:**
   [No Answer ⬍]

5. **Who did you vote for in 1996?**
   [No Answer ⬍]

6. **Even though not all of these candidates are necessarily running, if the presidential election were held today, who would you vote for?**
   [No Answer ⬍]

7. **Of the following TWO potential presidential candidates, who would you vote for?**
   [No Answer ⬍]

8. **Of the following presidential candidates, who would you vote for?**
   [No Answer ⬍]

9. **Do you consider yourself...**
   [No Answer ⬍]

10. What political concepts do you agree with? *(check all that apply)*

    ☐ - We need less government regulation in general
    ☐ - We need more responsible government regulation
    ☐ - States should have more responsibility than the Federal Gov.

    ☐ - The government should NOT mandate moral standards
    ☐ - The government SHOULD mandate moral standards

    ☐ - Tax breaks are more important than reducing the deficit
    ☐ - Reducing the deficit is more important than tax breaks

    ☐ - Unions are destroying American productivity
    ☐ - Unions protect the worker

    ☐ - The economy is more important than the environment
    ☐ - The environment is more important than the economy

11. In your opinion, what is the worst problem with our society?
    [No Answer ▲▼]

12. Of those items listed, what should be our next President's highest priority?
    [No Answer ▲▼]

13. Without turning this into a partisan/rhetorical argument, who do you want to see for president in 2000 and why? (*Limit this to one or two sentences*)
    [▲▼]

Thanks very much for participating in the survey!

To submit your survey choices, select:
[SUBMIT SURVEY]

or [Reset survey settings]

You can view the latest survey results after you submit your answers.

We hope you will also participate in other surveys online as well. Please note that you should only complete each survey <u>once</u>.

by phone and providing those who agree to participate with whatever equipment they lack. Since this approach considerably increases the cost of the survey, it is normally used as part of creating the panel of respondents who agree to be contacted for multiple surveys over time. The start-up costs can then be spread across many surveys. Knowledge Networks is a company that received funding from the U.S. National Science Foundation to create such a web-based survey panel. CentER Data in the Netherlands also uses this panel approach (Couper & Miller 2008:832–833). Another approach to reducing coverage bias in web-based surveys is to recruit a volunteer panel of Internet users and then weight the resulting sample to make it comparable to the general population in terms of demographics like gender, race, age, and education. This is the method many market research organizations adopt (Couper & Miller 2008:832–833).

When they are appropriate for a particular population, web surveys have some unique advantages (Selm & Jankowski 2006). Questionnaires completed on the web can elicit more honest reports about socially undesirable behavior or experiences, including illicit behavior and victimization in the general population and failing course grades among college students, when compared to results with phone interviews (Kreuter, Presser, & Tourangeau 2008; Parks, Pardi, & Bradizza 2006). Web-based surveys are relatively easy to complete, as respondents simply click on response boxes and the survey can be programmed to move respondents easily through sets of questions, not presenting questions that do not apply to the respondent, thus leading to higher rates of item completion (Kreuter et al. 2008). Pictures, sounds, and animation can be used as a focus of particular questions and graphic and typographic variation can be used to enhance visual survey appeal. Definitions of terms can also pop up when respondents scroll over them (Dillman 2007:458–459). In these ways, a skilled web programmer can generate a survey layout with many attractive features to make it more likely that respondents will give their answers—and have a clear understanding of the question (Smyth, Dillman, Christian, & Stern 2004:4–5). Responses can quickly be checked to make sure they fall within the allowable range. Because answers are recorded directly in the researcher's database, data entry errors are almost eliminated and results can be reported quickly. By taking advantage of these features, Titus K. L. Schleyer and Jane L. Forrest (2000:420) achieved a 74% response rate in a survey of dental professionals who were already Internet users.

In spite of some clear advantages of some types of web-based surveys, researchers who use this method must be aware of some important disadvantages. Coverage bias is the single biggest problem with web-based surveys of the general population and of segments of the population without a high level of Internet access, and none of the different web-based survey methods fully overcome this problem. Weighting web-based survey panels of Internet users by demographic and other characteristics does not result in similar responses on many questions to those that are obtained with a mailed survey to a sample of the larger population (Rookey, Hanway, & Dillman 2008). Although providing Internet access to all who agree to participate in a web-based survey panel reduces coverage bias, many potential respondents do not agree to participate in such surveys: The rate of agreement to participate was 57% in one Knowledge Networks survey and just 41.5% in a survey of students at the University of Michigan (Couper 2000:485–489). Only about one-third of Internet users contacted in phone surveys agree to provide an e-mail address for a web-based survey and then only one-third of those actually complete the survey (Couper 2000:488). Surveys by phone continue to elicit higher rates of response (Kreuter et al. 2008). Some researchers have found that when people are sent a mailed survey that also provides a link to a web-based survey alternative, they overwhelmingly choose the paper survey (Couper 2000:488).

Remember, regardless of your sampling work, there's zero chance of a noncomputer user responding to an electronic survey. But there's another, almost opposite problem with web surveys: Because they are so easy and cheap to set up, you can find hundreds of web surveys on a wide range of topics and for many different purposes. Among Internet users, almost anyone can participate in many of these web surveys. The large number of respondents such an uncontrolled method can generate should not cause you to forget the importance of a representative sample. Uncontrolled web surveys are guaranteed to produce, instead, a very biased sample (Dillman 2000:355).

When the population to be surveyed has a high rate of Internet use and access is controlled, however, the web makes possible fast and effective surveys (Dillman 2000:354–355). Many corporations use web surveys for gathering information and attitude profiles of their own employees and get response rates of 80% or more. Under proper conditions, electronic surveys are an excellent tool.

A Comparison of Survey Designs

Which survey design should you use for a study? Let's compare the four major survey designs: (1) mailed surveys, (2) phone surveys, (3) in-person surveys, and (4) electronic surveys. (Group-administered surveys are similar in most respects to mailed surveys except that they require the unusual circumstance of having access to the sample in a group setting.) Exhibit 7.8 summarizes these strong and weak points.

The most important difference among these four methods is their varying response rates. Because of the low response rates of *mailed surveys*, they are weakest from a sampling standpoint. However, researchers with limited time, money, and staff may still prefer a mailed survey. Mailed surveys can be useful in asking sensitive questions (e.g., questions about marital difficulties or financial situation), because respondents won't be embarrassed by answering in front of an interviewer.

Video Link 7.3
Watch a clip on survey designs and data.

Contracting with an established survey research organization for a *phone survey* is often the best alternative to a mailed survey. The persistent follow-up attempts that are necessary to secure an adequate response rate are much easier over the phone than in person, although you must be careful about the cell phone sampling and response problem. A phone survey limits the length and complexity of the questionnaire but offers the possibility of very carefully monitoring interviewers (Fowler 1988:61–73).

In-person surveys can be long and complex, and the interviewer can easily monitor the conditions (the room, noise and other distractions, etc.). Although interviewers may themselves distort results, either by changing the wording of questions or failing to record answers properly, this problem can be lessened by careful training and monitoring of interviewers and by tape-recording the answers.

The advantages and disadvantages of *electronic surveys* depend on the populations to be surveyed. Too many people do not have Internet connections for general use of Internet surveying. But when your entire sample has access and ability (e.g., college students, corporate employees), web-based surveys can be very effective.

So overall, in-person interviews are the strongest design and are generally preferable when sufficient resources and a trained interview staff are available; telephone surveys have many of the advantages of in-person interviews at much less cost, but coverage response rates are an increasing problem. Any decision about the best survey design for a particular study must take into account the particular features and goals of the study.

Ethical Issues in Survey Research

Survey research designs usually pose fewer ethical dilemmas than do experimental or field research designs. Potential respondents to a survey can easily refuse to participate, and a cover letter or introductory statement that identifies the sponsors of and motivations for the survey gives them the information required to make this

Exhibit 7.8	Advantages and Disadvantages of Four Survey Designs

Characteristics of Design	Mail Survey	Phone Survey	In-Person Survey	Web Survey
Representative sample				
Opportunity for inclusion is known				
For completely listed populations	High	High	High	Medium
For incompletely listed populations	Medium	Medium	High	Low
Selection within sampling units is controlled (e.g., specific family members must respond)	Medium	High	High	Low
Respondents are likely to be located				
If samples are heterogeneous	Medium	High	High	Low
If samples are homogeneous and specialized	High	High	High	High
Questionnaire construction and question design				
Allowable length of questionnaire Ability to include	Medium	Medium	High	Medium
Complex questions	Medium	Low	High	High
Open questions	Low	High	High	Medium
Screening questions	Low	High	High	High
Tedious, boring questions	Low	High	High	Low
Ability to control question sequence	Low	High	High	High
Ability to ensure questionnaire completion	Medium	High	High	Low
Distortion of answers				
Odds of avoiding social desirability bias	High	Medium	Low	High
Odds of avoiding interviewer distortion	High	Medium	Low	High
Odds of avoiding contamination by others	Medium	High	Medium	Medium
Administrative goals				
Odds of meeting personnel requirements	High	High	Low	Medium
Odds of implementing quickly	Low	High	Low	High
Odds of keeping costs low	High	Medium	Low	High

decision. Little is concealed from the respondents, and the methods of data collection are quite obvious. Only in group-administered survey designs might the respondents (such as students or employees) be, in effect, a captive audience, so they require special attention to ensure that participation is truly voluntary. (Those who do not wish to participate may be told they can just hand in a blank form.)

Sometimes, political or marketing surveys are used unscrupulously to sway opinion under the guise of asking for it. So-called push polls are sometimes employed in political campaigns to distort an opponent's image ("If you knew Congressman Jones was cheating on his wife, would you consider him fit for high office?"). Advertisers can use surveys that pretend to collect opinions or "register" a purchase for warranty purposes, but often they are really trying to collate information about where you live, your phone numbers, your buying habits, and the like.

Confidentiality is most often the primary focus of ethical concern in survey research. Many surveys include questions that might prove damaging to the subjects if their answers were disclosed. When a survey of employees asks, "Do you think management here, especially your boss, is doing a good job?" or when student course evaluations ask, "On a scale of 1 to 5, how fair would you say the professor is?" respondents may well hesitate; if the boss or professor saw the results, workers or students could be hurt.

To prevent any disclosure of such information, it is critical to preserve subject confidentiality. Only research personnel should have access to information that could be used to link respondents to their responses, and even that access should be limited to what is necessary for specific research purposes. Only numbers should be used to identify respondents on their questionnaires, and the researcher should keep the names that correspond to these numbers in a safe, private location, unavailable to staff and others who might come across them. Trustworthy assistants under close supervision should carry out follow-up mailings or contact attempts that require linking the ID numbers with names and addresses. If an electronic survey is used, encryption technology should be used to make information that is provided over the Internet secure from unauthorized people. Usually confidentiality can be protected readily; the key is to be aware of the issue. Don't allow bosses to collect workers' surveys or professors to pick up course evaluations. Be aware of your respondents' concerns and be even a little more careful than you need to be.

Few surveys can provide true **anonymity**, where no identifying information is ever recorded to link respondents with their responses. The main problem with anonymous surveys is that they preclude follow-up attempts to contact nonrespondents and they prevent panel designs, which measure change through repeated surveys of the same individuals. In-person surveys rarely can be anonymous, because an interviewer must, in almost all cases, know the name and address of the interviewee. However, phone surveys that are meant only to sample opinion at one point in time, as in political polls, can safely be completely anonymous. When no future follow-up is desired, group-administered surveys also can be anonymous. To provide anonymity in a mail survey, the researcher should omit identifying codes from the questionnaire but may include a self-addressed, stamped postcard, so the respondent can notify the researcher that the questionnaire has been returned without creating any linkage to the questionnaire itself (Mangione 1995:69).

Encyclopedia Link 7.2
Read about survey-specific research ethics.

Anonymity: Provided by research in which no identifying information is recorded that could be used to link respondents to their responses.

🖾 Conclusion

Survey research is an exceptionally efficient and productive method for investigating a wide array of social research questions. In addition to the potential benefits for social science, considerations of time and expense frequently make a survey the preferred data collection method. One or more of the five survey designs reviewed in this chapter can be applied to almost any research question. It is no wonder that surveys have become the most popular research method in sociology and that they frequently inform discussion and planning about important social and political questions. As use of the Internet increases, survey research should become even more efficient and popular.

The relative ease of conducting at least some types of survey research leads many people to imagine that no particular training or systematic procedures are required. Nothing could be further from the truth. But as a result of this widespread misconception, you will encounter a great many nearly worthless survey results. You must be prepared to examine carefully the procedures used in any survey before accepting its findings as credible. And if you decide to conduct a survey, you must be prepared to invest the time and effort required by proper procedures.

Key Terms

Anonymity   149
Cognitive interview   134
Computer-assisted personal interview
   (CAPI)   141
Context effects   135
Contingent question   132
Cover letter   137
Double-barreled question   130
Double negative   130

Electronic survey   142
E-mail survey   142
Fence-sitters   132
Filter question   132
Floaters   132
Group-administered survey   137
In-person interview   141
Interactive voice response (IVR)   140
Interpretive questions   134

Interview schedule   133
Mailed (self-administered)
   survey   137
Omnibus survey   129
Phone survey   138
Questionnaire   133
Skip pattern   132
Survey research   129
Web survey   143

Highlights

- Surveys are the most popular form of social research because of their versatility, efficiency, and generalizability. Many survey datasets, like the General Social Survey, are available for social scientists to use in teaching and research.

- Omnibus surveys cover a range of topics of interest and generate data useful to multiple sponsors.

- Questions must be worded carefully to avoid confusing respondents, encouraging less-than-honest responses, or triggering biases. Inclusion of "Don't know" choices and neutral responses may help, but the presence of such options also affects the distribution of answers. Open-ended questions can be used to determine the meaning that respondents attach to their answers. Answers to any survey questions may be affected by the questions that precede them in a questionnaire or interview schedule.

- Questions can be tested and improved through review by experts, focus group discussions, cognitive interviews, and/ or pilot testing. Every questionnaire and interview schedule should be pretested on a small sample that is like the sample to be surveyed.

- The cover letter for a mailed questionnaire should be credible, personalized, interesting, and responsible.

- Response rates in mailed surveys are typically well below 70%, unless multiple mailings are made to nonrespondents and the questionnaire and cover letter are attractive, interesting, and carefully planned. Response rates for group-administered surveys are usually much higher than for mailed surveys.

- Phone interviews using random digit dialing (RDD) allow fast turnaround and efficient sampling. Multiple callbacks are often required, and the rate of nonresponse to phone interviews is rising. Phone interviews should be limited in length to about 30 to 45 minutes. In-person interviews have several advantages over other types of surveys: They allow longer and more complex interview schedules, monitoring of the conditions when the questions are answered, probing for respondents' understanding of the questions, and high response rates. However, the interviewer must balance the need to establish rapport with the respondent with the need to adhere to a standardized format.

- Electronic surveys may be e-mailed or posted on the web. Interactive voice response systems using the telephone are another option. At this time, use of the Internet is not sufficiently widespread to allow e-mail or web surveys of the general population, but these approaches can be fast and efficient for populations with high rates of computer use.

- The decision to use a particular survey design must take into account the unique features and goals of the study. In general, in-person interviews are the strongest but most expensive survey design.

- Most survey research poses few ethical problems because respondents can decline to participate—an option that should be stated clearly in the cover letter or introductory statement. Special care must be taken when questionnaires are administered in group settings (to "captive audiences") and when sensitive personal questions are to be asked; subject confidentiality should always be preserved.

STUDENT STUDY SITE

The Student Study Site, available at **www.sagepub.com/chambliss4e,** includes useful study materials including web exercises with accompanying links, eFlashcards, videos, audio resources, journal articles, and encyclopedia articles, many of which are represented by the media links throughout the text. The site also features Interactive Exercises—represented by the green icon here—to help you understand the concepts in this book.

Exercises

Discussing Research

1. Response rates to phone surveys are declining, even as phone usage increases. Part of the problem is that lists of cell phone numbers are not available and wireless service providers do not allow outside access to their networks. Cell phone users may also have to pay for incoming calls. Do you think regulations should be passed to increase the ability of survey researchers to include cell phones in their random digit dialing surveys? How would you feel about receiving survey calls on your cell phone? What problems might result from "improving" phone survey capabilities in this way?

2. In-person interviews have for many years been the "gold standard" in survey research, because the presence of an interviewer increases the response rate, allows better rapport with the interviewee, facilitates clarification of questions and instructions, and provides feedback about the interviewee's situation. However, researchers who design in-person interviewing projects are now making increasing use of technology to ensure consistent questioning of respondents and to provide greater privacy while answering questions. But having a respondent answer questions on a laptop while the interviewer waits is a very different social process than asking the questions verbally. Which approach would you favor in survey research? What tradeoffs can you suggest there might be in terms of quality of information collected, rapport building, and interviewee satisfaction?

Finding Research

1. What resources are available for survey researchers? This question can be answered in part through careful inspection of a website maintained by the Survey Research Laboratory at the University of Illinois at Chicago (www.srl.uic.edu/srllink/srllink.htm#Organizations). Spend some time reviewing these resources and write a brief summary of them.

2. Go to the Research Triangle Institute site at www.rti.org. Click on "Survey Research & Services" then "Innovations." Read about their methods for computer-assisted interviewing and their cognitive laboratory methods for refining questions. What does this add to my treatment of these topics in this chapter? Give specific examples.

Critiquing Research

1. Read one of the original articles that reported one of the surveys described in this chapter. Critique the article using the questions presented in Exhibit 12.2 on page 261 as your guide but focus particular attention on sampling, measurement, and survey design.

2. Each of the following questions was used in a survey that we received at some time in the past. Evaluate each question and its response choices using the guidelines for question writing presented in this chapter. What errors do you find? Try to rewrite each question to avoid such errors and improve question wording.

   a. The first question in an *Info World* (computer publication) "product evaluation survey":

   How interested are you in PostScript Level 2 printers?

   _____Very _____Somewhat _____Not at all

   b. From the Greenpeace National Marine Mammal Survey:

   Do you support Greenpeace's nonviolent direct action to intercept whaling ships, tuna fleets, and other commercial fishermen in order to stop their wanton destruction of thousands of magnificent marine mammals?

   _____Yes _____No _____Undecided

   c. From a U.S. Department of Education survey of college faculty:

   How satisfied or dissatisfied are you with each of the following aspects of your instructional duties at this institution?

	Very Dissat.	Somewhat Dissat.	Somewhat Satisf.	Very Satisf.
i. The authority I have to make decisions about what courses I teach	1	2	3	4
ii. Time available for working with students as advisor, mentor	1	2	3	4

   d. From a survey about affordable housing in a Massachusetts community:

   Higher than single-family density is acceptable to make housing affordable.

Strongly Agree	Undecided	Disagree	Strongly Agree	Disagree
1	2	3	4	5

   e. From a survey of faculty experience with ethical problems in research:

   Are you reasonably familiar with the codes of ethics of any of the following professional associations?

	Very Familiar	Familiar	Not Too Familiar
American Sociological Association	1	2	0
Society for the Study of Social Problems	1	2	0
American Society of Criminology	1	2	0

   If you are familiar with any of the above codes of ethics, to what extent do you agree with them?

   Strongly Agree   Agree   No opinion   Disagree   Strongly Disagree

   Some researchers have avoided using a *professional code of ethics* as a guide for the following reason. Which responses, if any, best describe your reasons for not using all or any of parts of the codes?

	Yes	No
1. Vagueness	1	0
2. Political pressures	1	0
3. Codes protect only individuals, not groups	1	0

   f. From a survey of faculty perceptions:

   Of the students you have observed while teaching college courses, please indicate the percentage who significantly improve their performance in the following areas.

   Reading____%

   Organization____%

   Abstraction____%

   g. From a University of Massachusetts, Boston, student survey:

   A person has a responsibility to stop a friend or relative from driving when drunk.

   Strongly Agree_____ Agree_____ Disagree_____ Strongly Disagree_____

   Even if I wanted to, I would probably not be able to stop most people from driving drunk.

   Strongly Agree_____ Agree_____ Disagree_____ Strongly Disagree_____

3. We received in a university mailbox some years ago a two-page questionnaire that began with the following "cover letter" at the top of the first page:

Faculty Questionnaire

This survey seeks information on faculty perception of the learning process and student performance in their undergraduate careers. Surveys have been distributed in nine universities in the Northeast through random deposit in mailboxes of selected departments. This survey is being conducted by graduate students affiliated with the School of Education and the Sociology Department. We greatly appreciate your time and effort in helping us with our study.

Critique this cover letter and then draft a more persuasive one.

4. Go to the Centre for Applied Social Surveys Question Bank (http://surveynet.ac.uk/sqb/). Click on the link for one of the listed surveys. Review 10 questions used in the survey and critique them in terms of the principles for question writing that you have learned. Do you find any question features that might be attributed to the use of British English?

Doing Research

1. Write 10 questions for a one-page questionnaire that concerns a possible research question. Your questions should operationalize at least three of the variables on which you have focused, including at least one independent and one dependent variable. (You may have multiple questions to measure some variables.) Make all but one of your questions closed ended.

2. Conduct a preliminary pretest of the questionnaire by conducting cognitive interviews with two students or other persons like those to whom the survey is directed. Follow up the closed-ended questions with open-ended probes that ask the respondents what they meant by each response or what came to mind when they were asked each question. Take account of the feedback you receive when you revise your questions.

3. Polish up the organization and layout of the questionnaire, following the guidelines in this chapter. Prepare a rationale for the order of questions in your questionnaire. Write a cover letter directed to the appropriate population that contains appropriate statements about research ethics (human subject issues).

Ethics Questions

1. Group-administered surveys are easier to conduct than other types of surveys, but they always raise an ethical dilemma. If a teacher allows a social research survey to be distributed in class, or if an employer allows employees to complete a survey on company time, is the survey truly voluntary? Is it sufficient to read a statement to the group stating that their participation is entirely up to them? How would you react to a survey in your class? What general guidelines should be followed in such situations?

2. Tjaden and Thoennes (2000) sampled adults with random digit dialing to study violent victimization from a nationally representative sample of adults. What ethical dilemmas do you see in reporting victimizations that are identified in a survey? What about when the survey respondents are under the age of 18? What about children under the age of 12?

CHAPTER 8

Elementary Quantitative Data Analysis

Research|Social Impact Link 8.1

Read more about quantitative analysis and society.

"Show me the data," says your boss. Presented with a research conclusion, most people—not just bosses—want evidence to support it; presented with piles of data, you the researcher need to uncover what it all means. To handle the data gathered by your research, you need to use straightforward methods of data analysis.

In this chapter, we introduce several common statistics used in social research and explain how they can be used to make sense of the "raw" data gathered in your research. Such **quantitative data analysis**, using numbers to discover and describe patterns in your data, is the most elementary use of social statistics.

Why Do Statistics?

A **statistic**, in ordinary language usage, is a numerical description of a population, usually based on a sample of that population. (In the technical language of mathematics, a *parameter* describes a population, and a *statistic* specifically describes a sample.) Some statistics are useful for describing the results of measuring single variables or for constructing and evaluating multi-item scales. These statistics include frequency distributions, graphs, measures of central tendency and variation, and reliability tests. Other statistics are used primarily to describe the association among variables and to control for other variables, and thus, to enhance the causal validity of our conclusions. Cross-tabulation, for example, is one simple technique for measuring association and controlling other variables; it is introduced in this chapter. All of these statistics are termed **descriptive statistics**, because they describe the distribution of and relationship among variables. Statisticians also use **inferential statistics** to estimate the degree of confidence that can be placed in generalizations from a sample to the population from which the sample was selected.

> **Quantitative data analysis:** Statistical techniques used to describe and analyze variation in quantitative measures.

Video Link 8.1
Watch a clip about research and social problems.

> **Statistic:** A numerical description of some feature of a variable or variables in a sample from a larger population.
>
> **Descriptive statistics:** Statistics used to describe the distribution of and relationship among variables.
>
> **Inferential statistics:** Statistics used to estimate how likely it is that a statistical result based on data from a random sample is representative of the population from which the sample is assumed to have been selected.

Case Study: The Likelihood of Voting

In this chapter, we use for examples some data from the 2010 General Social Survey (GSS) on voting and other forms of political participation. What influences the likelihood of voting? Prior research on voting in both national and local settings provides a great deal of support for one hypothesis: The likelihood of voting increases with social status (Milbrath & Goel 1977:92–95; Salisbury 1975:326; Verba & Nie 1972:126). We will find out whether this hypothesis was supported in the 2010 GSS and examine some related issues.

The variables we use from the 2010 GSS are listed in Exhibit 8.1. We use these variables to illustrate particular statistics throughout this chapter.

How to Prepare Data for Analysis

Our analysis of voting in this chapter is an example of what is called **secondary data analysis**. It is secondary because we received the data secondhand. A great many high-quality datasets are available for reanalysis from the Inter-University Consortium for Political and Social Research at the University of Michigan (1996), and many others can be obtained from the government, individual researchers, and other research organizations (see Appendix C).

If you have conducted your own survey or experiment, your quantitative data must be prepared in a format suitable for computer entry. Questionnaires or other

> **Secondary data analysis:** Analysis of data collected by someone other than the researcher or the researcher's assistants.

Exhibit 8.1　List of GSS 2010 Variables for Analysis of Voting

Variable[a]	SPSS Variable Name	Description
Social Status		
Family income	INCOME4R	Family income (in categories)
Education	EDUCR6	Years of education completed (6 categories)
	EDUC4	Years of education completed (4 categories)
	EDUC3	Years of education, trichotomized
Age	AGE4	Years old (categories)
Gender	SEX	Sex
Marital status	MARITAL	Married, never married, widowed, divorced
Race	RACED	White, minority
Politics	PARTYID3	Political party affiliation
Voting	VOTE08D	Voted in 2004 presidential election (yes/no)
Political views	POLVIEWS3	Liberal, moderate, conservative
Interpersonal trust	TRUSTD	Believe other people can be trusted

a. Some variables recoded.

data entry forms can be designed to facilitate this process (Exhibit 8.2). Data from such a form can be entered online, directly into a database, or first on a paper form and then typed or even scanned into a computer database. Whatever data entry method is used, the data must be checked carefully for errors—a process called **data cleaning**. Most survey research organizations now use a database management program to monitor data entry so that invalid codes can be corrected immediately. After data are entered, a computer program must be written to "define the data." A data definition program identifies the variables that are coded in each column or range of columns, attaches meaningful labels to the codes, and distinguishes values representing missing data. The procedures vary depending on the specific statistical package used.

> **Data cleaning:** The process of checking data for errors after the data have been entered in a computer file.

What Are the Options for Displaying Distributions?

The first step in data analysis is usually to discover the variation in each variable of interest. How many people in the sample are married? What is their typical income? Did most of them complete high school? Graphs and frequency distributions are the two most popular display formats. Whatever format is used, the primary

---

| Exhibit 8.2 | Data Entry Procedures |

OMB Control No: 6691-0001
Expiration Date: 04/30/07

Bureau of Economic Analysis
Customer Satisfaction Survey

1. Which data products do you use?	Frequently (every week)	Often (every month)	Infrequently	Rarely	Never	Don't know or not applicable
GENERAL DATA PRODUCTS	(On a scale of 1-5, please circle the appropriate answer.)					
Survey of Current Business	5	4	3	2	1	N/A
CD-ROMs	5	4	3	2	1	N/A
BEA website (www.bea.gov)	5	4	3	2	1	N/A
STAT-USA website (www.stat-usa.gov)	5	4	3	2	1	N/A
Telephone access to staff	5	4	3	2	1	N/A
E-Mail access to staff	5	4	3	2	1	N/A
INDUSTRY DATA PRODUCTS						
Gross Product by Industry	5	4	3	2	1	N/A
Input-Output Tables	5	4	3	2	1	N/A
Satellite Accounts	5	4	3	2	1	N/A
INTERNATIONAL DATA PRODUCTS						
U.S. International Transactions .... (Balance of Payments)	5	4	3	2	1	N/A
U.S. Exports and Imports of Private Services ..	5	4	3	2	1	N/A
U.S. Direct Investment Abroad	5	4	3	2	1	N/A
Foreign Direct Investment in the United States ..	5	4	3	2	1	N/A
U.S. International Investment Position	5	4	3	2	1	N/A
NATIONAL DATA PRODUCTS						
National Income and Product Accounts (GDP)	5	4	3	2	1	N/A
NIPA Underlying Detail Data	5	4	3	2	1	N/A
Capital Stock (Wealth) and Investment .... by Industry	5	4	3	2	1	N/A
REGIONAL DATA PRODUCTS						
State Personal Income	5	4	3	2	1	N/A
Local Area Personal Income	5	4	3	2	1	N/A
Gross State Product by Industry	5	4	3	2	1	N/A
RIMS II Regional Multipliers	5	4	3	2	1	N/A

Central tendency: The most common value (for variables measured at the nominal level) or the value around which cases tend to center (for a quantitative variable).

Variability: The extent to which cases are spread out through the distribution or clustered around just one value.

Skewness: The extent to which cases are clustered more at one or the other end of the distribution of a quantitative variable rather than in a symmetric pattern around its center. Skew can be positive (a right skew), with the number of cases tapering off in the positive direction, or negative (a left skew), with the number of cases tapering off in the negative direction.

concern of the analyst is to display accurately the distribution's shape; that is, to show how cases are distributed across the values of the variable.

Three features are important in describing the shape of the distribution: (1) central tendency, (2) variability, and (3) skewness (lack of symmetry). All three features can be represented in a graph or in a frequency distribution.

We now examine graphs and frequency distributions that illustrate the three features of shape. Several summary statistics used to measure specific aspects of central tendency and variability are presented in a separate section.

Graphs

There are many types of graphs, but the most common and most useful for the statistician are bar charts, histograms, and frequency polygons. Each has two axes, the vertical axis (the *y*-axis) and the horizontal axis (the *x*-axis), and labels to identify the variables and the values, with tick marks showing where each indicated value falls along each axis.

A **bar chart** contains solid bars separated by spaces. It is a good tool for displaying the distribution of variables measured in discrete categories (e.g., nominal variables such as religion or marital status), because such categories don't blend into each other. The bar chart of marital status in Exhibit 8.3 indicates that about half of adult Americans were married at the time of the survey. Smaller percentages

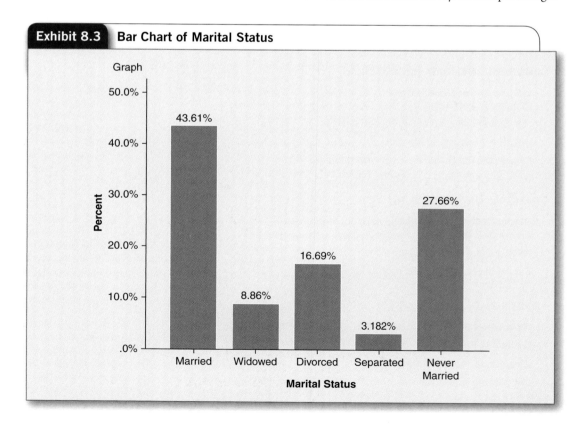

Exhibit 8.3   Bar Chart of Marital Status

were divorced, separated, widowed, or never married. The most common value in the distribution is *married*. There is a moderate amount of variability in the distribution, because the half who are not married are spread across the categories of widowed, divorced, separated, and never married. Because marital status is not a quantitative variable, the order in which the categories are presented is arbitrary, and there is no need to discuss skewness.

Histograms, in which the bars are adjacent, are used to display the distribution of quantitative variables that vary along a continuum that has no necessary gaps. Exhibit 8.4 shows a histogram of years of education from the 2010 GSS data. The distribution has a clump of cases centered at 12 years. The distribution is skewed because there are more cases just above the central point than below it.

In a **frequency polygon**, a continuous line connects the points representing the number or percentage of cases with each value. It is easy to see in the frequency polygon of years of education in Exhibit 8.5 that the most common value is 12 years (high school completion) and that this value also seems to be the center of the distribution. There is moderate variability in the distribution, with many cases having more than 12 years of education and almost one-third having completed at least 4 years of college (16 years). The distribution is highly skewed in the negative direction, with few respondents reporting less than 10 years of education.

> **Bar chart:** A graphic for qualitative variables in which the variable's distribution is displayed with solid bars separated by spaces.

> **Histogram:** A graphic for quantitative variables in which the variable's distribution is displayed with adjacent bars.

> **Frequency polygon:** A graphic for quantitative variables in which a continuous line connects data points representing the variable's distribution.

| **Exhibit 8.4** | **Histogram of Years of Education** |

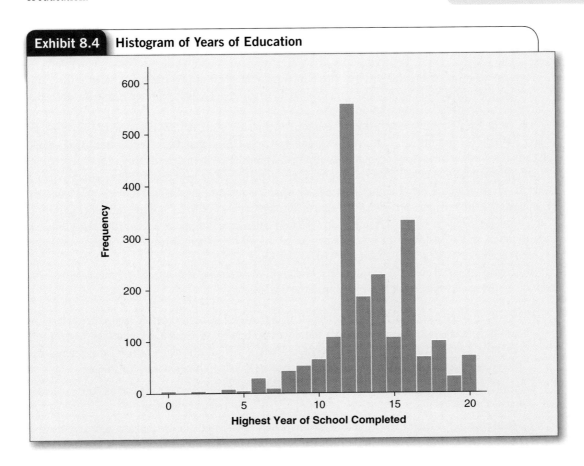

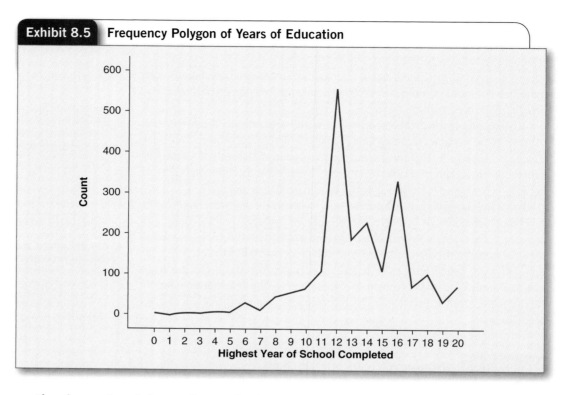

Exhibit 8.5 **Frequency Polygon of Years of Education**

Video Link 8.2
Watch for information
on data visualization.

If graphs are misused, they can distort rather than display the shape of a distribution. Compare, for example, the two graphs in Exhibit 8.6. The first graph shows that high school seniors reported relatively stable rates of lifetime use of cocaine between 1980 and 1985.

The second graph, using exactly the same numbers, appeared in a 1986 *Newsweek* article on "the coke plague" (Orcutt & Turner 1993). To look at this graph, you would think that the rate of cocaine usage among high school seniors had increased dramatically during this period. But, in fact, the difference between the two graphs is due simply to changes in how the graphs were drawn. In the *Newsweek* graph, the percentage scale on the vertical axis begins at 15 rather than at 0, making what was about a 1 percentage point increase look very big indeed. In addition, omission from this graph of the more rapid increase in reported usage between 1975 and 1980 makes it look as if the tiny increase in 1985 were a new, and thus more newsworthy, crisis. Finally, these numbers report "lifetime use," not current or recent use; such numbers can drop only when anyone who has used cocaine dies. The graph is, in total, grossly misleading.

Adherence to several guidelines (Tufte 1983; Wallgren, Wallgren, Persson, Jorner, & Haaland 1996) will help you to spot such problems and to avoid them in your own work:

- Begin the graph of a quantitative variable at 0 on both axes. The difference between bars can be misleadingly exaggerated by cutting off the bottom of the vertical axis and displaying less than the full height of the bars. It may at times be reasonable to violate this guideline, as when an age distribution is presented for a sample of adults; but in this case, be sure to mark the break clearly on the axis.

- Always use bars of equal width. Bars of unequal width, including pictures instead of bars, can make particular values look as if they carry more weight than their frequency warrants.

- Ensure that the two axes, usually, are of approximately equal length. Either shortening or lengthening the vertical axis will obscure or accentuate the differences in the number of cases between values.

- Avoid "chart junk"—a lot of verbiage or excessive marks, lines, lots of cross-hatching, and the like. It can confuse the reader and obscure the shape of the distribution.

Exhibit 8.6 Two Graphs of Cocaine Usage

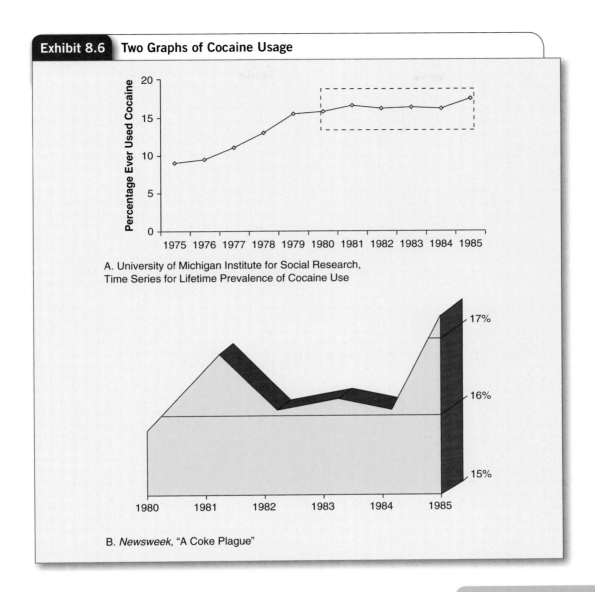

A. University of Michigan Institute for Social Research,
Time Series for Lifetime Prevalence of Cocaine Use

B. *Newsweek*, "A Coke Plague"

Frequency Distributions

Another good way to present a univariate (one-variable) distribution is with a **frequency distribution**. A frequency distribution displays the number, **percentage** (the relative frequencies), or both corresponding to each of a variable's values. A frequency distribution will usually be labeled with a title, a stub (labels for the values), a caption, and perhaps the number of missing cases. If percentages are presented rather than frequencies (sometimes both are included), the total number of cases in the distribution (the **base number N**) should be indicated (Exhibit 8.7).

Constructing and reading frequency distributions for variables with few values is not difficult. The frequency distribution of voting in Exhibit 8.7, for example, shows that 68.7% of the respondents eligible to vote said they voted and that 25.9% reported they did not vote. The total number of respondents to this question was 2,023, although 2,044 actually were interviewed. The rest were ineligible to vote,

Frequency distribution: Numerical display showing the number of cases, and usually the percentage of cases (the relative frequencies), corresponding to each value or group of values of a variable.

Percentage: The relative frequency, computed by dividing the frequency of cases in a particular category by the total number of cases and multiplying by 100.

Base number (N): The total number of cases in a distribution.

Exhibit 8.7	Frequency Distribution of Voting in the 2008 Presidential Election	

Value	Frequency	Valid Percentage
Voted	1390	72.4
Did not vote	530	27.6
Ineligible	103	
Don't know	12	
No answer	9	
Total	2044	100.0
N		(1920)

just refused to answer the question, said they did not know whether they had voted, or gave no answer.

When the distributions of variables with many values (for instance, age) are to be presented, the values must first be grouped. Exhibit 8.8 shows both an ungrouped and a grouped frequency distribution of age. You can see why it is so important to group the values, but we have to be sure that in doing so, we do not distort the distribution. Follow these two rules, and you'll avoid problems:

1. Categories should be logically defensible and should preserve the shape of the distribution.

2. Categories should be mutually exclusive and exhaustive so that every case is classifiable in one and only one category.

What Are the Options for Summarizing Distributions?

Summary statistics describe particular features of a distribution and facilitate comparison among distributions. We can, for instance, show that average income is higher in Connecticut than in Mississippi and higher in New York than in Louisiana. But if we just use one number to represent a distribution, we lose information about other aspects of the distribution's shape. For example, a measure of central tendency (such as the mean or average) would miss the point entirely for an analysis about differences in income inequality among states. A high average income could as easily be found in a state with little income inequality as in one with much income inequality; the average says nothing about the distribution of incomes. For this reason, analysts who report summary measures of central tendency usually also report a summary measure of variability or present the distributions themselves to indicate skewness.

Research in the News

GENERAL SOCIAL SURVEY SHOWS INFIDELITY ON THE RISE

In the **News**

Since 1972, about 12% of married men and 7% of married women have said each year that they have had sex outside their marriage. However, the lifetime rate of infidelity for men over age 60 increased from 20% in 1991 to 28% in 2006, while for women in this age group it increased from 5% to 15%. Infidelity has also increased among those under age 35: from 15% to 20% among young married men and from 12% to 15% among young married women. On the other hand, couples appear to be spending slightly more time with each other.

Source: Parker-Pope, Tara. 2008. Love, sex, and the changing landscape of infidelity. *The New York Times,* October 28:D1.

Exhibit 8.8 Grouped Versus Ungrouped Frequency Distributions

Ungrouped		Grouped	
Age	Percentage	Age	Percentage
18	0.5%	18–19	1.7%
19	1.2	20–29	16.7
20	1.2	30–39	17.8
21	1.7	40–49	18.0
22	.9	50–59	18.5
23	1.9	60–69	14.8
24	1.5	70–79	7.1
25	2.4	80–89	5.4
26	1.4		100.0%
27	2.1		(2041)
28	1.5		
29	2.2		
30	2.0		
31	2.1		
32	1.6		
33	1.8		
34	1.6		
35	2.2		
36	1.5		
37	1.9		
38	1.5		
39	1.7		
40	1.8		
41	1.9		
42	1.7		
43	2.1		
44	1.9		
45	1.8		
46	1.9		
. . .	. . .		

Encyclopedia Link 8.1
Read about when to
use measures of central
tendency.

Measures of Central Tendency

Central tendency is usually summarized with one of three statistics: the mode, the median, or the mean. For any particular application, one of these statistics may be preferable, but each has a role to play in data analysis. To choose an appropriate measure of central tendency, the analyst must consider a variable's level of measurement, the skewness of a quantitative variable's distribution, and the purpose for which the statistic is used.

Mode

The **mode** is the most frequent value in a distribution. In a distribution of Americans' religious affiliations, Protestant Christian is the most frequently occurring value—the largest single group. In an age distribution of college students, 18- to 22-year-olds are by far the largest group and, therefore, the mode. One silly, but easy, way to remember the definition of the *mode* is to think of apple pie *á la mode,* which means pie with a big blob of vanilla ice cream on top. Just remember, the mode is where the big blob is—the largest collection of cases.

> **Mode (probability average):** The most frequent value in a distribution; also termed the probability average.

The mode is also sometimes termed the **probability average**, because being the most frequent value, it is the most probable. For example, if you were to pick a case at random from the distribution of age (Exhibit 8.8), the probability of the case being in his or her 50s would be 18.5%—the most probable value in the distribution.

The mode is used much less often than the other two measures of central tendency, because it can so easily give a misleading impression of a distribution's central tendency. One problem with the mode occurs when a distribution is **bimodal**, in contrast to being **unimodal**. A bimodal distribution has two categories with a roughly equal number of cases and clearly more cases than the other categories. In this situation, there is no single mode.

> **Bimodal:** A distribution in which two nonadjacent categories have about the same number of cases and these categories have more cases than any others.
>
> **Unimodal:** A distribution of a variable in which only one value is the most frequent.

Nevertheless, there are occasions when the mode is very appropriate. The mode is the only measure of central tendency that can be used to characterize the central tendency of variables measured at the nominal level. In addition, because it is the most probable value, it can be used to answer questions such as which ethnic group is most common in a given school.

Median

The **median** is the position average, or the point that divides the distribution in half (the 50th percentile). Think of the median of a highway—it divides the road exactly in two parts. To determine the median, we simply array a distribution's values in numerical order and find the value of the case that has an equal number of cases above and below it. If the median point falls between two cases (which happens if the distribution has an even number of cases), the median is defined as the average of the two middle values and is computed by adding the values of the two middle cases and dividing by 2. The median is not appropriate for variables that are measured at the nominal level; their values cannot be put in order, so there is no meaningful middle position.

> **Median:** The position average, or the point, that divides a distribution in half (the 50th percentile).

The median in a frequency distribution is determined by identifying the value corresponding to a cumulative percentage of 50. Starting at the top of the years of education distribution in Exhibit 8.9, for example, and adding up the percentages, we find that we reach 44.2% in the 12-years category and then 69.8% in the 13- to 15-years category. The median is therefore 13 to 15.

> **Mean:** The arithmetic, or weighted, average computed by adding up the value of all the cases and dividing by the total number of cases.

Mean

The **mean** is just the arithmetic average. (Many people, you'll notice, use the word *average* a bit more generally to designate everything we've called central tendency.)

In calculating a mean, any higher numbers pull it up, and any lower numbers pull it down. Therefore, it takes into account the values of each case in a distribution—it is a weighted average. (The median, by contrast, only depends on whether the numbers are higher or lower compared to the middle, not *how* high or low.)

The mean is computed by adding up the values of all the cases and dividing the result by the total number of cases, thereby taking into account the value of each case in the distribution:

Mean = Sum of value of cases / Number of cases

In algebraic notation, the equation is $X = \sum x_i / N$. For example, to calculate the mean value of 8 cases, we add the values of all the cases ($\sum x_i$ and divide by the number of cases (N):

$$(28 + 117 + 42 + 10 + 77 + 51 + 64 + 55) / 8 = 55.5$$

Exhibit 8.9	Years of Education Completed

Years of Education	Percentage
Less than 8	5.7%
8–11	11.2
12	27.3
13–15	25.6
16	16.3
17 or more	13.8
	100.0
	(2044)

Computing the mean obviously requires adding up the values of the cases. So it makes sense to compute a mean only if the values of the cases can be treated as actual quantities—that is, if they reflect an interval or ratio level of measurement—or if we assume that an ordinal measure can be treated as an interval (which is a fairly common practice). It makes no sense to calculate the mean of a qualitative (nominal) variable such as religion, for example. Imagine a group of four people in which there were 2 Protestants, 1 Catholic, and 1 Jew. To calculate the mean, you would need to solve the equation (Protestant + Protestant + Catholic + Jew) / 4 = ?. Even if you decide that Protestant = 1, Catholic = 2, and Jew = 3 for data entry purposes, it still doesn't make sense to add these numbers because they don't represent quantities of religion. In general, certain statistics (such as the mean) can apply only if there is a high enough level of measurement.

Median or Mean?

Because the mean is based on adding the value of all the cases, it will be pulled in the direction of exceptionally high (or low) values. In a positively skewed distribution, the value of the mean is larger than the median—more so the more extreme the skew. For instance, in Seattle, the presence of Microsoft owner Bill Gates—possibly the world's richest person—probably pulls the mean wealth number up quite a bit. One extreme case can have a disproportionate effect on the mean.

This differential impact of skewness on the median and mean is illustrated in Exhibit 8.10. On the first balance beam, the cases (bags) are spread out equally, and the median and mean are in the same location. On the second balance beam, the median corresponds to the value of the middle case, but the mean is pulled slightly upward toward the value of the one case with an unusually high value. On the third beam, the mean is clearly pulled up toward an unusual value. In some distributions the two measures will have markedly different values, and in such instances usually the median is preferred. (Income is a very common variable that is best measured by the median, for instance.)

Measures of Variation

Central tendency is only one aspect of the shape of a distribution—the most important aspect for many purposes but still just a piece of the total picture. The distribution, we have seen, also matters. It is important to

Research|Social Impact Link 8.2
Read about measures of variation.

Exhibit 8.10 The Mean as a Balance Point

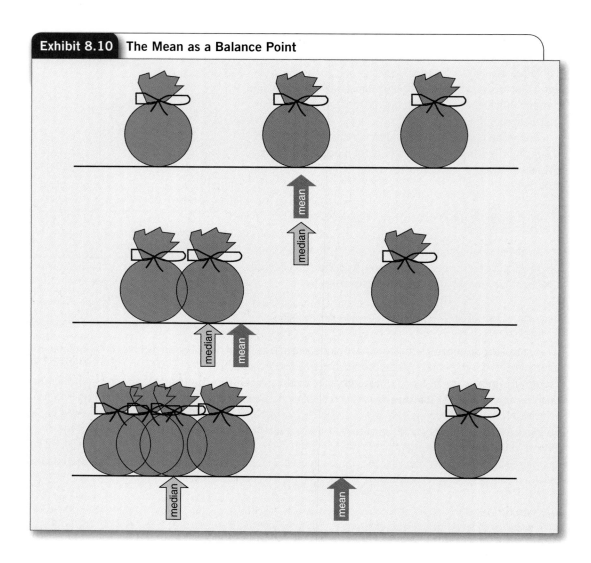

know that the median household income in the United States is a bit over $50,000 a year, but if the variation in income isn't known—the fact that incomes range from zero up to hundreds of millions of dollars—we haven't really learned much. Measures of variation capture how widely and densely spread income (for instance) is. Four popular measures of variation for quantitative variables are the range, the interquartile range, the variance, and the standard deviation (which is the single most popular measure of variability). Each conveys a certain kind of information, with strengths and weaknesses. Statistical measures of variation are used infrequently with qualitative variables and are not presented here.

Range: The true upper limit in a distribution minus the true lower limit (or the highest rounded value minus the lowest rounded value, plus 1).

Range

The **range** is the simplest measure of variation, calculated as the highest value in a distribution minus the lowest value, plus 1:

Range = Highest value − Lowest value + 1

It often is important to report the range of a distribution—to identify the whole range of possible values that might be encountered. However, because the range can be altered drastically by just one exceptionally high or low value—termed an **outlier**—it's not a good summary measure for most purposes.

Interquartile Range

The **interquartile range** avoids the problem outliers create by showing the range where most cases lie. **Quartiles** are the points in a distribution that correspond to the first 25% of the cases, the first 50% of the cases, and the first 75% of the cases. You already know how to determine the 2nd quartile, corresponding to the point in the distribution covering half of the cases—it is another name for the median. The interquartile range is the difference between the 1st quartile and the 3rd quartile (plus 1).

Variance

Variance, in its statistical definition, is the average squared deviation of each case from the mean; you take each case's distance from the mean, square that number, and take the average of all such numbers. Thus, variance takes into account the amount by which each case differs from the mean. The variance is mainly useful for computing the standard deviation, which comes next in our list here. An example of how to calculate the variance, using the following formula, appears in Exhibit 8.11:

$$\sigma^2 = \frac{\sum (X_i - \overline{X})^2}{N}$$

Symbol key: $\overline{X}$ = mean; N = number of cases; Σ = sum over all cases; X_i = value of case i on variable X.

The variance is used in many other statistics, although it is more conventional to measure variability with the closely related standard deviation than with the variance.

Standard Deviation

Very roughly, the **standard deviation** is the distance from the mean that covers a clear majority of cases (about two-thirds). More precisely, the standard deviation is simply the square root of the variance. It is the square root of the average squared deviation of each case from the mean:

$$\sigma^2 = \sqrt{\frac{\sum (X_i - \overline{X})^2}{N}}$$

Outlier: An exceptionally high or low value in a distribution.

Interquartile range: The range in a distribution between the end of the 1st quartile and the beginning of the 3rd quartile.

Quartiles: The points in a distribution corresponding to the first 25% of the cases, the first 50% of the cases, and the first 75% of the cases.

Variance: A statistic that measures the variability of a distribution as the average squared deviation of each case from the mean.

Exhibit 8.11		Calculation of the Variance	
Case #	**Score (X_i)**	**$X_i - \overline{X}$**	**$(X_i - \overline{X})^2$**
1	21	−3.27	10.69
2	30	5.73	32.83
3	15	−9.27	85.93
4	18	−6.27	39.31
5	25	0.73	0.53
6	32	7.73	59.75
7	19	−5.27	27.77
8	21	−3.27	10.69
9	23	−1.27	1.61
10	37	12.73	162.05
11	26	1.73	2.99
			434.15

Mean: $\overline{X}$ = 267/11 = 24.27

Sum of squared deviations = 434.15

Variance: σ^2 = 434.15/11 = 39.47

Variance: A statistic that measures the variability of a distribution as the average squared deviation of each case from the mean.

Standard deviation: The square root of the average squared deviation of each case from the mean.

Normal distribution: A symmetric distribution shaped like a bell and centered around the population mean, with the number of cases tapering off in a predictable pattern on both sides of the mean.

Symbol key: $\bar{X}$ = mean; N = number of cases; Σ = sum over all cases; X_i = value of case on i variable X; $\sqrt{}$ = square root.

The standard deviation has mathematical properties that make it the preferred measure of variability in many cases, particularly when a variable is normally distributed. A graph of a **normal distribution** looks like a bell, with one "hump" in the middle, centered around the population mean, and the number of cases tapering off on both sides of the mean (Exhibit 8.12). A normal distribution is symmetric: If you were to fold the distribution in half at its center (at the population mean), the two halves would match perfectly. If a variable is normally distributed, 68% of the cases (almost exactly two-thirds) will lie between ±1 standard deviation from the distribution's mean, and 95% of the cases will lie between 1.96 standard deviations above and below the mean.

So the standard deviation, in a single number, tells you quickly about how wide the variation is of any set of cases, or the range in which most cases will fall. It's very useful.

How Can We Tell Whether Two Variables Are Related?

Audio Link 8.1

Listen to an example of normal distribution.

Univariate distributions are nice, but they don't say how variables relate to each other—for instance, if religion affects education or if marital status is related to income. To establish cause, of course, one's first task is to show an association between independent and dependent variables (cause and effect). **Cross-tabulation** is a simple, easily understandable first step in such quantitative data analysis. Cross-tabulation displays the distribution of one variable within each category of another variable; it can also be termed a *bivariate distribution*, since it shows two variables at the same time. Exhibit 8.13 displays the cross-tabulation of voting by income so that we can see if the likelihood of voting increases as income goes up.

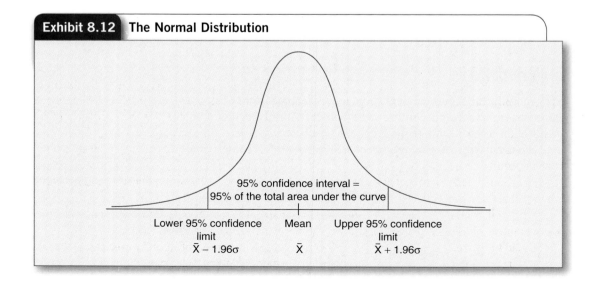

Exhibit 8.12   The Normal Distribution

95% confidence interval = 95% of the total area under the curve

Lower 95% confidence limit
$\bar{X} - 1.96\sigma$

Mean
$\bar{X}$

Upper 95% confidence limit
$\bar{X} + 1.96\sigma$

| Exhibit 8.13 | Cross-Tabulation of Voting in 2008 by Family Income: Cell Counts and Percentages |

Voting	<$20,000	$20,000–$39,999	$40,000–$74,999	$75,000+
Cell Counts	**Family Income**			
Voted	228	266	334	404
Did not vote	187	124	109	56
Total (n)	(415)	(390)	(443)	(460)
Percentages				
Voted	55	68	75	88
Did not vote	45	32	25	12
Total	100	100	100	100

The "crosstab" table is presented first (the upper part) with frequencies and then again (the lower part) with percentages. The *cells* of the table are where row and column values intersect; for instance, the first cell is where < $20,000 meets Voted; 228 is the value. Each cell represents cases with a unique combination of values of the two variables. The independent variable is usually the column variable, listed across the top; the dependent variable, then, is usually the row variable. This format isn't necessary, but social scientists typically use it.

> **Cross-tabulation (crosstab):** In the simplest case, a bivariate (two-variable) distribution showing the distribution of one variable for each category of another variable; can also be elaborated using three or more variables.

Reading the Table

The first (upper) table in Exhibit 8.13 shows the raw number of cases with each combination of values of voting and family income. It is hard to look at the table in this form and determine whether there is a relationship between the two variables. What we really want to know is the likelihood, for any level of income, that someone voted. So we need to convert the cell frequencies into percentages. Percentages show the likelihood per 100 (*per cent* in Latin) that something occurs. The second table, then, presents the data as percentages within the categories of the independent variable (the column variable, in this case). In other words, the cell frequencies have been converted into percentages of the column totals (the *N* in each column). For example, in Exhibit 8.13, the number of people earning less than $20,000 who voted is 228 out of 415, or 55%. Because the cell frequencies have been converted to percentages of the column totals, the numbers add up to 100 in each column but not across the rows.

Note carefully: You must *always* calculate percentages within levels of the independent variable—adding numbers down the columns in our standard format. In this example, we want to know the chance that a person with an income of less than $20,000 voted, so we calculate what percentage of those people voted. Then we *compare* that to the chance that people of other income levels voted. Calculating percentages across the table, by contrast, will not show the effect of the independent variable on voting. To repeat, *always* calculate percentages within levels of the independent variable (think: with**in** the **in**dependent variable).

To read the percentage table, compare the percentage distribution of voting/not voting across the columns. Start with the lowest income category (in the left column). Move slowly from left to right, looking at each distribution down the columns. As income increases, you will see that the percentage who voted also increases, from 55% of those with annual incomes under $20,000 (in the first cell in the first column) up to 88% of those with incomes of $75,000 or more (the last cell in the body of the table in the first row). This result is consistent with the hypothesis: It seems that higher income is moderately associated with a greater likelihood of voting.

Now look at Exhibit 8.14, which relates gender (as the independent variable) to voting (the dependent variable). The independent variable is listed across the top, and the percentages have been calculated, correctly, down the columns with values of the independent variable. Does gender affect voting? As you look down the first column, you see that 68.6% of men voted; then, in the second column, 75.3% of women voted. Gender did, in this table, have some effect on voting. Women were more likely to vote.

Some standard practices should be followed in formatting percentage tables (crosstabs): When a table is converted to percentages, usually just the percentages in each cell should be presented, and not the number of cases in each cell. Include 100% at the bottom of each column (if the independent variable is the column variable) to indicate that the percentages add up to 100, as well as the base number (N) for each column (in parentheses). If the percentages add up to 99 or 101 due to rounding error, just indicate so in a footnote. As noted already, there is no requirement that the independent variable always be the column variable, although consistency within a report or paper is a must. If the independent variable is the row variable, we calculate percentages in the cells of the table on the row totals (the N in each row), and the percentages add up to 100 across the rows.

Exhibit 8.15 shows two different tables. The top half shows voting by education—that is, the likelihood that a person with a given level of education voted in 2008. Look first at the voting distribution for high school graduation: The percent voting has jumped to over 68%—a significant change from the percentage for grade school completers. As you move across to the numbers for some college, then college graduates, it becomes obvious that education has a major effect on a person's likelihood of voting.

Now try looking at the lower table, which is a bit more complex, since it shows several levels of the dependent variable, family income. Try to see the effect that education has on income. Among the 283 grade school graduates surveyed (the first column on the left), you can see that 52.7%—more than half—have incomes under $20,000 a year. Shifting to the high school graduates, the number in that lowest-income category has clearly fallen: The distribution has shifted some toward the higher income results. With some college, that trend continues; and for college graduates, you can see that 50.3% of them—more than half!—are making over $75,000 a year. That's more than double (50.3 to 24.0) the percent of people who only did some college.

So, education has a powerful effect on a person's chances for making a high income—which may be why many of you are reading this book right now!

When you read research reports and journal articles, you will find that social scientists usually judge the strength of association on the basis of more statistics than just a cross-tabulation table. A **measure of association** is a descriptive statistic used to summarize the strength of an association. One measure of association in

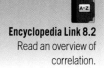

Encyclopedia Link 8.2
Read an overview of correlation.

Exhibit 8.14	Voting in 2008 by Gender	

Voting	Gender	
	Male	**Female**
Voted	68.6%	75.3%
Did not vote	31.4	24.7
Total	100%	100%
(n)	(829)	(1,091)

Exhibit 8.15	Voting in 2008 by Education and Income by Education

Voting by Education

Voting	Education			
	Grade School	**High School**	**Some College**	**College Graduate**
Voted	45.0%	68.2%	76.0%	87.6%
Did not vote	55.0	31.8	24.0	12.4
Total	100%	100%	100%	100%
(n)	(307)	(529)	(499)	(582)

Family Income by Education

Family Income	Education			
	Grade School	**High School**	**Some College**	**College Graduate**
<$20,000	52.7%	26.6%	23.6%	10.1%
$20,000–$39,999	30.7	30.1	22.3	12.8
$40,000–$74,999	12.4	27.8	30.1	26.8
$75,000+	4.2	15.5	24.0	50.3
Total	100%	100%	100%	100%
(n)	(283)	(489)	(475)	(555)

cross-tabular analyses with ordinal variables is called **gamma**. The value of gamma ranges from −1 to +1. The closer a gamma value is to −1 or +1, the stronger the relationship between the two variables; a gamma of zero indicates that there is no relationship between the variables. Inferential statistics go further, addressing whether an association exists in the larger population from which the (random) sample was drawn. Even when the empirical association between two variables supports the researcher's hypothesis, it is possible that the association was just due to the vagaries of random sampling. In a crosstab, estimation of this probability can be based on the inferential statistic, **chi-square**. The probability is customarily reported in a summary form such as $p < .05$, which can be translated as "The probability that the association was due to chance is less than 5 out of 100 (5%)."

When the analyst feels reasonably confident (at least 95% confident, or $p < .05$) that an association was not due to chance, it is said that the association is statistically significant. **Statistical significance** basically means we conclude that the relationship is actually there; it's not a chance occurrence. Convention (and the desire to avoid concluding that an association exists in the population when it doesn't) dictates that the criterion be a probability of less than 5%. Statistical significance, though, doesn't equal substantive significance. That is, while the relationship is really occurring, not just happening accidentally, it may still not matter very much. It may be a minor part of what's happening.

Measure of association: A type of descriptive statistic that summarizes the strength of an association.

Gamma: A measure of association that is sometimes used in cross-tabular analysis.

Chi-square: An inferential statistic used to test hypotheses about relationships between two or more variables in a cross-tabulation.

Statistical significance: The mathematical likelihood that an association is not due to chance, judged by a criterion the analyst sets (often that the probability is less than 5 out of 100, or $p < .05$).

Extraneous variable: A variable that influences both the independent and dependent variables so as to create a spurious association between them that disappears when the extraneous variable is controlled.

Elaboration analysis: The process of introducing a third variable into an analysis to better understand—to elaborate—the bivariate (two-variable) relationship under consideration. Additional control variables also can be introduced.

Controlling for a Third Variable

Cross-tabulation also can be used to study the relationship between three or more variables. The single most important reason for introducing a third variable is to see whether a bivariate relationship is spurious. A third, **extraneous variable**, for instance, may influence both the independent and dependent variables, creating an association between them that disappears when the extraneous variable is controlled. Ruling out possible extraneous variables helps to strengthen considerably the conclusion that the relationship between the independent and dependent variables is causal—that it is nonspurious. In general, adding variables is termed **elaboration analysis**: the process of introducing control or intervening variables into a bivariate relationship to better understand the relationship (Davis 1985; Rosenberg 1968).

For example, we have seen a positive association between incomes and the likelihood of voting; people with higher incomes are more likely to vote. But perhaps that association only exists because education influences both income and likelihood of voting; maybe when we control for education—that is, when we hold the value of education constant—we will find that there is no longer an association between income and voting. This possibility is represented by the hypothetical three-variable causal model in Exhibit 8.16, in which the arrows show that education influences both income and voting, thereby creating a relationship between the two. To test whether there is such an effect of education, we create the trivariate table in Exhibit 8.17, showing the bivariate crosstabs for various levels of education separately.

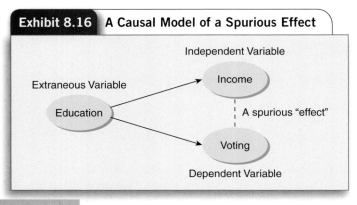

Exhibit 8.16 **A Causal Model of a Spurious Effect**

Independent Variable

Extraneous Variable

Income

Education

A spurious "effect"

Voting

Dependent Variable

This allows us to see if the income/voting relationship still exists after we hold education constant.

The trivariate cross-tabulation in Exhibit 8.17 shows that the relationship between voting and income is *not* spurious due to the effect of education. The association between voting and income occurs in all three subtables. So our original hypothesis—that income as a social status indicator has an effect on voting—is not weakened.

Our goal in introducing you to cross-tabulation has been to help you think about the association among variables and to give you a relatively easy tool for describing association. To read most statistical reports and to conduct more sophisticated analyses of social data, you will have to extend your statistical knowledge, at least to include the technique of *regression* or *correlation analysis.* These statistics have many advantages over cross-tabulation—as well as some disadvantages. You will need to take a course in social statistics to become proficient in the use of statistics based on regression and correlation.

Journal Link 8.1
Read an article where authors control for guilt.

Journal Link 8.2
Read about work assignment and career outcomes when controlling for work environment.

🔲 Analyzing Data Ethically: How Not to Lie With Statistics

Using statistics ethically means first and foremost being honest and open. Findings should be reported honestly, and the researcher should be open about the thinking that guided the decision to use particular

Exhibit 8.17 Voting in 2008 by Income and Education

Voting	Family Income			
	<$20,000	**$20,000–$39,999**	**$40,000–$74,999**	**$75,000+**
Education = Grade school				
Voted	38.0%	40.5%	55.2%	55.6%
Did not vote	62.0	59.5	44.8	44.4
Total	100%	100%	100%	100%
(n)	(137)	(79)	(29)	(9)
Education = High school				
Voted	58.4%	67.4%	65.6%	85.1%
Did not vote	41.6	32.6	34.4	14.9
Total	100%	100%	100%	100%
(n)	(125)	(138)	(131)	(74)
Education = Some college				
Voted	63.5%	80.6%	78.3%	81.5%
Did not vote	36.5	19.4	21.7	18.5
Total	100%	100%	100%	100%
(n)	(104)	(103)	(138)	(108)
Education = College graduate				
Voted	75.5%	82.6%	86.1%	92.2%
Did not vote	24.5	17.4	13.9	7.8
Total	100%	100%	100%	100%
(n)	(49)	(69)	(144)	(269)

statistics. Although this section has a mildly humorous title (after Darrell Huff's 1954 little classic, *How to Lie With Statistics),* make no mistake about the intent: It is possible to distort social reality with statistics, and it is unethical to do so knowingly, even when the error is due more to carelessness than to deceptive intent. There are a few basic rules to keep in mind:

- Inspect the shape of any distribution for which you report summary statistics to ensure that the statistics do not mislead your readers because of an unusual degree of skewness.

- When you create graphs, be sure to consider how the axes you choose may change the distribution's apparent shape; don't deceive your readers. You have already seen that it is possible to distort the shape of a distribution by manipulating the scale of axes, clustering categories inappropriately, and the like.

Journal Link 8.3
Read about research on alcohol and partner violence.

- Whenever you need to group data in a frequency distribution or graph, inspect the ungrouped distribution and then use a grouping procedure that does not distort the distribution's basic shape.

- Test hypotheses formulated in advance of data collection as they were originally stated. When evaluating associations between variables, it becomes very tempting to search around in the data until something interesting emerges. Social scientists sometimes call this a "fishing expedition." Although it's not wrong to examine data for unanticipated relationships, inevitably some relationships between variables will appear just on the basis of chance association alone. Exploratory analyses must be labeled in research reports as such.

- Be honest about the limitations of using survey data to test causal hypotheses. Finding that a hypothesized relationship is not altered by controlling for some other variables does not establish that the relationship is causal. There is always a possibility that some other variable that we did not think to control, or that was not even measured in the survey, has produced a spurious relationship between the independent and dependent variables in our hypothesis (Lieberson 1985). We have to think about the possibilities and be cautious in our causal conclusions.

回 Conclusion

Audio Link 8.2
Listen to more information about quantitative studies.

With some simple statistics (means, standard deviations, and the like), a researcher can describe social phenomena, identify relationships among them, explore the reasons for these relationships (especially through elaboration), and test hypotheses about them. Statistics—carefully constructed numbers that describe an entire population of data—are amazingly helpful in giving a simple summation of complex situations. Statistics provide a remarkably useful tool for developing our understanding of the social world, a tool that we can use both to test our ideas and to generate new ones.

Unfortunately, to the uninitiated, the use of statistics can seem to end debate right there—one can't argue with the numbers. But you now know better. Numbers are worthless if the methods used to generate the data are not valid, and numbers can be misleading if they are not used appropriately, taking into account the type of data to which they are applied. In a very poor town with one wealthy family, the mean income may be fairly high—but grossly misleading. And even assuming valid methods and proper use of statistics, there's one more critical step, because the numbers do not speak for themselves. Ultimately, how we interpret and report statistics determines their usefulness.

Key Terms

Bar chart   158
Base number (*N*)   161
Bimodal   164
Central tendency   158
Chi-square   171
Cross-tabulation
  (crosstab)   168
Data cleaning   156
Descriptive statistics   155

Elaboration analysis   172
Extraneous variable   172
Frequency
  distribution   161
Frequency polygon   159
Gamma   171
Histogram   159
Inferential statistics   155
Interquartile range   167

Mean   164
Measure of association   170
Median   164
Mode (probability average)   164
Normal distribution   168
Outlier   167
Percentage   161
Quantitative data analysis   154
Quartiles   167

Highlights

- Data entry options include direct collection of data through a computer, use of scannable data entry forms, and use of data entry software. All data should be cleaned during the data entry process.

- Use of secondary data can save considerable time and resources but may limit data analysis possibilities.

- Bar charts, histograms, and frequency polygons are useful for describing the shape of distributions. Care must be taken with graphic displays to avoid distorting a distribution's apparent shape.

- Frequency distributions display variation in a form that can be easily inspected and described. Values should be grouped in frequency distributions in a way that does not alter the shape of the distribution. Following several guidelines can reduce the risk of problems.

- Summary statistics are often used to describe the central tendency and variability of distributions. The appropriateness of the mode, mean, and median vary with a variable's level of measurement, the distribution's shape, and the purpose of the summary.

- The variance and standard deviation summarize variability around the mean. The interquartile range is usually preferable to the range to indicate the interval spanned by cases due to the effect of outliers on the range. The degree of skewness of a distribution is usually described in words rather than with a summary statistic.

- Cell frequencies in cross-tabulation should normally be converted to percentages within the categories of the independent variable. A cross-tabulation can be used to determine the existence, strength, direction, and pattern of an association.

- Elaboration analysis can be used in cross-tabular analysis to test for spurious relationships.

- Inferential statistics are used with sample-based data to estimate the confidence that can be placed in a statistical estimate of a population parameter. Estimates of the probability that an association between variables may have occurred on the basis of chance are also based on inferential statistics.

- Honesty and openness are the key ethical principles that should guide data summaries.

STUDENT STUDY SITE

The Student Study Site, available at **www.sagepub.com/chambliss4e,** includes useful study materials including web exercises with accompanying links, eFlashcards, videos, audio resources, journal articles, and encyclopedia articles, many of which are represented by the media links throughout the text. The site also features Interactive Exercises—represented by the green icon here—to help you understand the concepts in this book.

Exercises

Discussing Research

1. We presented in this chapter several examples of bivariate and trivariate cross-tabulations involving voting in the 2008 presidential election. What additional influences would you recommend examining to explain voting in elections? Suggest some additional independent variables for bivariate analyses with voting, as well as several additional control variables to be used in three-variable crosstabs.

2. When should we control just to be honest? Should social researchers be expected to investigate alternative explanations for their findings? Should they be expected to check to see if the associations they find occur for different subgroups in their samples? Justify your answers.

Finding Research

1. Do a web search for information on a social science subject in which you are interested. How much of the information you find relies on statistics as a tool for understanding the subject? How do statistics allow researchers to test their ideas about the subject and generate new ideas? Write your findings in a brief report, referring to the websites on which you relied.

2. The National Bureau of Economic Research provides many graphs and numeric tables about current economic conditions (www.nber.org/). Review some of these presentations. Which displays are most effective in conveying information? Summarize what you can learn from this site about economic conditions.

Critiquing Research

1. Become a media critic. For the next week, scan a newspaper or some magazines for statistics. How many articles can you find that use frequency distributions, graphs, and the summary statistics introduced in this chapter? Are these statistics used appropriately and interpreted correctly? Would any other statistics have been preferable or useful in addition to those presented?

Doing Research

1. Create frequency distributions from lists in U.S. Census Bureau reports on the characteristics of states, cities, or counties or any similar listing of data for at least 100 cases (http://factfinder2 .census.gov/faces/nav/jsf/pages/index.xhtml0). You will have to decide on a grouping scheme for the distribution of variables, such as average age and population size; how to deal with outliers in the frequency distribution; and how to categorize qualitative variables, such as the predominant occupation. Decide what summary statistics to use for each variable. How well were the features of each distribution represented by the summary statistics? Describe the shape of each distribution. Propose a hypothesis involving two of these variables and develop a crosstab to evaluate the support for this hypothesis. Describe each relationship in terms of the four aspects of an association after converting cell frequencies to percentages in each table within the categories of the independent variable. Does the hypothesis appear to have been supported?

2. Exhibit 8.18 is a three-variable table created with survey data from 355 employees hired during the previous year at a large telecommunications company. Employees were asked if the presence of on-site child care at the company's offices was important in their decision to join the company.

   Reading the table:

   a. Does gender affect attitudes?

   b. Does marital status affect attitudes?

   c. Which of the preceding two variables matters more?

   d. Does being married affect men's attitudes more than women's?

3. If you have access to the SPSS statistical program, you can analyze data contained in the 2010 General Social Survey (GSS) file on the Study Site for this text. See Appendix C for instructions on using SPSS.

   Develop a description of the basic social and demographic characteristics of the U.S. population in 2010. Examine each characteristic with three statistical techniques: a graph, a frequency distribution, and a measure of central tendency (and a measure of variation, if appropriate).

   a. From the menu, select Graphs and then Legacy Dialogs and Bar. Select Simple Define [Marital—Category Axis]. Bars represent % of cases. Select Options (do not display groups defined by missing values). Finally, select Histogram for each of the variables [EDUC, EARNRS, TVHOURS, ATTEND].

   b. Describe the distribution of each variable.

   c. Generate frequency distributions and descriptive statistics for these variables. From the menu, select Analyze/ Descriptive Statistics/Frequencies. From the Frequencies window, set MARITAL, EDUC, EARNRS, TVHOURS, ATTEND. For the Statistics, choose the mean, median, range, and standard deviation.

   d. Collapse the categories for each distribution. Be sure to adhere to the guidelines given in the section "Grouped Data." Does the general shape of any of the distributions change as a result of changing the categories?

| Exhibit 8.18 | Is Child Care Important? By Gender and Marital Status |

	MEN		WOMEN	
	Single	**Married**	**Single**	**Married**
Not important	54%	48%	33%	12%
Somewhat important	24%	30%	45%	31%
Very important	22%	22%	22%	57%
	100%	100%	100%	100%
n =	(125)	(218)	(51)	(161)

e. Which statistics are appropriate to summarize the central tendency and variation of each variable? Do the values of any of these statistics surprise you?

4. Try describing relationships with support for capital punishment by using graphs. Select two relationships you identified in previous exercises and represent them in graphic form. Try drawing the graphs on lined paper (graph paper is preferable).

Ethics Questions

1. Review the frequency distributions and graphs in this chapter. Change one of these data displays so that you are "lying with statistics." (You might consider using the graphic technique discussed by Orcutt & Turner, 1993.)

2. Consider the relationship between voting and income that is presented in Exhibit 8.13. What third variables do you think should be controlled in the analysis to understand better the basis for this relationship? How might social policies be affected by finding out that this relationship was due to differences in neighborhood of residence rather than to income itself?

CHAPTER 9

Qualitative Methods

Observing, Participating, Listening

Q ualitative research goes straight to where people live—and die:

We see what those poor bastards go through. Seriously, when [a dying medical patient has] been resuscitated nine or ten times and their chest looks like raw meat, they've been fried from being defibrillated, they've had their chest pumped on, they've got a flat chest because their ribs are no more connected to their sternum . . . You know this guy doesn't have a chance in hell. I mean, he's already blown out, squash, herniated his brain, he doesn't have any spontaneous respirations, he's flat EEGs. You take care of him for eight hours, you know that this person is not viable, and you feel for him and you feel for the family . . . When you're resuscitating somebody and they get no response going into the code for an hour, and now has no EKG, no heart tracing, pupils are blown, fixed, no spontaneous respiration, blood gases are out in the ozone . . . you are the one that's going to turn to the resident and say, "Don't you think this is about it, don't you think we should call this?" (interview, as cited in Chambliss 1996:164)

Throughout this chapter, you will learn that some of our greatest insights into social processes can result from what appear to be very ordinary activities: observing, participating, listening, and talking. But you will also learn that qualitative research is much more than just doing what comes naturally: Qualitative researchers must observe keenly, take notes systematically, question respondents strategically, and prepare to spend more time and invest more of their whole selves than often occurs with experiments or surveys.

We begin with an overview of the major features of qualitative research. The next section discusses participant observation research, which is the most distinctive qualitative method. We then discuss intensive interviewing—a type of interviewing that qualifies as qualitative rather than quantitative research—and focus groups, an increasingly popular qualitative method. The last two sections discuss how to analyze qualitative data and make ethical decisions in qualitative research.

🔲 What Are *Qualitative* Methods?

Qualitative methods refer to several distinct research activities: participant observation, intensive interviewing, and focus groups.

Although these three qualitative designs differ in many respects, they share several features, in addition to the collection of qualitative data itself, that distinguish them from experimental and survey research designs (Denzin & Lincoln 1994; Maxwell 1996; Wolcott 1995):

- Qualitative researchers typically begin with *an exploratory research question* about what people think and how they act, and why, in some social setting. Their research approach is primarily inductive.

- The designs *focus on previously unstudied processes and unanticipated phenomena,* because previously unstudied attitudes and actions can't adequately be understood with a structured set of questions or within a highly controlled experiment.

- They have an *orientation to social context,* to the interconnections between social phenomena rather than to their discrete features.

- They *focus on human subjectivity,* on the meanings that participants attach to events and that people give to their lives.

- They have a *sensitivity to the subjective role of the researcher.* Qualitative researchers consider themselves as necessarily part of the social process being studied and, therefore, keep track of their own actions in, and reactions to, that social process.

> **Qualitative methods:** Methods, such as participant observation, intensive interviewing, and focus groups, that are designed to capture social life as participants experience it rather than in categories the researcher predetermines. These methods typically involve exploratory research questions, inductive reasoning, an orientation to social context, and a focus on human subjectivity and the meanings participants attach to events and to their lives.

Research|Social Impact Link 9.1
Read more about qualitative methods.

Case Study: Beyond Caring

In preparing to write his 1996 book *Beyond Caring: Hospitals, Nurses, and the Social Organization of Ethics,* Dan Chambliss spent many months, spread over 12 years, studying hospital nurses at work. Observing in several different hospitals, in different regions of the United States, Chambliss watched countless operations and emergency room crises, but he also sat up nights chatting with nurses on geriatric floors (specializing in the care of old people) and quietly watched for hours at a time while nurses did postoperative care; bathed patients; helped patients walk down the hall; or just met with each other and with doctors, technicians, and aides to discuss the day's work. He also conducted more than 100 formal interviews, averaging 1.5 hours or more each; he attended birthday parties and softball games and saw nurses in social situations as well as at professional conferences. This project exemplifies **field research**, which combines various forms of qualitative research.

> **Field research:** Research in which natural social processes are studied as they happen and left relatively undisturbed.

The resulting data are nothing like the clean list of responses given to a survey questionnaire. Instead, Chambliss (1996) wrote his book from boxes full of notes on his observations, such as these:

> [Today I witnessed] the needle injection of local anesthetic into a newborn (3 weeks) baby's skull, so they could remove a shunt. The two residents doing it discussed whether a local anesthetic would be sufficient; a general [anesthetic] would be dangerous. One said, "I can do it if you can." This exchange was carried out a couple of times. A nurse (man) stroked the infant's hand, talked softly to it, and calmed it immediately as they were setting up, putting in the IVs—hard to do, the veins are so small.
>
> The resident injected the local anesthetic. Everyone around was affected by the immediate widening of the baby's eyes as the needle first went in, and then the screaming. The resident doing it, though, was absolutely concentrated on the task. At one point the female resident mentioned her concern, saying something about the whole point of anesthetic is to lessen pain, not to increase it. The baby was put in pain, couldn't have known any reason for it, was helpless to resist. [Field Notes] (pp. 135–136)

Journal Link 9.1
Read how field research was used to examine social behavior after Hurricane Katrina.

So fieldwork involves, at its simplest, spending time with people in their own settings, watching them do what they do. Gary Allen Fine, a prominent field researcher, has studied Little League baseball, restaurant kitchens, high school debate teams, and people who hunt for mushrooms, to name a few settings. Chambliss had complete access to the working (and sometimes personal) lives of the nurses he studied.

Such research obviously requires a huge investment of time. Chambliss moved his residence several times during his research, living in apartments near the medical centers that he studied. He built his entire schedule, for months on end, around the opportunities for seeing often unseen things—emergency resuscitations, hidden malpractice, even the boredom of some nursing work.

But the investment can be worth the cost. Chambliss's (1996) early research on nurses primarily relied on tape-recorded interviews:

> These [interviews] produced many dramatic stories and often confirmed theories I already held, but as I began to spend more time in hospitals I began to doubt the veracity of interviews. I began to see how the interviews were a reflection of my interests as much as of my subjects' lives. The stories told were more exciting than the ordinary drudgery I saw; the nurses described in stories seemed more committed and courageous than some of those I actually watched. Interviewees told what they noticed and remembered, which I discovered to be a highly selective version of what actually occurred. Much of life, I found, consists precisely in not noticing what one does all the time. "There aren't any ethical problems here I can think of," said a pediatric research nurse mentioned earlier; "You should talk with people on the ethics committee," said nurses gathered outside the room of an AIDS patient. (Pp. 194)

Chambliss wanted to learn about nurses, so in a sense he did the obvious: He worked and talked with nurses, many of them, over a long period of time. But he also took care to study a variety of hospitals and different services within hospitals; he also "sampled" different times of the day and night and different kinds of patients. True, such research is inductive, and the researcher is open to surprises; Chambliss couldn't run controlled experiments or easily isolate independent and dependent variables. But even the most unstructured kind of research still adheres to the basic discipline of scientific method.

There are many different qualitative methods. Here we first focus attention on three qualitative methods that illustrate the flexibility of this approach: ethnography, netnography, and ethnomethodology. We then discuss how to collect data using three different qualitative strategies: participant observation, intensive interviewing, and focus groups. In Chapter 10, you will learn how researchers analyze data collected with these methods.

Video Link 9.1
Watch a clip about ethnography.

Ethnography

Field research borrows heavily from a long-standing traditional method of anthropological studies called **ethnography**. Ethnography is the study of a culture or cultures that some group of people share (Van Maanen 1995:4). As a method, it usually refers to participant observation by a single investigator immersing in the group for a long period of time (often 1 or more years). Ethnographic research can also be termed *naturalistic,* because it seeks to describe and understand the natural social world as it really is, in all its richness and detail. Anthropological field research has traditionally been ethnographic, and much sociological fieldwork shares these same characteristics. But there are no particular methodological techniques associated with ethnography other than just "being there." The analytic process relies on the thoroughness and insight of the researcher to "tell us like it is" in the setting, as she experienced it.

> **Ethnography:** The study and systematic recording of human cultures.

Code of the Street, Elijah Anderson's (2000:11) award-winning study of Philadelphia's inner city, captures the flavor of this approach:

> My primary aim in this work is to render ethnographically the social and cultural dynamics of the interpersonal violence that is currently undermining the quality of life of too many urban neighborhoods . . . How do the people of the setting perceive their situation? What assumptions do they bring to their decision making?

Anderson's methods are described in the book's preface: participant observation, including direct observation and in-depth interviews; impressionistic materials drawn from various social settings around the city; and interviews with a wide variety of people. Like most traditional ethnographers, Anderson (2000) describes his concern with being "as objective as possible" and using his training, as other ethnographers do, "to look for and to recognize underlying assumptions, their own and those of their subjects, and to try to override the former and uncover the latter" (p. 11).

From analysis of the data obtained in these ways, a rich description emerges of life in the inner city. Although we often do not "hear" the residents speak, we feel the community's pain in Anderson's (2000) description of "the aftermath of death":

> When a young life is cut down, almost everyone goes into mourning. The first thing that happens is that a crowd gathers about the site of the shooting or the incident. The police then arrive, drawing more of a crowd. Since such a death often occurs close to the victim's house, his mother or his close relatives and friends may be on the scene of the killing. When they arrive, the women and girls often wail and moan, crying out their grief for all to hear, while the young men simply look on, in studied silence. . . . Soon the ambulance arrives. (p. 138)

Researcher Interview Link 9.1
Watch a researcher describe her ethnographic research.

Anderson (2000) uses these descriptions as a foundation on which he develops the key concepts in his analysis, such as "code of the street":

> The "code of the street" is not the goal or product of any individual's actions but is the fabric of everyday life, a vivid and pressing milieu within which all local residents must shape their personal routines, income strategies, and orientations to schooling, as well as their mating, parenting, and neighbor relations. (p. 326)

Anderson's (2003) report on his Jelly's Bar study illustrates how an ethnographic analysis deepened as he became more socially integrated into the Jelly's Bar group. He thus became more successful at "blending the local knowledge one has learned with what we already know sociologically about such settings" (p. 39).

> I engaged the denizens of the corner and wrote detailed field notes about my experiences, and from time to time looked for patterns and relationships in my notes. In this way, an understanding of the setting came to me in time, especially as I participated more fully in the life of the corner and wrote my field notes about my experiences; as my notes accumulated, and as I reviewed them occasionally and supplemented them with conceptual memos to myself, their meanings became more clear, while even more questions emerged. (p. 15)

Some qualitative analysts are abandoning this rich ethnographic tradition, however. Many doubt that social scientists can perceive the social world objectively or receive impressions from people that are unaffected by their being studied (Van Maanen 2002).

Netnography

As you know from social media like Facebook, *communities* now refer not only to people in a common physical location, but also to relationships that develop online. Online communities may be formed by persons with similar interests or backgrounds, perhaps to create new social relationships that location or schedules did not permit, or to supplement relationships that emerge in a course of work or school or other ongoing social activities. Like communities of people who interact face-to-face, online communities can develop a culture and become sources of identification and attachment (Kozinets 2010:14–15). And like physical communities, researchers can study online communities through immersion in the group for an extended period. **Netnography**, also termed *cyberethnography* and *virtual ethnography* (James & Busher 2009:34–35), is the use of ethnographic methods to study online communities.

In some respects, netnography is similar to traditional ethnography. The researcher prepares to enter the field by becoming familiar with online communities and their language and customs, formulating an exploratory research question about social processes or orientation in that setting, selecting an appropriate community to study. Unlike in-person ethnographies, netnographies can focus on communities whose members are physically distant and dispersed. The selected community should be relevant to the research question, involve frequent communication among actively engaged members, and have a number of participants who, as a result, generate a rich body of textual data (Kozinets 2010:89).

The netnographer's self-introduction should be clear and friendly. Robert Kozinets (2010) provides the following example written about the online discussion space, alt.coffee:

> I've been lurking here for a while, studying online coffee culture on alt.coffee, learning a lot, and enjoying it very much . . . I just wanted to pop out of lurker status to let you know I am here . . . I will be wanting to quote some of the great posts that have appeared here, and I will contact the individuals

Journal Link 9.2
Read about an ethnographic study exploring community identity and the 9/11 attacks.

Netnography (cyberethnography and virtual ethnography): The use of ethnographic methods to study online communities.

Video Link 9.2
Watch an explanation of netnography as a modern adaptation of an ethnography.

by personal e-mail who posted them to ask their permission to quote them. I also will be making the document on coffee culture available to any interested members of the newsgroup for their perusal and comments—to make sure I get things right. (p. 93)

A netnographer must keep both observational and reflective field notes, but unlike a traditional ethnographer can return to review the original data—the posted text—long after it was produced. The data can then be coded, annotated with the researcher's interpretations, checked against new data to evaluate the persistence of social patterns, and used to develop a theory that is grounded in the data.

Research in the News

READERS' ONLINE FEEDBACK CAN BE VICIOUS

After a woman published an article in an online magazine about postpartum post–traumatic stress disorder, following a traumatic delivery experience with her baby boy, the nasty comments started to pour in to the area reserved for reader responses. She was told not to have any more babies and that she would be a bad mother. In a similar incident, an uninsured woman who had written of her inability to function after a car accident was told to "Get a Minnie Mouse bandage and go to sleep." Why do some people get so vicious on the Internet? One social scientist suggested that it is because of the lack of face-to-face interaction, which provides constant feedback about others' feelings through body language and gestures.

Source: Brodesser-Akner, Taffy. 2010. E-playgrounds can get vicious (online feedback from readers)." *The New York Times*, April 22:E8.

Ethnomethodology

Ethnomethodology, a notable variation of fieldwork, studies the way that participants construct the social world in which they live—how they "create reality"—rather than trying to describe the social world objectively. In fact, ethnomethodologists do not necessarily believe that we can find an objective reality; instead, how participants come to create and sustain a sense of "reality" is the focus of study. In the words of Jaber F. Gubrium and James A. Holstein (1997), in ethnomethodology, as compared to the naturalistic orientation of ethnography,

> **Ethnomethodology:** A qualitative research method focused on the way that participants in a social setting create and sustain a sense of reality.

> The focus shifts from the scenic features of everyday life onto the ways through which the world comes to be experienced as real, concrete, factual, and "out there." An interest in members' methods of constituting their worlds supersedes the naturalistic project of describing members' worlds as they know them. (p. 41)

Unlike the ethnographic analyst, who seeks to describe the social world as the participants see it, the ethnomethodological analyst seeks to maintain some distance from that world. The ethnomethodologist views a "code" of conduct, like that described by Elijah Anderson (2003), not as a description of a real normative force that constrains social action but as the way that people in the setting create a sense of order and social structure (Gubrium & Holstein 1997:44–45). The ethnomethodologist focuses on how reality is constructed, not on what it is.

How Does Participant Observation Become a Research Method?

Dan Chambliss used **participant observation** (or *fieldwork* or *field research*) to study nurses because it leaves natural social processes, in their natural setting, relatively undisturbed. It is a means for seeing the social world as the research subjects see it, in its totality, and for understanding subjects' interpretations of that world (Wolcott 1995:66). Participant observers seek to avoid the artificiality of experimental designs and the unnatural structured questioning of survey research (Koegel 1987:8). This method encourages consideration of the context in which social interaction occurs, of the complex and interconnected nature of social relations, and of the sequencing of events (Bogdewic 1999:49). Through it, we can understand the *mechanisms* (one of the criteria for establishing cause) of social life.

Participant observation: A qualitative method for gathering data that involves developing a sustained relationship with people while they go about their normal activities.

In his study of nursing homes, Timothy Diamond (1992) explained how his exploratory research question led him to adopt the method of participant observation:

> How does the work of caretaking become defined and get reproduced day in and day out as a business? . . . The everyday world of Ina and Aileen and their co-workers, and that of the people they tend . . . I wanted to collect stories and to experience situations like those Ina and Aileen had begun to describe. I decided that . . . I would go inside to experience the work myself. (p. 5)

The term *participant observer* actually represents a continuum of roles (Exhibit 9.1), ranging from being a complete observer who does not participate in group activities and is publicly defined as a researcher to being a covert participant who acts just like other group members and does not disclose his research role. Many field researchers develop a role between these extremes, publicly acknowledging being a researcher but nonetheless participating in group activities.

Choosing a Role

The first concern of all participant observers is deciding what balance to strike between observing and participating and whether to reveal their roles as researchers. These decisions must take into account the specifics of the social situation being studied, the researcher's own background and personality, the larger sociopolitical context, and ethical concerns. Which balance of participating and observing is most appropriate also changes during most projects—often many times.

Complete Observation

Complete observation: A role in participant observation in which the researcher does not participate in group activities and is publicly defined as a researcher.

In **complete observation**, researchers try to see things as they happen, without actively participating in these events. Chambliss watched nurses closely, but he never bathed a patient, changed a dressing, started an intravenous line, or told a family that their loved one had died. Once during an emergency surgery for a ruptured ectopic pregnancy—a drastic, immediately life-threatening event—a surgeon ordered him to "put in a Foley" (a urinary catheter), but a nurse quickly said, "He's a researcher, I'll do it." Of course, at the same time as observing a setting,

Exhibit 9.1 The Observational Continuum

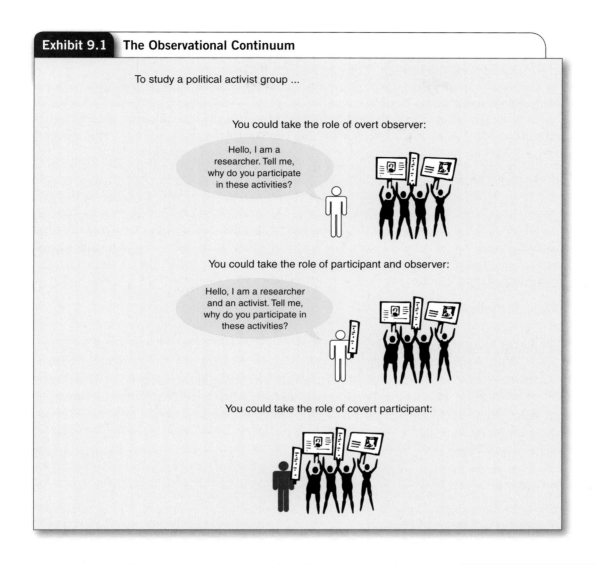

To study a political activist group ...

You could take the role of overt observer:

Hello, I am a researcher. Tell me, why do you participate in these activities?

You could take the role of participant and observer:

Hello, I am a researcher and an activist. Tell me, why do you participate in these activities?

You could take the role of covert participant:

researchers must take into account the ways in which their presence as observers itself alters the social situation being observed. Such **reactive effects** occur because it is not "natural" for someone to be present, recording observations for research and publication purposes (Thorne 1993:20).

Reactive effects: The changes in an individual or group behavior that are due to being observed or otherwise studied.

Mixed Participation or Observation

Most field researchers adopt a role that involves some active participation in the setting. Usually they inform at least some group members of their research interests, but then they participate in enough group activities to develop rapport with members and to gain a direct sense of what group members experience. This is not an easy balancing act. In his massive, 10-year study of gangs in urban America, Martin Sanchez Jankowski (1991:13) participated in nearly all the things they did. "I ate where they ate, I slept where they slept, I stayed with their families, I traveled where they went, and . . . I fought with them. The only things that I did not participate in were those activities that were illegal . . . (including taking drugs)."

And Jankowski (1991) says that although, for instance, the fights he was in "often left bruises, I was never seriously hurt. Quite remarkably, in the more than 10 years during which I conducted this research, I was only seriously injured twice" (p. 12).

A strategy of mixed participation and observation has two clear ethical advantages. Because group members know the researcher's real role in the group, they can choose to keep some information or attitudes hidden. By the same token, a researcher such as Jankowski can decline to participate in unethical or dangerous activities. Most field researchers get the feeling that, after they have become known and at least somewhat trusted figures in the group, their presence does not have any palpable effect on members' actions.

One especially interesting example of a mixed strategy is Chambliss's work on Olympic-level competitive swimmers. While working as a pure observer with a large number of world-class swimmers and teams, Chambliss himself coached, for 6 years, a small, local team in New York State. Here he tried to apply what he had learned through his years of research about what produces Olympic athletes. If his theories were correct, he reasoned, he should be able to make his *own* team much better. And, in fact, his swimmers improved dramatically, from being a rather poor local team to producing some state champions and even a few national-class athletes (Chambliss 1989). His written reports thus include a very unusual mix of observations, theorizing, and practical field experimentation to test his theory.

Complete Participation

Some field researchers adopt a **complete participation** role in which one operates as a fully functioning member of the setting. Most often, such research is also **covert**, or secret—other members don't know that the researcher is doing research. In one famous covert study, Laud Humphreys (1970) served as a "watch queen" so that he could learn about men engaging in homosexual acts in a public restroom. In another case, Randall Alfred (1976) joined a group of Satanists to investigate group members and their interaction. And Erving Goffman (1961) worked as a state mental hospital attendant while studying the treatment of psychiatric patients.

> **Complete (covert) participation:** A role in field research in which the researcher does not reveal his or her identity as a researcher to those who are observed.

Covert participants don't disrupt their settings, but they do face other problems. They must write up notes from memory and must do so when it would be natural for them to be away from group members. Researchers often run to the bathroom to scribble their notes, jot reminders on napkins to expand on later, or whisper into hand-held recorders when they are out of the room. Researchers' spontaneous reactions to every event are unlikely to be consistent with those of the regular participants (Mitchell 1993), because they are not "really" interested in washroom sex, Satanists, or psychiatric ward attendants. When Timothy Diamond (1992) did covert research as an aide in a nursing home, his economic resources showed:

> "There's one thing I learned when I came to the States," [said a Haitian nursing assistant]. "Here you can't make it on just one job." She tilted her head, looked at me curiously, then asked, "You know, Tim, there's just one thing I don't understand about you. How do you make it on just one job?" (pp. 47–48)

Ethical issues have been at the forefront of the debate over the strategy of covert participation. Some covert observers may become so wrapped up in the role they are playing that they adopt not just the mannerisms but also the perspectives and goals of the regular participants—they "go native"—and so may end up "going along to get along" with group activities that are themselves unethical. Kai Erikson (1967) argues that covert participation is, therefore, by its very nature unethical and should not be allowed except in public settings. If others suspect the researcher's identity or if the researcher contributes to, or impedes, group action, these consequences can be adverse. Covert researchers cannot anticipate the unintended consequences of their actions for research subjects or even for other researchers; covert research may, for instance, increase public distrust of all social scientists.

Video Link 9.3
Watch a clip about qualitative accounts that help to inform research questions.

Of the thirty-seven gangs studied, thirteen were in the Los Angeles area, twenty were in the New York City area, and four were in the Boston area. Various ethnic groups are represented in the sample, which includes gangs composed of Irish, African-American, Puerto Rican, Chicano, Dominican, Jamaican, and Central American members. The sample also involves gangs of varying size. The smallest had thirty-four members; the largest had more than one thousand . . . Within this sample, stratified by ethnicity, I randomly selected ten in each city.

It was my intention to study African-American gangs, Latino gangs, Asian gangs, and white gangs, and so gangs representing each of these ethnic groups were chosen. Because I wanted to include gangs of varying membership sizes, I randomly selected gangs from my ethnically stratified list until I obtained a sample representing gangs of different sizes. Since my overall strategy was to study five gangs in Los Angeles and five in New York for two years, then add more, and finally add several Boston gangs, I selected five of the original ten chosen and began my effort to secure their participation. (pp. 6–7)

Exhibit 9.2 **Sampling Plan for Participant Observation in Schools**

Information Source*	Type of Information to Be Obtained				
	Collegiality	Goals and Community	Action Expectations	Knowledge Orientation	Base
SETTINGS					
Public places (halls, main offices)					
Teacher's lounge	X	X		X	X
Classrooms		X	X	X	X
Meeting rooms	X		X	X	
Gymnasium or locker room		X			
EVENTS					
Faculty meetings	X		X		X
Lunch hour	X				X
Teaching		X	X	X	X
PEOPLE					
Principal		X	X	X	X
Teachers	X	X	X	X	X
Students		X	X	X	
ARTIFACTS					
Newspapers		X	X		X
Decorations		X			

*Selected examples in each category.

Exhibit 9.3 Theoretical Sampling

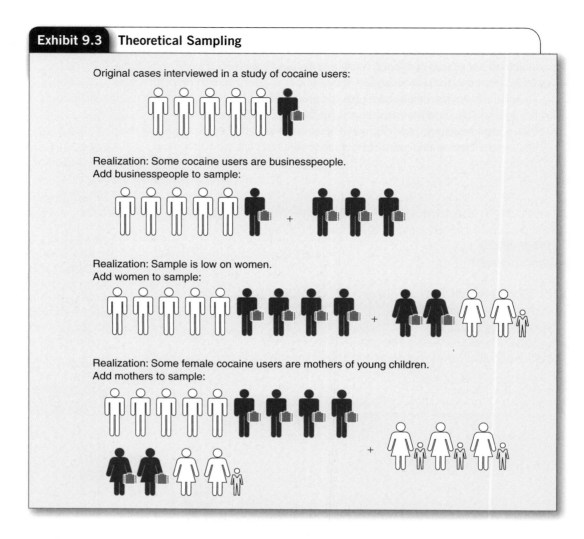

Original cases interviewed in a study of cocaine users:

Realization: Some cocaine users are businesspeople.
Add businesspeople to sample:

Realization: Sample is low on women.
Add women to sample:

Realization: Some female cocaine users are mothers of young children.
Add mothers to sample:

Taking Notes

Notes are the primary means of recording participant observation data (Emerson, Fretz, & Shaw 1995). It is almost always a mistake to try to take comprehensive notes while engaged in the field—the process of writing extensively is just too disruptive. The usual procedure is to jot down brief notes about highlights of the observation period. These brief notes then serve as memory joggers when writing the actual **field notes** later. It also helps to maintain a daily log in which each day's activities are recorded (Bogdewic 1999: 58–67). With the aid of the **jottings** and some practice, researchers usually remember a great deal of what happened—as long as the comprehensive field notes are written immediately afterward or at least within the next 24 hours, and before they have been discussed with anyone else.

Usually writing up notes takes much longer—at least three times longer—than the observing did. Field notes must be as complete, detailed, and true to what was observed and heard as possible. Direct quotes should be distinguished clearly from paraphrased quotes, and both should be set off from the researcher's observations and reflections. The surrounding context should receive as much attention as possible, and a map of the setting should be included, with indications of where individuals were at different times.

Field notes: Notes that describe what has been observed, heard, or otherwise experienced in a participant observation study. These notes usually are written after the observational session.

Jottings: Brief notes written in the field about highlights of an observation period.

Careful note taking yields a big payoff. On page after page, field notes will suggest new concepts, causal connections, and theoretical propositions. Notes also should include descriptions of the methodology and a record of the researcher's feelings and thoughts while observing. Exhibit 9.4 illustrates these techniques with notes from the Chambliss study.

Encyclopedia Link 9.1
Read about the importance of field notes in qualitative research.

Exhibit 9.4   Sample Field Notes From the Chambliss Nursing Study

Note: Original field notes, either written on site or typed later that day. Identifying information has been blacked out. "ISCU" stands for "Infant Special Care Unit," where premature infants are cared for. The first sentence reads, "Don't observe us tonight," "we're short [staffed]," a quotation from a nurse in the unit.

Managing the Personal Dimensions

Field researchers cannot help but be affected on a personal, emotional level by social processes in the social situation they are studying. At the same time, those being studied react to researchers not just as researchers but as personal acquaintances—and often as friends. Managing and learning from this personal side of field research is an important part of any project.

The researcher, like his informants, is a social animal. He has a role to play, and he has his own personality needs that must be met in some degree if he is to function successfully. Where the researcher operates out of a university, just going into the field for a few hours at a time, he can keep his personal social life separate from field activity. His problem of role is not quite so complicated. If, on the other hand, the researcher is living for an extended period in the community she is studying, her personal life is inextricably mixed with her research. (Whyte 1955:279)

Barrie Thorne (1993), a sociologist known for her research on gender roles among children, wondered whether "my moments of remembering, the times when I felt like a ten-year-old girl, [were] a source of distortion or insight?" (p. 26). She concluded they were both: "Memory, like observing, is a way of knowing and can be a rich resource." But "when my own responses . . . were driven by emotions like envy or aversion, they clearly obscured my ability to grasp the full social situation" (p. 26).

There is no formula for successfully managing the personal dimension of field research. It is much more art than science and flows more from the researcher's own personality and natural approach to other people than from formal training. But novice field researchers often neglect to consider how they will manage personal relationships when they plan and carry out their projects. Attention to a few guidelines based on our personal experience with field research, provided in Exhibit 9.5, should maximize the likelihood of a project's success.

Systematic Observation

Observations can be made in a more systematic, quantitative design that allows systematic comparisons and more confident generalizations. A researcher using systematic observation develops a standard form on which to record variation within the observed setting in terms of variables of interest. Such variables might include the frequency of some behavior(s), the particular people observed, the weather or other environmental conditions, and the number and state of repair of physical structures. In some systematic observation studies, records will be obtained from a random sample of places or times.

Robert Sampson and Stephen Raudenbush's (1999) study of disorder and crime in urban neighborhoods provides an excellent example of systematic observation methods. In this section we explore their use of systematic social observation to learn about these neighborhoods. A systematic observational strategy increases the reliability of observational data by using explicit rules that standardize coding practices across observers (Reiss 1971). It is a method particularly well suited to overcome one of the limitations of survey research on crime and disorder: Residents who are fearful of crime perceive more neighborhood disorder than do residents who are less fearful, even though both are observing the same neighborhood (Sampson & Raudenbush 1999:606).

This ambitious multiple-methods investigation combined observational research, survey research, and archival research. The observational component involved a stratified probability (random) sample of 196 Chicago census tracts. A specially equipped sport-utility vehicle was driven down each street in these tracts at the rate of 5 miles per hour. Two video recorders taped the blocks on both sides of the street, while two observers peered out of the vehicle's windows and recorded their observations in the logs. The result was an observational record of 23,816 face blocks (the block on one side of the street is a face block). The observers recorded in their logs codes that indicated land use, traffic, physical conditions, and evidence of

---

Exhibit 9.5   **Nine Steps to Successful Field Research**

1. *Have a simple, one-sentence explanation of your project.* "I want to learn about the problems nurses face in their work," or "I want to learn what makes a great swimming team." People will ask what you're doing, but no one cares to hear all your theories.

2. *Be yourself.* Don't lie about who you are. First, it's wrong. Second, you'll get caught and ruin the trust you're trying to build. (Yes, there are exceptions, but very few.)

3. *Don't interfere.* They got along just fine before you came along, and they can do it again. Don't be a pest.

4. *Listen, actively.* Be genuinely interested in what they say. Movie stars, politicians, and other celebrities are used to having other people listen to what they say, but that's not true for most people. If you really care to listen, they'll tell you everything.

5. *Show up,* at every opportunity—3:00 in the morning, or if you have to walk 5 miles. Go to their parties and their funerals. Make a 5-hour trip for a 15-minute interview, and they'll notice—and give you everything you want.

6. *Pay attention to everything,* especially when you're bored. That's when the important stuff is happening, the stuff *no one else* notices.

7. *Protect your sources,* more than is necessary. When word gets around that you can be trusted, you won't believe what people will tell you.

8. *Write everything down, that day.* By tomorrow, you'll forget 90% of the best material, and then it's gone forever.

9. *Always remember: It's not about you, it's about them.* Don't try to be smart, or savvy, or hip; don't try to be the center of attention. Stop thinking about yourself all the time. Pay attention to other people.

---

physical disorder (Exhibit 9.6). The videotapes were sampled and then coded for 126 variables, including housing characteristics, businesses, and social interactions. Physical disorder was measured by counting such features as cigarettes or cigars in the street, garbage, empty beer bottles, graffiti, condoms, and syringes. Indicators of social disorder included adults loitering, drinking alcohol in public, fighting, and selling drugs. To check for reliability, a different set of coders recoded the videos for 10% of the blocks. The repeat codes achieved 98% agreement with the original codes.

Sampson and Raudenbush also measured crime levels with data from police records, census tract socio-economic characteristics with census data, and resident attitudes and behavior with a survey. The combination of data from these sources allowed a test of the relative impact on the crime rate of residents' informal social control efforts and of the appearance of social and physical disorder.

Pierre St. Jean (2006) extended the research of Sampson and Raudenbush with a mixed-methods study of high crime areas that used resident surveys, participant observation, in-depth interviews with residents and offenders, and also systematic social observation. St. Jean recorded neighborhood physical and social appearances with video cameras mounted in a van that was driven along neighborhood streets. Pictures were then coded for the presence of neighborhood disorder (Exhibit 9.7).

This study illustrates both the value of multiple methods and the technique of recording observations in a form from which quantitative data can be obtained. The systematic observations give us much greater confidence in the measurement of relative neighborhood disorder than we would have from unstructured descriptive reports or from responses of residents to survey questions. Interviews with residents and participant observation helped to identify the reasons that offenders chose particular locations when deciding where to commit crimes.

Video Link 9.4
Watch Pierre St. Jean's video.

Exhibit 9.6   Neighborhood Disorder Indicators Used in Systematic Observation Log

Variable	Category	Frequency
Physical Disorder		
Cigarettes, cigars on street or gutter	no	6,815
	yes	16,758
Garbage, litter on street or sidewalk	no	11,680
	yes	11,925
Empty beer bottles visible in street	no	17,653
	yes	5,870
Tagging graffiti	no	12,859
	yes	2,252
Graffiti painted over	no	13,390
	yes	1,721
Gang graffiti	no	14,138
	yes	973
Abandoned cars	no	22,782
	yes	806
Condoms on sidewalk	no	23,331
	yes	231
Needles/syringes on sidewalk	no	23,392
	yes	173
Political message graffiti	no	15,097
	yes	14
Social Disorder		
Adults loitering or congregating	no	14,250
	yes	861
People drinking alcohol	no	15,075
	yes	36
Peer group, gang indicators present	no	15,091
	yes	20
People intoxicated	no	15,093
	yes	18
Adults fighting or hostilely arguing	no	15,099
	yes	12
Prostitutes on street	no	15,100
	yes	11
People selling drugs	no	15,099
	yes	12

Exhibit 9.7 One Building in St. Jean's (2007) Study

▣ How Do You Conduct Intensive Interviews?

Participant observation can provide a wonderfully rich view, then, of the social world. But it remains a *view,* seen by the observer. Often we wonder what individuals think or feel or how they see their world. For this purpose, one can use intensive interviews.

Unlike the more structured interviewing that may be used in survey research (discussed in Chapter 7), **intensive**, or **depth**, **interviewing** relies on open-ended questions to develop a comprehensive picture of the interviewee's background, attitudes, and actions—to "listen to people as they describe how they understand the worlds in which they live and work" (Rubin & Rubin 1995:3).

For instance,

> We had two or three patients, and they were terminally ill with cancer. We would give the patients, every two or three hours around the clock toward the end, morphine sulphate, intramuscular.
>
> I was really worried about giving them a morphine injection because the morphine depresses the respiration. I thought, well, is this injection going to do them in?

Intensive (depth) interviewing: A qualitative method that involves open-ended, relatively unstructured questioning in which the interviewer seeks in-depth information on the interviewee's feelings, experiences, and perceptions.

If I don't give the injection, they will linger on longer, but they might also have more pain. If I do give the injection, the end result of death is going to occur faster. Am I playing God?" [Interview] (Chambliss 1996: 171)

Audio Link 9.1

Listen to more information on conducting interviews.

The key to eliciting such a response is *active listening*—which is not the same as just being quiet. Instead, you must actively question, ask for explanations, and show a genuine deep curiosity about the subject's views and feelings. Your own opinions are not important here; you must suspend all judgment of what the respondent is saying, even if you regard the person's opinions as obnoxious or even immoral. Remember, the goal is to learn what the *respondent* thinks, not to express what you think.

Therefore, depth interviews may be highly unstructured. Rather than asking standard questions in a fixed order, a researcher conducting intensive interviews may allow the specific content and order of questions to vary from one interviewee to another. Like participant observation studies, intensive interviewing engages researchers actively with subjects. The researchers must listen to lengthy explanations, ask follow-up questions tailored to the preceding answers, and seek to learn about interrelated belief systems or personal approaches to things rather than measure a limited set of variables. As a result, intensive interviews are often much longer than standardized interviews, sometimes as long as 15 hours, conducted in several different sessions.

The intensive interview can become more like a conversation between partners than between a researcher and a subject (Kaufman 1986:22–23). Some call it "a conversation with a purpose" (Rossman & Rallis 1998:126). Robert Bellah, Richard Madsen, William Sullivan, Ann Swidler, and Steven Tipton (1985) elaborate on this aspect of intensive interviewing in a methodological appendix to their national best seller about American individualism, *Habits of the Heart*:

Research|Social Impact Link 9.2

Read about how qualitative studies help create bodies of research for other research questions.

We did not, as in some scientific version of "Candid Camera," seek to capture their beliefs and actions without our subjects being aware of us. Rather, we sought to bring our preconceptions and questions into the conversation and to understand the answers we were receiving not only in terms of the language but also, so far as we could discover, in the lives of those we were talking with. Though we did not seek to impose our ideas on those with whom we talked. . . . we did attempt to uncover assumptions, to make explicit what the person we were talking to might rather have left implicit. The interview as we employed it was active, Socratic. (p. 304)

Saturation point: The point at which subject selection is ended in intensive interviewing because new interviews seem to yield little additional information.

Random selection is rarely used to select respondents for intensive interviews, but the selection method still must be considered carefully. Researchers should try to select interviewees who are knowledgeable about the subject of the interview, who are open to talking, and who represent a range of perspectives (Rubin & Rubin 1995:65–92). Selection of new interviewees should continue, if possible, at least until the **saturation point** is reached, the point when new interviews seem to yield little additional information (Exhibit 9.8).

Establishing and Maintaining a Partnership

Because intensive interviewing does not engage researchers as participants in subjects' daily affairs, the problems of entering the field are much reduced. However, the logistics of arranging long periods for personal interviews can still be pretty complicated. It also is important to establish rapport with subjects by considering in advance how they will react to the interview arrangements and by developing an approach that does not violate their standards for social behavior. Interviewees should be treated with respect, as knowledgeable partners whose time is valued (in other words, don't be late for your appointments). A commitment to confidentiality should be stated and honored (Rubin & Rubin 1995).

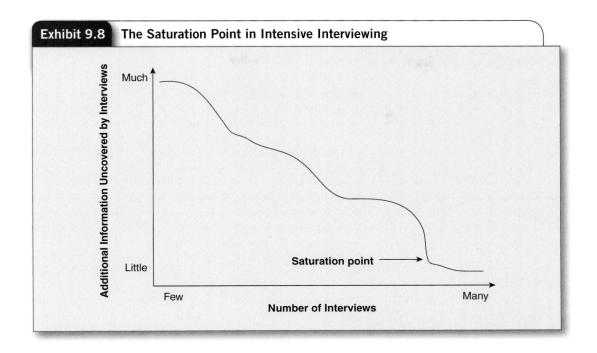

Exhibit 9.8 The Saturation Point in Intensive Interviewing

Asking Questions and Recording Answers

Intensive interviewers must plan their main questions around an outline of the interview topic. The questions generally should be short and to the point. More details can then be elicited through nondirective probes (such as "Can you tell me more about that?" or "Uh-huh," echoing the respondent's comment, or just maintaining a moment of silence). Follow-up questions can then be tailored to answers to the main questions.

Interviewers should strategize throughout an interview about how best to achieve their objectives while taking into account interviewees' answers. *Habits of the Heart* (Bellah et al., 1985) again provides a useful illustration:

[Coinvestigator Steven] Tipton, in interviewing Margaret Oldham [a pseudonym], tried to discover at what point she would take responsibility for another human being:

Q: So what are you responsible for?

A: I'm responsible for my acts and for what I do.

Q: Does that mean you're responsible for others, too?

A: No.

Q: Are you your sister's keeper?

A: No.

Q: Your brother's keeper?

A: No.

Q: Are you responsible for your husband?

A: I'm not. He makes his own decisions. He is his own person. He acts his own acts. I can agree with them, or I can disagree with them. If I ever find them nauseous enough, I have a responsibility to leave and not deal with it any more.

Q: What about children?

A: I . . . I would say I have a legal responsibility for them, but in a sense I think they in turn are responsible for their own acts. (p. 304)

Do you see how the interviewer actively encouraged the subject to *explain* what she meant by "responsibility"? This sort of active questioning undoubtedly did a better job of clarifying the interviewee's concept of responsibility than a fixed set of questions would have.

Tape recorders commonly are used for recording intensive interviews and focus group interviews. They do not inhibit most interviewees and, in fact, are routinely ignored. Occasionally respondents are very concerned with their public image and may therefore speak "for the tape recorder," but such individuals are unlikely to speak frankly in any research interview. In any case, constant note taking during an interview prevents adequate displays of interest and is distracting. Sometimes, though, the very act of visibly turning off the recorder may free a respondent to tell that one great secret being kept—and interviewers often use that very technique.

▣ How Do You Run Focus Groups?

Finally, for quick, emotionally resonant answers, **focus groups** can be the qualitative researcher's best friend. Long favored by advertisers, marketing researchers, and political consultants who want to see "what message pushes their buttons," focus groups are collections of unrelated individuals, convened by a researcher and then led in group discussion of a topic for 1 to 2 hours. The researcher asks specific questions and guides the discussion, but the resulting information is qualitative and relatively unstructured. Focus groups need not involve representative samples; instead, a few individuals are recruited for the group who have the time to participate, have some knowledge pertinent to the focus group topic, and share key characteristics with the target population. Throughout the Mellon Project on liberal arts education at Hamilton College, focus groups—of dean's list students, minority students, or study abroad participants, for instance—have been used to assess major problem areas in various programs rapidly and to develop areas for more systematic investigation.

> **Focus groups:** A qualitative method that involves unstructured group interviews in which the focus group leader actively encourages discussion among participants on the topics of interest.

Focus group research typically proceeds like this: The researcher convenes a series of groups, each including 7 to 10 people, for the discussions. Sometimes the groups are heterogeneous, with many dissimilar people (old and young, boss and employees, Democrats and Republicans); this can stimulate a broader array of opinions. But usually groups are, by design, homogeneous by categories one wishes to compare. For instance, a business might run eight focus groups, four from the sales offices and four from service offices, to learn how these different functions see their customers. Or a college could run focus groups of freshmen and sophomores to learn about the different ways these groups approach course registration. It's generally best (though not always possible) to have group members be strangers so that personal relationships don't affect their answers, and it's crucial to avoid power differentials—no bosses with subordinates, teachers with students, or parents with their children. Such combinations will prevent open and honest opinion from emerging (Krueger & Casey 2000).

Once completed, focus group discussions are relatively easy to analyze: Just compare the responses, on each question, from one kind of group (say, salespeople) to responses for the same question by another kind of group (say, service representatives).

Richard Krueger (1988) provides a good example of a situation in which focus groups were used effectively:

> [A] University recently launched a $100 million fund drive. The key aspect of the drive was a film depicting science and research efforts. The film was shown in over two dozen focus groups of alumni, with surprising results to University officials. Alumni simply did not like the film and instead were more attracted to supporting undergraduate humanistic education. (pp. 33–37)

Focus group methods share with other field research techniques an emphasis on discovering unanticipated findings and exploring hidden meanings. Although weak in developing reliable, generalizable results (the strength of survey research), focus groups can be indispensable for developing hypotheses and survey questions, for investigating the meaning of survey results, and for quickly assessing the range of opinion about an issue. Exhibit 9.9 presents guidelines for running focus groups.

Audio Link 9.2
Listen to a focus group in action.

Exhibit 9.9   Keys to Running Focus Groups

- A great moderator—Is neutral and genuinely respects the participants and is a great listener who can draw people out.
- Main questions—These ask what you really want to know, can be answered by participants, are clear and understandable to the participants, and provide useful answers.
- Participants—Are homogeneous by relevant category for comparisons, with no power differentials within the group.
- Sampling—Is purposeful, representing the entire range of responses, and is random within the pools meeting criteria. Ideally, participants in any group should be strangers to each other. Use reminders to attend with incentives.
- Recording—Audio recording, with an assistant taking notes, is best.
- Analysis—Compare answers of different groups to different questions (groups on differently colored paper, sorted by question, etc.).
- Reporting—You are speaking for the participants. Lead with the big insights and answer the questions that were asked of the study. Interesting quotations get attention!
- When in doubt—Ask the potential participants about food, setting, issues, moderator, etc.

Basically, good focus groups get honest answers, on important topics, from people who know.

🖳 Analyzing Qualitative Data

The data for a qualitative study most often are notes jotted down in the field or during an interview—from which the original comments, observations, and feelings are reconstructed—or text transcribed from audiotapes. What to do with all this material? Chapter 10 is a full discussion of this topic, but a few comments here provide an introduction.

Many field research projects have slowed to a halt because a novice researcher becomes overwhelmed by the quantity of information that has been collected. (A 1-hour interview can generate 20 to 25 pages of single-spaced text [Kvale 1996:169].) Boxes filled with those scribbled notes—old napkins with some words, flyers for events attended, e-mails from informants—can seem daunting, but a few simple steps can clarify the whole mess rather quickly.

The Phases of Analysis

Encyclopedia Link 9.2
Read more about
deductive qualitative
analysis.

Basically, there are two approaches to analyzing qualitative data: the deductive, or hypothesis-testing approach, and the inductive, or exploratory approach. Jankowski (1991), in his study of gangs, used a *deductive* method:

> I began the analysis by establishing topics that would need to be covered in a book about gangs, such as gang recruitment, gang organization, violence, and so on. I then proceeded to read each of my notes (daily notes, daily summaries, weekly summaries) and place them in stacks having to do with each topic I wanted to cover.
>
> When notes pertained to more than one topic, I photocopied them and placed each under the additional topics.
>
> My analysis began by taking a topic and reviewing what other researchers had found concerning gangs. Their findings would be written down in hypothesis form and then I would read my notes to determine what my evidence suggested. (p. 16)

Encyclopedia Link 9.3
Read an overview of
inductive qualitative analysis.

Jankowski created a series of topics and sorted his notes into those categories, which were created in advance. He then used the material in each category to test hypotheses from the literature on gangs.

An *inductive* approach, by contrast, allows themes and topics to emerge from the data themselves. In his nursing study, Chambliss first spent several weeks reading through all of his notes and skimming transcripts, taking notes on the topics covered; then he read through the notes themselves, organizing them into a handful of topics (e.g., routines of work, ethical problems, the role of the nurse) that seemed to arise repeatedly; and finally he resorted all of his "notes and quotes" into huge piles, one for each of those main topics. (As he did this work, his notes covered the entire living room floor of his apartment.)

As observation, interviewing, and reflection continue, researchers refine their definitions of problems, concepts, and select indicators. They can then check the frequency and distribution of phenomena. How many people made a particular type of comment? How often did social interaction lead to arguments? Hypotheses are modified as researchers gain experience in the setting. For the final analysis, the researchers check their models carefully against their notes and make a concerted attempt to discover negative evidence that might suggest the model is incorrect. Such an approach combines the inductive with the deductive.

🔲 Ethical Issues in Qualitative Research

Qualitative research can raise some complex ethical issues. No matter how hard the field researcher strives to study the social world naturally, leaving no traces, the very act of research itself imposes something "unnatural" on the situation. It is up to the researchers themselves to identify and take responsibility for the consequences of their involvement. Five main ethical issues arise:

1. *Voluntary participation*—Ensuring that subjects are participating in a study voluntarily is not often a problem with intensive interviewing and focus group research, but it is often a point of contention in participant observation studies. Few researchers or institutional review boards are willing to condone covert participation, because it does not offer any way to ensure that participation by the subjects is voluntary. Even when the researcher's role is more open, interpreting the standard of voluntary participation still can be difficult. Should the requirement of voluntary participation apply equally to every member of an organization being observed? What if the manager consents, the workers are ambivalent, and the union says no?

2. *Subject well-being*—Before beginning a project, every field researcher should consider carefully how to avoid harm to subjects. It is not possible to avoid every theoretical possibility of harm or to be sure that a project will cause no adverse consequences whatsoever to any individual, but direct harm to the reputations or feelings of particular individuals should be avoided at all costs. The risk of such harm can be minimized by maintaining the confidentiality of research subjects and by not adversely affecting the course of events while engaged in a setting. Whyte (1955:335–337) found himself regretting having recommended that a particular politician be allowed to speak to a social club he was observing, because the speech led to serious dissension in the club and strains between Whyte and some club members.

3. *Identity disclosure*—Current ethical standards require informed consent of research subjects, and most would argue that this standard cannot be met in any meaningful way if researchers do not disclose fully their identity. But how much disclosure about the study is necessary, and how hard should researchers try to make sure that their research purposes are understood? In field research on Codependents Anonymous, Leslie Irvine (1998) found that the emphasis on anonymity and the expectations for group discussion made it difficult for her to disclose her identity. Can a balance be struck between the disclosure of critical facts and a coherent research strategy?

4. *Confidentiality*—Field researchers normally use fictitious names for the characters in their reports, but doing so does not always guarantee confidentiality to their research subjects. In Chambliss's nursing book, reference to "the director of the medical center" might have identified that person, at least to other employees of the center who knew Chambliss did his research there. And anyone studying public figures or national leaders in a social movement must exercise special care, because their own followers or enemies can privately recognize such people. Researchers should thus make every effort to expunge any possible identifying material from published information and to alter unimportant aspects of a description when necessary to prevent identity disclosure. In any case, no field research project should begin if some participants clearly will suffer serious harm by being identified in project publications.

5. *Online research*—The large number of discussion groups and bulletin boards on the Internet has stimulated much interest in conducting research like that of Fox and Roberts (1999), who observed physicians' LISTSERVs in the United Kingdom. Such research can violate the principles of voluntary participation and identity disclosure when researchers participate in discussions and record and analyze text but do not identify themselves as researchers (Associated Press, 2000).

These ethical issues cannot be evaluated independently. The final decision to proceed must be made after weighing the relative benefits and risks to participants. Few qualitative research projects will be barred by consideration of these ethical issues, however, except for those involving covert participation. The more important concern for researchers is to identify the ethically troublesome aspects of their proposed research and resolve them before the project begins, as well as to act on new ethical issues as they come up during the project.

🖾 Conclusion

Qualitative research has both immediate and lasting attractions. Many of the classic works of social science, from Sigmund Freud's *Interpretation of Dreams* (1900/1999) and Margaret Mead's *Coming of Age in Samoa* (1928/2001) to Erving Goffman's *Presentation of Self in Everyday Life* (1959) and Kristin Luker's *Abortion and the Politics of Motherhood* (1985), rest on qualitative forms of social research. Telling true stories of real people, laying out their feelings and emotions, is qualitative research—interviews, fieldwork, and focus groups cut through the dry numbers and correlations, the abstract variables, and the hypotheses of contemporary quantitative social science. Qualitative research aims to go, as we said at the beginning of this chapter, where real people live. It thereby can become, at its best, a form of literature, beautifully teaching its readers the deeper truths of the human condition. More modestly, many students simply find reading reports of qualitative research to be far more interesting than the statistics used in survey analysis.

But "interesting" is not always the same as accurate, correct, or even representative. The juiciest stories that Dan Chambliss heard from his nurses were not, as it happens, what typically happened in their lives. Researchers love a good quote, but it may not represent the truth of a setting; fieldworkers love finding a key informant whose views may not be those of the average subject. Like journalists, even the best qualitative researchers may be drawn to the odd, the unusual, or the available—and all of those may be poor substitutes for representative sampling, standardized questions, and other more sober approaches to learning about social life. The statistics of survey analysis and the control groups of experiments force us to face reality with self-discipline; they make it harder to fool ourselves about what we see.

In the end, qualitative methods are one—and only one—excellent set of tools, complementary in purpose to the tools of surveys, experiments, and other methods. Each has its strengths and its weaknesses. When surveys find that college students complain about "social life" but also rejoice that they "made my best friends ever here," interviews can explain the (apparent) contradiction. When police statistics and crime surveys can't fathom the logic of gang life, Martin Sanchez Jankowski steps in and tells us the story in all its richness. And remember: No experiment, however carefully designed with an eye to protecting internal validity, could ever have uncovered what Sigmund Freud found by just sitting quietly next to a patient on a couch—and listening.

Key Terms

Complete observation   184
Complete (covert) participation   186
Ethnography   181
Ethnomethodology   183
Field notes   190
Field research   180

Focus groups   198
Gatekeeper   187
Intensive (depth)
   interviewing   195
Jottings   190
Key informant   188

Netnography   182
Participant observation   184
Qualitative methods   179
Reactive effects   185
Saturation point   196
Theoretical sampling   188

Highlights

- Qualitative methods are most useful in exploring new issues, in investigating hard-to-study groups, and in determining the meaning people give to their lives and actions. In addition, most social research projects can be improved in some respects by taking advantage of qualitative techniques.

- Qualitative researchers tend to develop ideas inductively; they try to understand the social context and sequential nature of attitudes and actions and explore the subjective meanings that participants attach to events. They rely primarily on participant observation, intensive interviewing, and, in recent years, focus groups.

- Participant observers may adopt one of several roles for a particular research project. Each role represents a different balance between observing and participating. Many field researchers prefer a moderate role, participating as well as observing in a group but acknowledging publicly the researcher role. Such a role avoids

the ethical issues posed by covert participation while still allowing the insights into the social world derived from participating directly in it. The role that the participant observer chooses should be based on an evaluation of the problems likely to arise from reactive effects and the ethical dilemmas of covert participation.

- Field researchers must develop strategies for entering the field, developing and maintaining relations in the field, sampling, and recording and analyzing data. Selection of sites or other units to study may reflect an emphasis on typical cases, deviant cases, or critical cases that can provide more information than others. Sampling techniques commonly used within sites or in selecting interviewees in field research include theoretical sampling.

- Recording and analyzing notes is a crucial step in field research. Jottings are used as brief reminders about events in the field, whereas daily logs are useful to chronicle the researcher's activities. Detailed field notes should be recorded daily. Periodic analysis of the notes can guide refinement of methods used in the field and of the concepts, indicators, and models developed to explain what has been observed.

- Intensive interviews involve open-ended questions and follow-up probes, with the specific question content and order varying from one interview to another.

- Focus groups combine elements of participant observation and intensive interviewing. They can increase the validity of attitude measurement by revealing what people say when presenting their opinions in a group context instead of the artificial one-on-one interview setting.

- Computer software is used increasingly for the analysis of qualitative, textual, and pictorial data. Users can record their notes, categorize observations, specify links between categories, and count occurrences.

- The four main ethical issues in field research concern voluntary participation, subject well-being, identity disclosure, and confidentiality.

STUDENT STUDY SITE

The Student Study Site, available at **www.sagepub.com/chambliss4e,** includes useful study materials including web exercises with accompanying links, eFlashcards, videos, audio resources, journal articles, and encyclopedia articles, many of which are represented by the media links throughout the text. The site also features Interactive Exercises—represented by the green icon here—to help you understand the concepts in this book.

Exercises

Discussing Research

1. Maurice Punch (1994) once opined that "the crux of the matter is that some deception, passive or active, enables you to get at data not obtainable by other means" (p. 91). What aspects of the social world would be difficult for participant observers to study without being covert? Might any situations require the use of covert observation to gain access? What might you do as a participant observer to lessen access problems while still acknowledging your role as a researcher?

2. Review the experiments and surveys described in previous chapters. Pick one and propose a field research design that would focus on the same research question but use participant observation techniques in a local setting. Propose the role that you would play in the setting, along the participant observation continuum, and explain why you would favor this role. Describe the stages of your field research study, including your plans for entering the field, developing and maintaining relationships, sampling, and recording and analyzing data. Then discuss what you would expect your study to add to the findings resulting from the study described in the book.

3. Intensive interviews are the core of many qualitative research designs. How do they differ from the structured survey procedures that you studied in the last chapter? What are their advantages and disadvantages over standardized interviewing? How does intensive interviewing differ from the qualitative method of participant observation? What are the advantages and disadvantages of these two methods?

Finding Research

1. Go to the *Annual Review of Sociology's* website (http://annualreviews.org). Search for articles that use qualitative methods as the primary method of gathering data on any one of the following subjects: child development/socialization, gender/sex roles, or aging/gerontology. Enter "Qualitative AND Methods" in the subject field to begin this search. Review at least five articles and report on the specific method of field research used in each.

2. Go to Intute's database of social sciences Internet resources (www.intute.ac.uk/socialsciences/). Choose "Research Tools and Methods" and then "Qualitative Methods." Now choose three or

four interesting sites to find out more about field research—either professional organizations of field researchers or journals that publish their work. Explore the sites to find out what information they provide regarding field research, what kinds of projects are being done that involve field research, and the purposes for which specific field research methods are being used.

3. You have been asked to do field research on the World Wide Web's impact on the socialization of children in today's world. The first part of the project involves your writing a compare-and-contrast report on the differences between how you and your generation were socialized as children and the way children today are being socialized. Collect your data by surfing the web "as if you were a kid." The web is your field, and you are the field researcher.

   Using any of the major search engines, explore the web within the "Kids" or "Children" subject heading, keeping field notes on what you observe.

   Write a brief report based on the data you have collected. How has the web impacted child socialization in comparison to when you were a child?

Critiquing Research

1. Read and summarize one of the qualitative studies discussed in this chapter or another classic study recommended by your instructor. Review and critique the study using the article review questions presented in Exhibit 12.2 on page 261. What questions are answered by the study? What questions are raised for further investigation?

2. Write a short critique of the ethics of Ellis's (1986) study (discussed in Chapter 2). Read the book ahead of time to clarify the details and then focus on each of the ethical guidelines presented in this chapter: voluntary participation, subject well-being, identity disclosure, and confidentiality. Conclude with a statement about the extent to which field researchers should be required to disclose their identities and the circumstances in which they should not be permitted to participate actively in the social life they study.

Doing Research

1. Conduct a brief observational study in a public location on campus where students congregate. A cafeteria, a building lobby, or a lounge would be ideal. You can sit and observe, taking occasional notes unobtrusively and without violating any expectations of privacy. Observe for 30 minutes. Write up field notes, being sure to include a description of the setting and a commentary on your own behavior and your reactions to what you observed.

2. Review the experiments and surveys described in previous chapters. Pick one and propose a field research design that would focus on the same research question but with participant observation techniques in a local setting. Propose the role along the participant observation continuum that you would play in the setting and explain why you would favor this role. Describe the stages of your field research study, including your plans for entering the field, developing and maintaining relationships, sampling, and recording and analyzing data. Then discuss what you would expect your study to add to the findings resulting from the study described in the book.

3. Develop an interview guide that focuses on a research question addressed in one of the studies in this book. Using this guide, conduct an intensive interview with one person who is involved with the topic in some way. Take only brief notes during the interview; then write up as complete a record of the interview as you can immediately afterward. Turn in an evaluation of your performance as an interviewer and note taker together with your notes.

Ethics Questions

1. Should covert observation ever be allowed in social science research? Do you believe that social scientists should simply avoid conducting research on groups or individuals who refuse to admit researchers into their lives? Some have argued that members of privileged groups do not need to be protected from covert research by social scientists—that this restriction should only apply to disadvantaged groups and individuals. Do you agree? Why or why not?

2. Should any requirements be imposed on researchers who seek to study other cultures to ensure that procedures are appropriate and interpretations are culturally sensitive? What practices would you suggest for cross-cultural researchers to ensure that ethical guidelines are followed? (Consider the wording of consent forms and the procedures for gaining voluntary cooperation.)

Qualitative Data Analysis

I was at lunch standing in line and he [another male student] came up to my face and started saying stuff and then he pushed me. I said . . . I'm cool with you, I'm your friend and then he push me again and calling me names. I told him to stop pushing me and then he push me hard and said something about my mom. And then he hit me, and I hit him back. After he fell I started kicking him.

—Morrill, Yalda, Adelman,
Musheno, and Bejarano (2000:521)

A real student writing an in-class essay about conflicts in which he had participated made this statement. It was written for a team of social scientists who were studying conflicts in high schools to better understand their origins and to inform prevention policies.

In qualitative data analysis, the raw data to be analyzed are text—words—rather than numbers. In the high school conflict study by Morrill et al. (2000), there were initially no variables or hypotheses. The use of text, not numbers, and the (initial) absence of variables are just two of the ways in which qualitative analysis differs from quantitative.

In this chapter, we present and illustrate the features that most qualitative analyses share. There is no one correct way to analyze textual data. To quote Michael Quinn Patton (2002), "Qualitative analysis transforms data into findings. No formula exists for that transformation. Guidance, yes. But no recipe. Direction can and will be offered, but the final destination remains unique for each inquirer, known only when—and if—arrived at" (p. 432).

We first discuss different types of qualitative analyses and then describe computer programs for qualitative data analysis. You will see that these increasingly popular programs are blurring the distinctions between quantitative and qualitative approaches to textual analysis.

What Is Distinctive About Qualitative Data Analysis?

Qualitative data analysis:
Techniques used to search and code textual, aural, and pictorial data and to explore relationships among the resulting categories.

The focus on text—on qualitative data rather than on numbers—is the most important feature of **qualitative data analysis**. The "text" that qualitative researchers analyze is most often transcripts of interviews or notes from participant observation sessions, but the term can also refer to pictures or other images that the researcher examines.

What can one learn from a text? There are two kinds of answers to this question. Some researchers view textual analysis as a way to understand what participants "really" thought or felt or did in some situation or at some point in time. The text becomes a way to get "behind the numbers" that are recorded in a quantitative analysis to see the richness of real social experience. In this approach, interviews or field studies can, for instance, illuminate what survey respondents really meant by their answers.

Other qualitative researchers, however, have adopted a *hermeneutic* perspective on texts, viewing interpretations as never totally true or false. The text has many possible interpretations (Patton 2002:114). The meaning of a text, then, is negotiated among a community of interpreters, and to the extent that some agreement is reached about meaning at a particular time and place, that meaning can only be based on consensual community validation. From the hermeneutic perspective, a researcher constructs a "reality" with his interpretations of a text provided by the subjects of research; other researchers with different backgrounds could come to markedly different conclusions.

Research|Social Impact
Link 10.1
Read more about using qualitative methods to study social change.

Qualitative and quantitative data analyses, then, differ in the priority given to the views of the subjects of the research versus those of the researcher. Qualitative data analysts seek to capture the setting or people who produced this text on their own terms rather than in terms of predefined (by researchers) measures and hypotheses. So, qualitative data analysis tends typically to be inductive—the analyst identifies important categories in the data, as well as patterns and relationships, through a process of discovery. There are often no predefined measures or hypotheses. Anthropologists term this an **emic focus**, which means representing the setting in terms of the participants, rather than an **etic focus**, in which the setting and its participants are represented in terms that the researcher brings to the study.

> **Emic focus:** Representing a setting with the participants' terms.
>
> **Etic focus:** Representing a setting with the researcher's terms.

Good qualitative data analyses focus on the interrelated aspects of the setting or group, or person, under investigation—the case—rather than breaking the whole up into separate parts. The whole is always understood to be greater than the sum of its parts, so the social context of events, thoughts, and actions becomes essential for interpretation. Within this framework, it doesn't really make sense to focus on two variables out of an interacting set of influences and test the relationship between just those two.

Qualitative data analysis is an iterative and reflexive process that begins as data are being collected rather than after data collection has ceased (Stake 1995). Next to her field notes or interview transcripts, the qualitative analyst jots down ideas about the meaning of the text and how it might relate to other issues. This process of reading through the data and interpreting it continues throughout the project. When it appears that additional concepts need to be investigated or new relationships explored, the analyst adjusts the data collection. This process is termed **progressive focusing** (Parlett & Hamilton 1976).

> **Progressive focusing:** The process by which a qualitative analyst interacts with the data and gradually refines his or her focus.

> We emphasize placing an interpreter in the field to observe the workings of the case, one who records objectively what is happening but simultaneously examines its meaning and redirects observation to refine or substantiate those meanings. Initial research questions may be modified or even replaced in mid-study by the case researcher. The aim is to thoroughly understand [the case]. If early questions are not working, if new issues become apparent, the design is changed. (Stake 1995:9)

Elijah Anderson (2003) describes the progressive focusing process in his memoir about his study of Jelly's Bar:

> I also wrote conceptual memos to myself to help me sort out my findings. Usually not more than a page long, they represented theoretical insights that emerged from my engagement with the data in my field notes. As I gained tenable hypotheses and propositions, I began to listen and observe selectively, focusing in on those events that I thought might bring me alive to my research interests and concerns. This method of dealing with the information I was receiving amounted to a kind of dialogue with the data, sifting out ideas, weighing new notions against the reality with which I [was] faced there on the streets and back at my desk. (pp. 235–236)

Following a few guidelines will help when a researcher starts analyzing qualitative data (Miller & Crabtree 1999):

- Know yourself—your biases and preconceptions.

- Know your question.

- Seek creative abundance. Consult others and keep looking for alternative interpretations.

- Be flexible.

- Exhaust the data. Try to account for all the data in the texts, then publicly acknowledge the unexplained and remember the next principle.

- Celebrate anomalies. They are the windows to insight.

- Get critical feedback. The solo analyst is a great danger to self and others.

- Be explicit. Share the details with yourself, your team members, and your audiences. (pp. 142–143)

Qualitative Data Analysis as an Art

If you miss the certainty of predefined measures and deductively derived hypotheses, you are beginning to understand the difference between quantitative and qualitative data analyses. Qualitative data analysis is even described by some as involving as much "art" as science—as a "dance." In the words of William Miller and Benjamin Crabtree (1999),

> Interpretation is a complex and dynamic craft, with as much creative artistry as technical exactitude, and it requires an abundance of patient plodding, fortitude, and discipline. There are many changing rhythms; multiple steps; moments of jubilation, revelation, and exasperation . . . The dance of interpretation is a dance for two, but those two are often multiple and frequently changing, and there is always an audience, even if it is not always visible. Two dancers are the interpreters and the texts. (pp. 138–139)

The "dance" of qualitative data analysis captures the alternation between immersion in the text to identify meanings and editing the text to create categories and codes. The process involves three steps in reading the text:

1. When the researcher reads the text literally, he or she is focused on its literal content and form; the text "leads" the dance.

2. Then the researcher reads the text reflexively, focusing on how his or her own orientation shapes interpretations and focus. Now, the researcher leads the dance.

3. Finally, the researcher reads the text interpretively; the researcher tries to construct his or her own interpretation of what the text means. (pp. 138–139)

In this artful way, analyzing text involves both inductive and deductive processes: The researcher generates concepts and linkages between them based on reading the text and also checks the text to see whether his concepts and interpretations are reflected in it.

Qualitative Compared to Quantitative Data Analysis

Video Link 10.1
Watch and compare
data analysis techniques.

With these points in mind, let's review the differences of the logic behind qualitative versus quantitative analysis (Denzin & Lincoln 2000:8–10; Patton 2002:13–14):

- A focus on meanings rather than on quantifiable phenomena

- Collection of much data on a few cases rather than little data on many cases

- Study in depth and detail, without predetermined categories or directions, rather than emphasis on analyses and categories determined in advance

- Conception of the researcher as an "instrument" rather than as the designer of objective instruments to measure particular variables

- Sensitivity to context, rather than seeking universal generalizations

- Attention to the impact of the researcher's and others' values on the course of the analysis, rather than presuming the possibility of value-free inquiry

- A goal of rich descriptions of the world rather than measurement of specific variables

Of course, even the most qualitative textual data can also be transposed to quantitative data through a process of categorization and counting. Some qualitative analysts also share with quantitative researchers a positivist goal of describing the world as it "really" is, but others have adopted a postmodern hermeneutic goal of trying to understand how different people see and make sense of the world, without believing that there is one uniquely correct description.

▣ What Techniques Do Qualitative Data Analysts Use?

Most approaches to qualitative data analysis take five steps:

1. Documentation of the data and data collection

2. Conceptualization and coding

3. Examining relationships to show how one concept may influence another

4. Authenticating conclusions by evaluating alternative explanations, disconfirming evidence, and searching for negative cases

5. Reflexivity

The analysis of qualitative research notes begins in the field at the time of observation and/or interviewing, as the researcher identifies problems and concepts that appear likely to help in understanding the situation. Simply reading the notes or transcripts is an important step in the analytic process. Researchers should make frequent notes in the margins to identify important statements and to propose ways of coding the data: "husband/wife conflict," perhaps, or "tension reduction strategy."

An interim stage may consist of listing the concepts developed in the notes and perhaps diagramming the relationships among concepts (Maxwell 1996:78–81). In large projects, regular team meetings are an important part of this process. In her study of neighborhood police officers, Susan Miller's (1999) research team met to go over their field notes and to resolve points of confusion, as well as to talk with other skilled researchers who helped to identify emerging concepts:

> The fieldwork team met weekly to talk about situations that were unclear and to troubleshoot any problems. We also made use of peer-debriefing techniques. Here, multiple colleagues, who were familiar with qualitative data analysis but not involved in our research, participated in preliminary analysis of our findings. (p. 233)

The back-and-forth of refining concepts usually continues throughout the entire qualitative research project. Let's examine each of the steps of qualitative analysis in more detail.

Documentation

The data for a qualitative study most often are notes jotted down in the field or during an interview or text transcribed from audiotapes. "The basic data are these observations and conversations, the actual words of people reproduced to the best of my ability from the field notes" (Diamond 1992:7). What to do with all this material? As mentioned in Chapter 9, many novice researchers have become overwhelmed by the quantity of information, and their research projects have ground to a halt as a result.

Analysis is less daunting, however, if the researcher maintains a disciplined transcription schedule:

> Usually, I wrote these notes immediately after spending time in the setting or the next day. Through the exercise of writing up my field notes, with attention to "who" the speakers and actors were, I became aware of the nature of certain social relationships and their positional arrangements within the peer group. (Anderson 2003:235)

You can see Anderson's analysis already emerging from the simple process of taking notes.

The first formal analytical step is documentation. The various contacts, interviews, written documents, and notes all need to be saved and catalogued in some fashion. Documentation is critical to qualitative research for several reasons: It is essential for keeping track of what will be a rapidly growing volume of notes, tapes, and documents; it provides a way of developing an outline for the analytic process; and it encourages ongoing conceptualizing and strategizing about the text.

Miles and Huberman (1994:53) provide a good example of a contact summary form that was used to keep track of observational sessions in a qualitative study of a new school curriculum (Exhibit 10.1).

Conceptualization, Coding, and Categorizing

Identifying and refining important concepts is a key part of the iterative process of qualitative research. Sometimes conceptualization begins with a simple observation that is interpreted directly, "pulled apart," and then put back together more meaningfully. Robert Stake provides an example (1995):

> When Adam ran a pushbroom into the feet of the children nearby, I jumped to conclusions about his interactions with other children: aggressive, teasing, arresting. Of course, just a few minutes earlier I had seen him block the children climbing the steps in a similar moment of smiling bombast. So I was aggregating, and testing my unrealized hypotheses about what kind of kid he was, not postponing my interpreting . . . My disposition was to keep my eyes on him. (p. 74)

Encyclopedia Link 10.1
Read about the processes and strategies of qualitative data coding.

The focus in this conceptualization "on the fly" is to provide a detailed description of what was observed and a sense of why it was important.

More often, analytic insights are tested against new observations; the initial statement of problems and concepts is refined; and the researcher then collects more data, interacts with it again, and the process continues. Elijah Anderson (2003) recounts how his conceptualization of social stratification at Jelly's Bar developed over a long period of time:

> I could see the social pyramid, how certain guys would group themselves and say in effect, "I'm here and you're there." I made sense of these crowds [initially] as the "respectables," the "non-respectables," and the "near-respectables." . . . Inside, such non-respectables might sit on the crates, but if a respectable came along and wanted to sit there, the lower status person would have to move. (pp. 18–19)

Exhibit 10.1   **Example of a Contact Summary Form**

Contact type: Site: Tindale_____

Visit   X_____   Contact date: 11/28-29/79____

Phone   _____   Today's date: 12/28/79_____

         (with whom)   Written by: BLT_____

1. What were the main issues or themes that struck you in this contact?

   Interplay between highly prescriptive, "teacher-proof" curriculum that is top-down imposed and the actual writing of the curriculum by the teachers themselves.

   Split between the "watchdogs" (administrators) and the "house masters" (dept. chairs & teachers) vis à vis job foci.

   District curric, coord'r as decision maker re school's acceptance of research relationship.

2. Summarize the information you got (or failed to get) on each of the target questions you had for this contact.

Question	Information
History of dev. of innov'n	Conceptualized by Curric., Coord'r, English Chairman & Assoc. Chairman; written by teachers in summer; revised by teachers following summer with field testing data
School's org'l structure	Principal & admin'rs responsible for discipline; dept chairs are educ'l leaders
Demographics	Racial conflicts in late 60's; 60% black stud. pop.; heavy emphasis on discipline & on keeping out non-district students slipping in from Chicago
Teachers' response to innov'n	Rigid, structured, etc. at first; now, they say they like it/ NEEDS EXPLORATION
Research access	Very good; only restriction: teachers not required to cooperate

3. Anything else that struck you as salient, interesting, illuminating or important in this contact?

   Thoroughness of the innov'n's development and training.

   Its embeddedness in the district's curriculum, as planned and executed by the district curriculum coordinator.

   The initial resistance to its high prescriptiveness (as reported by users) as contrasted with their current acceptance and approval of it (again, as reported by users).

4. What new (or remaining) target questions do you have in considering the next contact with this site?

   How do users really perceive the innov'n? If they do indeed embrace it, what accounts for the change from early resistance?

   Nature and amount of networking among users of innov'n.

   Information on "stubborn" math teachers whose ideas weren't heard initially—who are they? Situation particulars? Resolution?

   Follow-up on English teacher Reilly's "fall from the chairmanship."

   Follow a team through a day of rotation, planning, etc.

   CONCERN: The consequences of eating school cafeteria food two days per week for the next four or five months . . .

   Stop

But this initial conceptualization changed with experience as Anderson (2003:28) realized that the participants themselves used other terms to differentiate social status: *winehead, hoodlum,* and *regular*. What did they mean by these terms? "The 'regulars' basically valued 'decency.' They associated decency with conventionality but also with 'working for a living,' or having a 'visible means of support'" (29). In this way, Anderson progressively refined his concept as he gained experience in the setting.

Howard S. Becker (1958) provides another excellent illustration of this iterative process of conceptualization in his study of medical students:

When we first heard medical students apply the term "crock" to patients, we made an effort to learn precisely what they meant by it. We found, through interviewing students about cases both they and the observer had seen, that the term referred in a derogatory way to patients with many subjective symptoms but no discernible physical pathology. Subsequent observations indicated that this usage was a regular feature of student behavior and thus that we should attempt to incorporate this fact into our model of student-patient behavior. The derogatory character of the term suggested in particular that we investigate the reasons students disliked these patients. We found that this dislike was related to what we discovered to be the students' perspective on medical school: the view that they were in school to get experience in recognizing and treating those common diseases most likely to be encountered in general practice. "Crocks," presumably having no disease, could furnish no such experience. We were thus led to specify connections between the student-patient relationship and the student's view of the purpose of his professional education. Questions concerning the genesis of this perspective led to discoveries about the organization of the student body and communication among students, phenomena which we had been assigning to another [segment of the larger theoretical model being developed]. Since "crocks" were also disliked because they gave the student no opportunity to assume medical responsibility, we were able to connect this aspect of the student-patient relationship with still another tentative model of the value system and hierarchical organization of the school, in which medical responsibility plays an important role. (p. 658)

In this excerpt, the researcher was first alerted to a concept by observations in the field, then refined his understanding of this concept by investigating its meaning. By observing the concept's frequency of use, he came to realize its importance. Finally, he incorporated the concept into an explanatory model of student-patient relationships.

Matrix: A chart used to condense qualitative data into simple categories and provide a multidimensional summary that will facilitate subsequent, more intensive analysis.

A well-designed chart, or **matrix**, can facilitate the coding and categorization process. Exhibit 10.2 shows an example of a coding form designed by Miles and Huberman (1994:93–95) to represent the extent to which teachers and teachers' aides ("users") and administrators at a school gave evidence of various supporting conditions that indicated preparedness for a new reading program. The matrix condenses data into simple categories, reflects further analysis of the data to identify "degree" of support, and provides a multidimensional summary that will facilitate subsequent, more intensive analysis. Direct quotes still impart some of the flavor of the original text.

Examining Relationships and Displaying Data

Examining relationships is the centerpiece of the analytic process, because it allows the researcher to move from simple description of the people and settings to explanations of why things happened as they did with those people in that setting. A matrix can show how different concepts are related or, perhaps, what causes are linked with what effects.

Exhibit 10.2	Example of Checklist Matrix

Presence of Supporting Conditions		
Condition	**For Users**	**For Administrators**
Commitment	*Strong*—"wanted to make it work."	*Weak* at building level. Prime movers in central office committed; others not.
Understanding	"*Basic*" ("felt I could do it, but I just wasn't sure how.") for teacher. *Absent* for aide ("didn't understand how we were going to get all this.")	*Absent* at building level and among staff. *Basic* for 2 prime movers ("got all the help we needed from developer.") *Absent* for other central office staff.
Materials	*Inadequate:* ordered late, puzzling ("different from anything I ever used"), discarded.	N.A.
Front-end training	"*Sketchy*" for teacher ("it all happened so quickly"); no demo class. *None* for aide ("totally unprepared. I had to learn along with the children.")	Prime movers in central office had training at developer site; none for others.
Skills	*Weak-adequate* for teacher. "*None*" for aide.	One prime mover (Robeson) skilled in substance; others unskilled.
Ongoing inservice	*None*, except for monthly committee meeting; no substitute funds.	*None*
Planning, coordination time	*None*: both users on other tasks during day; lab tightly scheduled, no free time.	*None*
Provisions for debugging	*None* systematized; spontaneous work done by users during summer.	*None*
School admin. support	*Adequate*	N.A.
Central admin. support	*Very strong* on part of prime movers.	Building admin. only acting on basis of central office commitment.
Relevant prior experience	*Strong* and useful in both cases: had done individualized instruction, worked with low achievers. But [the] aide [had] no diagnostic experience.	*Present* and useful in central office, esp. Robeson (specialist).

In Exhibit 10.3, a matrix relates stakeholders' stake in a new program with the researcher's estimate of their attitude toward the program. Each cell of the matrix was to be filled in with a summary of an illustrative case study. In other matrix analyses, quotes might be included in the cells to represent the opinions of these different stakeholders, or the number of cases of each type might appear in the cells. The possibilities are almost endless. Keeping this approach in mind will generate many fruitful ideas for structuring a qualitative data analysis.

Exhibit 10.3 Coding Form for Relationships: Stakeholders' Stakes

Estimate of Various Stakeholders' Inclination Toward the Program

How high are the stakes for various primary stakeholders?	Favorable	Neutral or Unknown	Antagonistic
High			
Moderate			
Low			

Note: Construct illustrative case studies for each cell based on fieldwork.

Research|Social Impact Link 10.2
Read more about examining relationships.

The simple relationships that are identified with a matrix like that shown in Exhibit 10.3 can be examined and then extended to create a more complex causal model. Such a model can represent the multiple relationships among the important explanatory constructs. A great deal of analysis must precede the construction of such a model with careful attention to identification of important variables and the evidence that suggests connections between them. Exhibit 10.4 provides an example from a study of the implementation of a school program.

Exhibit 10.4 Example of a Causal Network Model

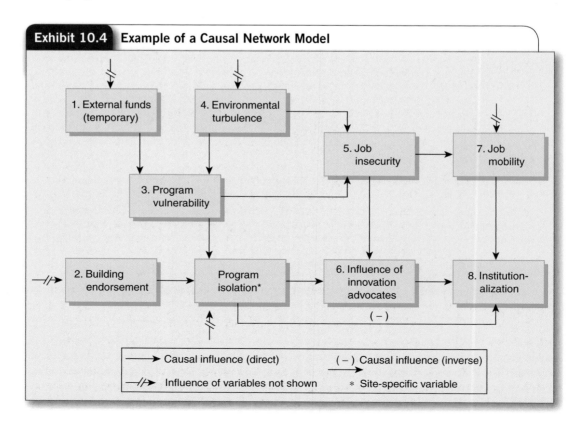

Authenticating Conclusions

No set standards exist for evaluating the validity or *authenticity* of conclusions in a qualitative study, but the need to consider carefully the evidence and methods on which conclusions are based is just as great as with other types of research. Individual items of information can be assessed in terms of at least three criteria (Becker 1958):

1. How credible was the informant? Were statements made by someone with whom the researcher had a relationship of trust or by someone the researcher had just met? Did the informant have reason to lie? If the statements do not seem to be trustworthy as indicators of actual events, can they at least be used to help understand the informant's perspective?

2. Were statements made in response to the researcher's questions, or were they spontaneous? Spontaneous statements are more likely to indicate what would have been said had the researcher not been present.

3. How does the presence or absence of the researcher or the researcher's informant influence the actions and statements of other group members? Reactivity to being observed can never be ruled out as a possible explanation for some directly observed social phenomenon. However, if the researcher carefully compares what the informant says goes on when the researcher is not present, what the researcher observes directly, and what other group members say about their normal practices, the extent of reactivity can be assessed to some extent.

A qualitative researcher's conclusions should also be judged by their ability to explain credibly some aspect of social life. Explanations should capture group members' tacit knowledge of the social processes that were observed, not just their verbal statements about these processes. **Tacit knowledge**—"the largely unarticulated, contextual understanding that is often manifested in nods, silences, humor, and naughty nuances"—is reflected in participants' actions as well as their words and in what they fail to state but nonetheless feel deeply and even take for granted (Altheide & Johnson 1994:492–493). These features are evident in Whyte's (1955) analysis of Cornerville social patterns:

Tacit knowledge: In field research, a credible sense of understanding of social processes that reflects the researcher's awareness of participants' actions, as well as their words, and of what they fail to state, feel deeply, and take for granted.

> The corner-gang structure arises out of the habitual association of the members over a long period of time. The nuclei of most gangs can be traced back to early boyhood. . . . Home plays a very small role in the group activities of the corner boy.
>
> . . . The life of the corner boy proceeds along regular and narrowly circumscribed Channels. . . . Out of [social interaction within the group] arises a system of mutual obligations which is fundamental to group cohesion. . . . The code of the corner boy requires him to help his friends when he can and to refrain from doing anything to harm them. When life in the group runs smoothly, the obligations binding members to one another are not explicitly recognized. (pp. 255–257)

Comparing conclusions from a qualitative research project to those other researchers obtained by conducting similar projects can also increase confidence in their authenticity. Miller's 1999 study of neighborhood police officers (NPOs) found striking parallels in the ways they defined their masculinity to processes reported in research about males in nursing and other traditionally female jobs (as cited in Bachman & Schutt 2007):

> In part, male NPOs construct an exaggerated masculinity so that they are not seen as feminine as they carry out the social-work functions of policing. Related to this is the almost defiant expression of heterosexuality, so that the men's sexual orientation can never truly be doubted even if their gender roles

are contested. Male patrol officers' language—such as their use of terms like "pansy police" to connote neighborhood police officers—served to affirm their own heterosexuality. In addition, the male officers, but not the women, deliberately wove their heterosexual status into conversations, explicitly mentioning their female domestic partner or spouse and their children. This finding is consistent with research conducted in the occupational field. The studies reveal that men in female-dominated occupations, such as teachers, librarians, and pediatricians, over-reference their heterosexual status to ensure that others will not think they are gay. (p. 307)

Reflexivity

Encyclopedia Link 10.2
Read about
the role of reflexivity in
qualitative data analysis.

Confidence in the conclusions from a field research study is also strengthened by an honest and informative account about how the researcher interacted with subjects in the field, what problems she encountered, and how these problems were or were not resolved. Such a "natural history" of the development of the evidence enables others to evaluate the findings. Such an account is important first and foremost because of the evolving and variable nature of field research: To an important extent, the researcher "makes up" the method in the context of a particular investigation rather than applying standard procedures that are specified before the investigation begins.

Barrie Thorne (1993) provides a good example of this final element of the analysis:

Many of my observations concern the workings of gender categories in social life. For example, I trace the evocation of gender in the organization of everyday interactions, and the shift from boys and girls as loose aggregations to "the boys" and "the girls" as self-aware, gender-based groups. In writing about these processes, I discovered that different angles of vision lurk within seemingly simple choices of language. How, for example, should one describe a group of children? A phrase like "six girls and three boys were chasing by the tires" already assumes the relevance of gender. An alternative description of the same event—"nine fourth-graders were chasing by the tires"—emphasizes age and downplays gender. Although I found no tidy solutions, I have tried to be thoughtful about such choices . . . After several months of observing at Oceanside, I realized that my field notes were peppered with the words "child" and "children," but that the children themselves rarely used the term. "What do they call themselves?" I badgered in an entry in my field notes. The answer it turned out, is that children use the same practices as adults. They refer to one another by using given names ("Sally," "Jack") or language specific to a given context ("that guy on first base"). They rarely have occasion to use age-generic terms. But when pressed to locate themselves in an age-based way, my informants used "kids" rather than "children." (pp. 8–9)

Qualitative data analysts, more often than quantitative researchers, display real sensitivity to how a social situation or process is interpreted from a particular background and set of values and not simply based on the situation itself (Altheide & Johnson 1994). Researchers are only human, after all, and must rely on their own senses and process all information through their own minds. By reporting how and why they think they did what they did, they can help others determine whether, or how, the researchers' perspectives influenced their conclusions.

Elijah Anderson's (2003) memoir about the Jelly's Bar research illustrates the type of "tracks" that an ethnographer makes, as well as how the ethnographer can describe those tracks. Anderson acknowledges that his tracks began as a child:

While growing up in the segregated black community of South Bend, from an early age, I was curious about the goings on in the neighborhood, but particularly streets, and more particularly, the corner taverns that my uncles and my dad would go to hang out and drink in. . . . Hence, my selection of Jelly's as a field setting was a matter of my background, intuition, reason, and with a little bit of luck. (pp. 1–2)

After starting to observe at Jelly's, Anderson's (2003) "tracks" led to Herman:

> After spending a couple of weeks at Jelly's, I met Herman and I felt that our meeting marked a big achievement. We would come to know each other well . . . [He was] something of an informal leader at Jelly's . . . We were becoming friends . . . He seemed to genuinely like me, and he was one person I could feel comfortable with. (p. 4)

Anderson's (2003) observations were shaped in part by Herman's perspective, but we also learn here that Anderson maintained some engagement with fellow students. This contact outside the bar helped to shape his analysis: "By relating my experiences to my fellow students, I began to develop a coherent perspective or a 'story' of the place which complemented the accounts that I had detailed in my accumulating field notes" (p. 6).

So, Anderson's analysis came in part from the way in which he "played his role" as a researcher and participant, not just from the setting itself.

Journal Link 10.2
Read about using qualitative interviewing to understand the role relationships play in the life of a female ex-offender.

What Are Some Alternatives in Qualitative Data Analysis?

The qualitative data analyst can choose from many interesting alternative approaches. Of course, the research question should determine the approach, but a researcher's preferences will also inevitably play a role as well. The alternative approaches we present here (narrative analysis, conversation analysis, and grounded theory) will give you a good sense of the possibilities (Patton 2002).

Narrative Analysis

Narrative "displays the goals and intentions of human actors; it makes individuals, cultures, societies, and historical epochs comprehensible as wholes" (Richardson 1995:200). **Narrative analysis** focuses on "the story itself" and seeks to preserve the integrity of personal biographies or a series of events that cannot adequately be understood in terms of their discrete elements (Riessman 2002:218). The coding for a narrative analysis is typically of the narratives as a whole rather than of the different elements within them. The coding strategy revolves around reading the stories and classifying them into general patterns.

For example, Calvin Morrill and his colleagues (2000) read through 254 conflict narratives written by ninth graders (mentioned at the beginning of this chapter) and found four different types of stories:

> **Narrative analysis:** A form of qualitative analysis in which the analyst focuses on how respondents impose order on the flow of experience in their lives and so make sense of events and actions in which they have participated.

1. Action tales, in which the author represents himself or herself and others as acting within the parameters of taken-for-granted assumptions about what is expected for particular roles among peers

2. Expressive tales, in which the author focuses on strong, negative emotional responses to someone who has wronged him or her

3. Moral tales, in which the author recounts explicit norms that shaped his or her behavior in the story and influenced the behavior of others

4. Rational tales, in which the author represents himself or herself as a rational decision maker navigating through the events of the story (p. 534)

Morrill et al. (2000:534–535) also classified the stories along four stylistic dimensions: (1) plot structure (such as whether the story unfolds sequentially), (2) dramatic tension (how the central conflict is represented), (3) dramatic resolution (how the central conflict is resolved), and (4) predominant outcomes (how the story ends). Coding reliability was checked through a discussion by the two primary coders, who found that their classifications agreed for a large percentage of the stories.

The excerpt that begins this chapter exemplifies what Morrill et al. (2000:536) termed an "action tale." Such tales

> unfold in matter-of-fact tones kindled by dramatic tensions that begin with a disruption of the quotidian order of everyday routines. A shove, a bump, a look . . . triggers a response . . . Authors of action tales typically organize their plots as linear streams of events as they move briskly through the story's scenes . . . This story's dramatic tension finally resolves through physical fighting, but . . . only after an attempted conciliation. (p. 356)

Audio Link 10.1
Learn more about narrative analysis.

You can contrast that "action tale," with the following narrative, which Morrill et al. (2000:545–546) classify as a "moral tale," in which the student authors "explicitly tell about their moral reasoning, often referring to how normative commitments shape their decision making":

> I . . . got into a fight because I wasn't allowed into the basketball game. I was being harassed by the captains that wouldn't pick me and also many of the players. The same type of things had happened almost every day where they called me bad words so I decided to teach the ring leader a lesson. I've never been in a fight before but I realized that sometimes you have to make a stand against the people that constantly hurt you, especially emotionally. I hit him in the face a couple of times and I got respect I finally deserved. (pp. 545–546)

Morrill et al. (2000:553) summarize their classification of the youth narratives in a simple table that highlights the frequency of each type of narrative and the characteristics associated with each of them (Exhibit 10.5). How does such an analysis contribute to our understanding of youth violence? Morrill et al.

Exhibit 10.5   Summary Comparison of Youth Narratives*

Representation of	Action Tales (N = 144)	Moral Tales (N = 51)	Expressive Tales (N = 35)	Rational Tales (N = 24)
Bases of everyday conflict	disruption of everyday routines & expectations	normative violation	emotional provocation	goal obstruction
Decision making	intuitive	principled stand	sensual	calculative choice
Conflict handling	confrontational	ritualistic	cathartic	deliberative
Physical violence†	in 44% (N = 67)	in 27% (N = 16)	in 49% (N = 20)	in 29% (N = 7)
Adults in youth conflict control	invisible or background	sources of rules	agents of repression	institutions of social control

*Total N = 254.

†Percentages based on the number of stories in each category.

first emphasize that their narratives "suggest that consciousness of conflict among youths—like that among adults—is not a singular entity but comprises a rich and diverse range of perspectives" (551).

Theorizing inductively, Morrill et al. (2000:553–554) then attempt to explain why action tales were much more common than the more adult-oriented normative, rational, or emotionally expressive tales. They say that one possibility is to be found in Gilligan's theory of moral development, which suggests that younger students are likely to limit themselves to the simpler action tales that "concentrate on taken-for-granted assumptions of their peer and wider cultures, rather than on more self-consciously reflective interpretation and evaluation" (pp. 553–554). More generally, Morrill et al. argue, "We can begin to think of the building blocks of cultures as different narrative styles in which various aspects of reality are accentuated, constituted, or challenged, just as others are deemphasized or silenced" (p. 556).

In this way, Morrill et al.'s narrative analysis allowed an understanding of youth conflict to emerge from the youths' own stories while also informing our understanding of broader social theories and processes.

Conversation Analysis

Conversation analysis is a specific qualitative method for analyzing ordinary conversation. Unlike narrative analysis, it focuses on the sequence and details of conversational interaction rather than on the "stories" that people are telling. Like ethnomethodology, from which it developed, conversation analysis focuses on how reality is constructed rather than on what it "is."

Three premises guide conversation analysis (Gubrium & Holstein 2000):

1. Interaction is sequentially organized, and talk can be analyzed in terms of the process of social interaction rather than in terms of motives or social status.

2. Talk, as a process of social interaction, is contextually oriented—it is both shaped by interaction and creates the social context of that interaction.

3. These processes are involved in all social interaction, so no interactive details are irrelevant to understanding it. (p. 492)

Consider these premises as you read the following dialogue between British researcher Ann Phoenix (2003) and a boy she called "Thomas" in her study of notions of masculinity, bullying, and academic performance among 11- to 14-year-old boys in 12 London schools:

Thomas: It's your attitude, but some people are bullied for no reason whatsoever just because other people are jealous of them . . .

Q: How do they get bullied?

Thomas: There's a boy in our year called James, and he's really clever and he's basically got no friends, and that's really sad because . . . he gets top marks in every test and everyone hates him. I mean, I like him . . . (p. 235)

Phoenix (2003) notes that here,

Thomas dealt with the dilemma that arose from attempting to present himself as both a boy and sympathetic to school achievement. He . . . distanced himself from . . . being one of those who bullies a boy just because they are jealous of his academic attainments . . . constructed for himself the position of being kind and morally responsible. (p. 235)

Note that Thomas was a boy talking to a woman. Do you imagine that his talk would have been quite different if his conversation had been with other boys?

An example of the very detailed data recorded in a formal conversation analysis appears in Exhibit 10.6. It is from David R. Gibson's (2005:1566) study of the effects of superior–subordinate and friendship interaction on the transitions that occur in the course of conversation—in this case, in meetings of managers. Every type of "participation-shift" (P-shift) is recorded and distinguished from every other type. Some shifts involve "turn claiming," in which one person (X) begins to talk after the first person (A) has addressed the group as a whole (0), without being prompted by the first speaker. Some shifts involve "turn receiving," in which the first person (A) addresses the second (B), who then responds. In "turn usurping," by contrast, the second person (X) speaks after the first person (A) has addressed a comment to a third person (B), who is thus prevented from responding. Examining this type of data can help us to see how authority is maintained or challenged in social groups.

Grounded Theory

Theory development occurs continually in qualitative data analysis (Coffey & Atkinson 1996:23). The goal of many qualitative researchers is to create **grounded theory**—that is, to build up inductively a systematic

Exhibit 10.6	Inventory of P-Shifts With Examples

P-Shift	Example
Turn claiming:	
AO-XA	John talks to the group, then Frank talks to John.
AO-XO	John talks to the group, then Frank talks to the group.
AO-XY	John talks to the group, then Frank talks to Mary.
Turn receiving:	
AB-BA	John talks to Mary, then Mary replies.
AB-BO	John talks to Mary, then Mary talks to the group.
AB-BY	John talks to Mary, then Mary talks to Irene.
Turn usurping:	
AE-XA	John talks to Mary, then Frank talks to John.
AB-XB	John talks to Mary, then Frank talks to Mary.
AB-XO	John talks to Mary, then Frank talks to the group.
AB-XY	John talks to Mary, then Frank talks to Irene.

Note: The initial speaker is denoted A and the initial target B, unless the group is addressed (or the target was ambiguous), in which case the target is O. Then, the P-shift is summarized in the form (speaker1) (target1)-(speaker2) (target2), with A or B appearing after the hyphen only if the initial speaker or target serves in one of these two positions in the second turn. When the speaker in the second turn is someone other than A or B, X is used; and when the target in the second turn is someone other than A, B, or the group O, Y is used.

theory that is "grounded" in, or based on, the observations. The observations are summarized into conceptual categories, which are tested directly in the research setting with more observations. Over time, as the conceptual categories are refined and linked, a theory evolves (Glaser & Strauss 1967; Huberman & Miles 1994:436).

As observation, interviewing, and reflection continue, researchers refine their definitions of problems and concepts and select indicators. They can then check the frequency and distribution of phenomena: How many people made a particular type of comment? How often did social interaction lead to arguments? Social system models may then be developed, which specify the relationships among different phenomena. These models are modified as researchers gain experience in the setting. For the final analysis, the researchers check their models carefully against their notes and make a concerted attempt to discover negative evidence that might suggest the model is incorrect.

> **Grounded theory:** Systematic theory developed inductively, based on observations that are summarized into conceptual categories, reevaluated in the research setting, and gradually refined and linked to other conceptual categories.

Mixed Methods

Different methods have different strengths and weaknesses; by combining them, or using *mixed methods,* you can often avoid the problems or misunderstandings that come from using just one.

Combining Qualitative Methods

After a devastating earthquake in Izmit, Turkey on August 17, 1999 killed 19,000 people, Elif Kale-Lostuvali (2007) conducted research using a combination of qualitative methodologies—including participant observation and intensive interviewing—to study citizen-state encounters in the region.

One important concept that emerged from Kale-Lostuvali's observations and interviews was a distinction locals made between a *mağdur* (sufferer) and a *deprenzade* (son of the earthquake). This was a critical distinction, because a *mağdur* was seen as deserving of government assistance, while a *deprenzade* was considered to be taking advantage of the situation for personal gain. Kale-Lostuvali (2007) drew on both interviews and participant observation to develop an understanding of this complex concept:

Video Link 10.2
Watch a researcher discuss when to use a mixed method approach.

> A prominent narrative frequently repeated in the disaster area elaborated the contrast between *mağdur* (sufferer, that is, the truly needy) and *depremzade* (sons of the earthquake). The *mağdur* (sufferers) were the deserving recipients of the aid that was being distributed. However, they (1) were in great pain and could not pursue what they needed; or (2) were proud and could not speak of their need; or (3) were humble, always grateful for the little they got, and were certainly not after material gains; or (4) were characterized by a combination of the preceding. And because of these characteristics, they had not been receiving their rightful share of the aid and resources. In contrast, *deprenzade* (sons of the earthquake) were people who took advantage of the situation. (p. 755)

The qualitative research by Spencer Moore and his colleagues (2004) on the social response to Hurricane Floyd demonstrates the interweaving of data from focus groups and from participant observation with the workers. Reports of heroic acts by rescuers, innumerable accounts of "neighbors helping neighbors," and the comments of HWATF [task force] participants suggest that residents, stranded motorists, relief workers, and rescuers worked and came together in remarkable ways during the relief and response phases of the disaster:

> Like people get along better . . . they can talk to each other. People who hadn't talked before, they talk now, a lot closer. That goes, not only for the neighborhood, job-wise, organization-wise, and all that . . . [Our] union sent some stuff for some of the families that were flooded out. (Focus Group #4) (pp. 210–211)

Combining Qualitative and Quantitative Methods

Conducting qualitative interviews can often enhance the value of a research design that uses primarily quantitative measurement techniques. Qualitative data can provide information about the quality of standardized case records and quantitative survey measures, as well as offer some insight into the meaning of particular fixed responses.

To help illustrate this combination, it makes sense to use official records to study the treatment of juveniles accused of illegal acts, because these records document the critical decisions to arrest, to convict, or to release (Dannefer & Schutt 1982). But research based on official records is only as good as the records themselves. In contrast to the controlled interview process in a research study, there is little guarantee that officials' acts and decisions were recorded in a careful and unbiased manner.

Case Study: Juvenile Court Records

Interviewing officials who create the records, or observing them while they record information, can strengthen research based on official records. A participant observation study of how probation officers screened cases in two New York juvenile court intake units shows how important such information can be (Needleman 1981). As indicated in Exhibit 10.7, Carolyn Needleman (1981) found that the concepts most researchers believe they are measuring with official records differ markedly from the meaning probation officers attach to those records.

Researchers assume that sending a juvenile case to court indicates a more severe disposition than retaining a case in the intake unit. But probation officers often diverted cases from court because they thought the courts would be too lenient. Researchers assume that probation officers evaluate juveniles as individuals, but in these settings, probation officers often based their decisions on juveniles' current social situation (e.g., whether they were living in a stable home), without learning anything about the individual juvenile. Perhaps most troubling for research using case records, Needleman (1981) found that probation officers decided how to handle cases first and then created an official record that appeared to justify their decisions.

Case Study: Mental Health System

The same observation can be made about the value of supplementing fixed-choice survey questions with more probing, open-ended questions. For example, Renee Anspach (1991) wondered about the use of standard surveys to study the effectiveness of mental health systems. Instead of drawing a large sample and asking a set of closed-ended questions, Anspach used snowball sampling techniques to select some administrators, case managers, clients, and family members in four community mental health systems, and then asked these respondents a series of open-ended questions. When asked whether their programs were effective, the interviewees were likely to respond in the affirmative. Their comments in response to other questions, however, pointed to many program failings. Anspach concluded that the respondents simply wanted the interviewer (and others) to believe in the program's effectiveness, for several reasons: Administrators wanted to maintain funding and employee morale; case managers wanted to ensure cooperation by talking up the program with clients and their families; and case managers also preferred to deflect blame for problems to clients, families, or system constraints.

Exhibit 10.7	Researchers' and Juvenile Court Workers' Discrepant Assumptions	
Researcher Assumptions	**Intake Worker Assumptions**
• Being sent to court is a harsher sanction than diversion from court. • Screening involves judgments about individual juveniles. • Official records accurately capture case facts.	• Being sent to court often results in more lenient and less effective treatment. • Screening centers on the juvenile's social situation. • Records are manipulated to achieve the desired outcome.

Case Study: Housing Loss in Group Homes

Ethnographic data complements quantitative data in Schutt's mixed-methods analysis of the value of group and independent living options for people who are homeless and have been diagnosed with severe mental illness. Exhibit 10.8 displays the quantitative association between lifetime substance abuse—a diagnosis recorded on a numerical scale that was made on the basis of an interview with a clinician—and housing loss—another quantitative indicator from service records (Schutt 2011:135).

The ethnographic notes recorded in the group homes reveal orientations and processes that help to explain the substance abuse-housing loss association (Schutt 2011):

> . . . the time has come where he has to decide once and for all to drink or not . . . Tom has been feeling "pinned to the bed" in the morning. He has enjoyed getting high with Sammy and Ben, although the next day is always bad. . . . Since he came back from the hospital Lisandro has been acting like he is taunting them to throw him out by not complying with rules and continuing to drink . . . (pp. 131, 133)

In this way, the analysis of the quantitative data reveals *what* happened, while Schutt's analysis of the ethnographic data helps to understand why.

Historical and Comparative Methods

Although the United States and several European nations have maintained democratic systems of governance for more than 100 years, democratic rule has more often been brief and unstable, when it has occurred at all. What explains the presence of democratic practices in one country and their absence in another? Are democratic politics a realistic option for every nation? What about Libya? Egypt? Iraq? Are there some prerequisites in historical experience, cultural values, or economic resources? (Markoff 2005:384–386). A diverse set of methodological tools allow us to investigate social processes at other times and in other places, when the

Researcher Interview Link 10.1
Watch a researcher describe immigration patterns using comparative research.

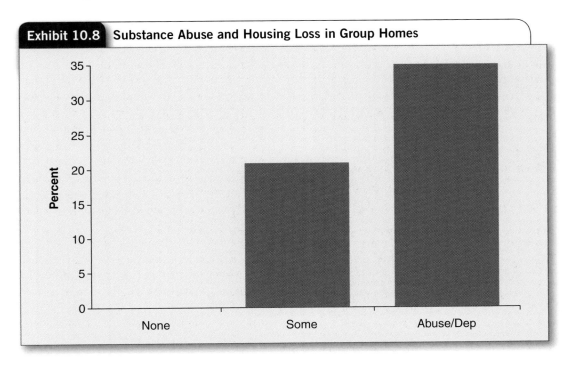

Exhibit 10.8 Substance Abuse and Housing Loss in Group Homes

actual participants in these processes are not available. The central insight behind historical and comparative research methods is that we can improve our understanding of social process when we make comparisons to other times and places. Max Weber's comparative study of world religions (Bendix 1962), Emile Durkheim's (1984) historical analysis of the division of labor, and Seymour Martin Lipset's (1990) contrast of U.S. and Canadian politics affirm the value of this insight.

Rueschemeyer, Stephens, and Stephens (1992) used a method such as this to explain why some nations in Latin America developed democratic politics, whereas others became authoritarian or bureaucratic-authoritarian states. First, Rueschemeyer et al. developed a theoretical framework that gave key attention to the power of social classes, state (government) power, and the interaction between social classes and the government. They then classified the political regimes in each nation over time (Exhibit 10.9). Next, they noted how each nation varied over time in terms of the variables they had identified as potentially important for successful democratization.

**Researcher Interview
Link 10.2**
Watch a researcher describe a specific study that used comparative research.

Exhibit 10.9 Classification of Regimes Over Time

	Constitutional Oligarchic	Authoritarian; Traditional, Populist, Military, or Corporatist	Restricted Democrat	Fully Democratic	Bureaucratic-Authoritarian
Argentina	before 1912	1930–46 1951–55 1955–58 1962–63	1958–62 1963–66	1912–30 1946–51 1973–76 1983–90	1966–73 1976–83
Brazil	before 1930	1930–45	1945–64 1985–90		1964–85
Bolivia	before 1930	1930–52 1964–82	1982–90	1952–64	
Chile	before 1920	1924–32	1920–24 1932–70 1990	1970–73	1973–89
Colombia	before 1936	1949–58	1936–49 1958–90		
Ecuador	1916–25	before 1916 1925–48 1961–78	1948–61 1978–90		
Mexico Paraguay		up to 1990 up to 1990			
Peru		before 1930 1930–39 1948–56 1962–63 1968–80	1939–48 1956–62 1963–68	1980–90	
Uruguay		before 1903 1933–42	1903–19	1919–33 1942–73 1984–90	1973–84
Venezuela		before 1935 1935–45	1958–68	1945–48 1968–90	

Their analysis identified several conditions for initial democratization: consolidation of state power (ending overt challenges to state authority); expansion of the export economy (reducing conflicts over resources); industrialization (increasing the size and interaction of middle and working classes); and some agent of political articulation of the subordinate classes (which could be the state, political parties, or mass movements).

Contemporary comparative research has helped to explain variation in a much more tightly focused phenomenon in democratic politics, voter turnout. Here is a critical conundrum in political science: Although free and competitive elections are a defining feature of democratic politics, elections themselves cannot orient governments to popular sentiment if citizens do not vote (LeDuc, Niemi, & Norris 1996). As a result, the low levels of voter participation in U.S. elections have long been a source of practical concern and research interest.

International data give our first clue for explaining voter turnout: The historic rate of voter participation in the United States (48.3%, on average) is much lower than it is in many other countries that have free, competitive elections; for example, Italy has a voter turnout of 92.5%, on average, since 1945 (Exhibit 10.10).

Is this variation due to differences among voters in knowledge and wealth? Do media and political party get-out-the-vote efforts matter? Mark Franklin's (1996:219–222) analysis of international voting data indicates that neither explanation accounts for much of the international variation in voter turnout. Instead, it is the structure of competition and the importance of issues that are influential. Voter turnout is maximized where structural features maximize competition: compulsory voting (including, in Exhibit 10.10, Austria, Belgium, Australia, and Greece), mail and Sunday voting (including the Netherlands and Germany), and multiday voting. Voter turnout also tends to be higher where the issues being voted on are important and where results are decided by proportional representation (as in Italy and Israel, in Exhibit 10.10) rather than on a winner-take-all basis (as in U.S. presidential elections)—so individual votes are more important.

Franklin (1996) concludes that it is these characteristics that explain the low level of voter turnout in the United States, not the characteristics of individual voters. The United States lacks the structural features that make voting easier, the proportional representation that increases the impact of individuals' votes, and, often, the sharp differences between candidates that are found in countries with higher turnout. Because these structural factors generally do not vary within nations, we would never realize their importance if our analysis was limited to data from individuals in one nation.

Visual Sociology

The analysis of the "text" of social life, then, can be conducted in a variety of ways. But words are not the only form) of qualitative data. For about 150 years, people have been recording the social world with photography. This creates the possibility of "observing" the social world through photographs and films and of interpreting the resulting images as a text. Visual sociology is a method both to learn how others "see" the social world and to create images of it for further study. As with written text, however, the visual sociologist must be sensitive to the way in which a photograph or film "constructs" the reality that it depicts.

An analysis by Eric Margolis (2004) of photographic representations of American Indian boarding schools gives you an idea of the value of analysis of photographs. On the left is a picture taken in 1886 of Chiricahua Apaches who had just arrived at the Carlisle Indian School in Carlisle, Pennsylvania (Exhibit 10.11). The school was run by a Captain Richard Pratt, who, like many Americans in that period, felt that tribal societies were communistic, indolent, dirty, and ignorant, while Western civilization was industrious and individualistic. So Captain Pratt set out to acculturate American Indians to the dominant culture. The second picture shows the result: the same group of Apaches looking like European, not Native, Americans, dressed in "standard" (per the dominant culture) uniforms with standard haircuts and with more standard posture.

Audio Link 10.2
Listen to how visual sociology is used.

Exhibit 10.10 Average Percentage of Voters Who Participated in Presidential or Parliamentary Elections, 1945–1998*

Country	Vote %	Country	Vote %
Italy	92.5	St. Kitts and Nevis	58.1
Cambodia	90.5	Morocco	57.6
Seychelles	96.1	Cameroon	56.3
Iceland	89.5	Paraguay	56.0
Indonesia	88.3	Bangladesh	56.0
New Zealand	86.2	Estonia	56.0
Uzbekistan	86.2	Gambia	55.8
Albania	85.3	Honduras	55.3
Austria	85.1	Russia	55.0
Belgium	84.9	Panama	53.4
Czech	84.8	Poland	52.3
Netherlands	84.8	Uganda	50.6
Australia	84.4	Antigua and Barbuda	50.2
Denmark	83.6	Burma/Myanmar	50.0
Sweden	83.5	Switzerland	49.3
Mauritius	82.8	USA	48.3
Portugal	82.4	Mexico	48.1
Mongolia	82.3	Peru	48.0
Tuvalu	81.9	Brazil	47.9
Western Samoa	81.9	Nigeria	47.6
Andorra	81.3	Thailand	47.4
Germany	80.9	Sierra Leone	46.8
Slovenia	80.6	Botswana	46.5
Aruba	80.4	Chile	45.9
Namibia	80.4	Senegal	45.6
Greece	80.3	Ecuador	44.7
Guyana	80.3	El Salvador	44.3
Israel	80.0	Haiti	42.9
Kuwait	79.6	Ghana	42.4
Norway	79.5	Pakistan	41.8
San Marino	79.1	Zambia	40.5
Finland	79.0	Burkina Faso	38.3
Suriname	77.7	Nauru	37.3
Malta	77.6	Yemen	36.8
Bulgaria	77.5	Colombia	36.2
Romania	77.2	Niger	35.6

*Based on entire voting-age population in countries that held at least two elections during these years. Only countries with highest and lowest averages are shown.

Exhibit 10.11   **Pictures of Chiricahua Apache Children Before and After Starting Carlisle Indian School, Carlisle, Pennsylvania, 1886**

Many other pictures display the same type of transformation. Are these pictures each "worth a thousand words"? They capture the ideology of the school management, but we can be less certain that they document accurately the "before and after" status of the students. Captain Pratt "consciously used photography to represent the boarding school mission as successful" (Margolis 2004:79). While he clearly tried to ensure a high degree of conformity, there were accusations that the contrasting images were exaggerated to overemphasize the change (Margolis:78). In these photographs, reality was being constructed, not just depicted.

With the widespread use of cell phone cameras and video recorders, visual sociology will certainly become an increasingly important aspect of qualitative analyses of social settings and the people in them. The result will be richer descriptions of the social world, but remember Darren Newbury's (2005) reminder to readers of his journal, *Visual Studies*: "Images cannot be simply taken of the world, but have to be made within it" (p. 1).

Research in the News

TAPING AND ANALYZING FAMILY LIFE

In the News

Researchers at the University of California, Los Angeles, recruited 32 local families and then, for 1 week, videotaped them almost continuously while they were awake and at home. A researcher roamed the houses with a handheld computer, recording at 10-minute intervals each family member's location and activities; family members also participated in in-depth interviews and completed questionnaires. Couples reported less stress if they had a more rigid division of labor, whether equal or not. Half the fathers spent at least as much time as their wives along with their children when they were at home, and were more likely to engage in physical activities, while their wives were more likely to watch television with their children. No one spent time in the yard.

Source: Carey, Benedict. 2010. Families' every hug and fuss, taped, analyzed and archived." *The New York Times*, May 23:A1.

How Can Computers Assist Qualitative Data Analysis?

Computer-assisted qualitative data analysis can dramatically accelerate the techniques used traditionally to analyze such text as notes, documents, or interview transcripts: preparation, coding, analysis, and reporting (Coffey & Atkinson 1996; Richards & Richards 1994). Two of the most popular programs can illustrate these steps: HyperRESEARCH and QSR NVivo. (You can link to a trial copy of HyperRESEARCH and tutorials about it on the book's Study Site at www.sagepub.com/chambliss4e.)

> **Computer-assisted qualitative data analysis:** Analysis of textual, aural, or pictorial data using a special computer program that facilitates searching and coding text.

Text preparation begins with typing or scanning text in a word processor or, with NVivo, directly into the program's rich text editor. NVivo will create or import a rich text file (*.rtf). HyperRESEARCH requires that your text be saved as a text file (as "ASCII" in most word processors, or *.txt) before you transfer it into the analysis program. HyperRESEARCH expects your text data to be stored in separate files corresponding to each unique case, such as an interview with one subject.

Coding the text involves categorizing particular text segments. This is the foundation of much qualitative analysis. Either program allows you to assign a code to any segment of text (in NVivo, you drag through the characters to select them; in HyperRESEARCH, you click on the first and last words to select text). You can either make up codes as you go through a document or assign codes that you have already developed to text segments. Exhibits 10.12a and 10.12b show the screens that appear in the two programs at the coding stage,

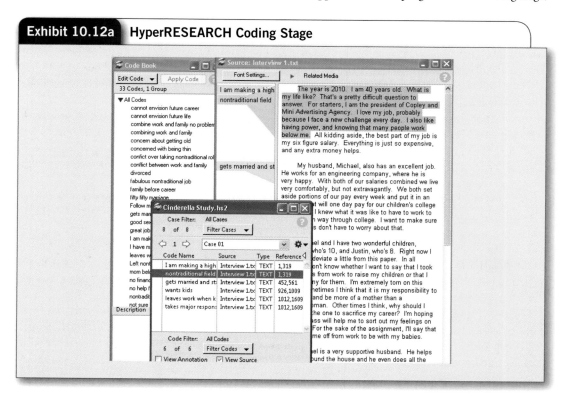

Exhibit 10.12a   HyperRESEARCH Coding Stage

Exhibit 10.12b NVivo Coding Stage

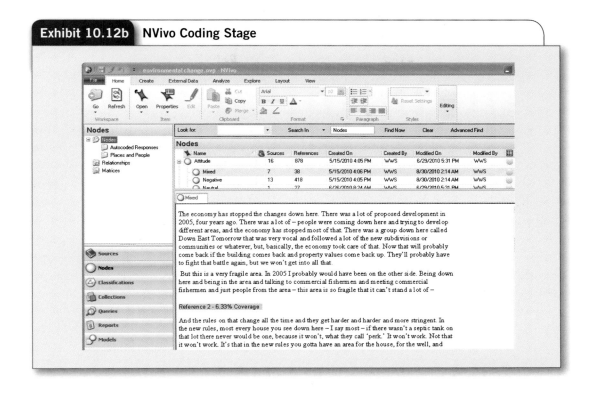

when a particular text segment is being labeled. You can also have the programs "autocode" text by identifying a word or phrase that should always receive the same code, or, in NVivo, by coding each section identified by the style of the rich text document—for example, each question or speaker. (Of course, you should check carefully the results of autocoding.) Both programs also let you examine the coded text "in context"—embedded in its place in the original document.

In qualitative data analysis, coding is not a one-time-only or one-code-only procedure. Both HyperRESEARCH and NVivo allow you to be inductive and holistic in your coding: You can revise codes as you go along, assign multiple codes to text segments, and link your own comments ("memos") to text segments. In NVivo, you can work "live" with the coded text to alter coding or create new, more subtle categories. You can also place hyperlinks to other documents in the project or any multimedia files outside it.

Analysis focuses on reviewing cases or text segments with similar codes and examining relationships among different codes. You may decide to combine codes into larger concepts. You may specify additional codes to capture more fully the variation among cases. You can test hypotheses about relationships among codes. NVivo allows development of an indexing system to facilitate thinking about the relationships among concepts and the overarching structure of these relationships. It also allows you to draw more free-form models (Exhibit 10.13). In HyperRESEARCH, you can specify combinations of codes that identify cases that you want to examine.

Reports from both programs can include text to illustrate the cases, codes, and relationships that you specify. You can also generate counts of code frequencies and then import these counts into a statistical program for quantitative analysis. However, the many types of analyses and reports that can be developed with qualitative analysis software do not lessen the need for a careful evaluation of the quality of the data on which conclusions are based.

Exhibit 10.13 A Free-Form Model in NVivo

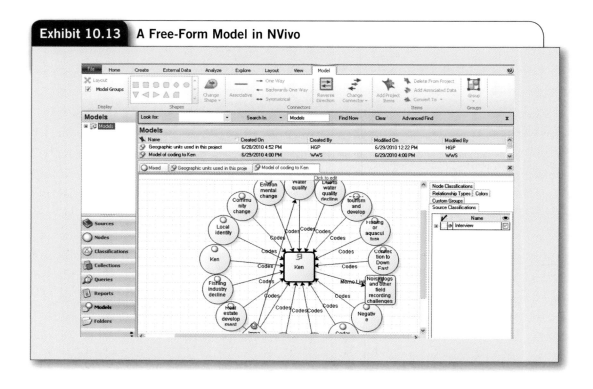

In practice, using these programs is not always as time saving as it may first appear (Bachman & Schutt 2007:319). Scott Decker and Barrik Van Winkle (1996:53–54) described the difficulty they faced in using a computer program to identify instances of "drug sales":

> The software we used is essentially a text retrieval package . . . One of the dilemmas faced in the use of such software is whether to employ a coding scheme within the interviews or simply to leave them as unmarked text. We chose the first alternative, embedding conceptual tags at the appropriate points in the text. An example illustrates this process. One of the activities we were concerned with was drug sales. Our first chore (after a thorough reading of all the transcripts) was to use the software to "isolate" all of the transcript sections dealing with drug sales. One way to do this would be to search the transcripts for every instance in which the word "drugs" was used. However, such a strategy would have the disadvantages of providing information of too general a character while often missing important statements about drugs. Searching on the word "drugs" would have produced a file including every time the word was used, whether it was in reference to drug sales, drug use, or drug availability, clearly more information than we were interested [in]. However, such a search would have failed to find all of the slang used to refer to drugs ("boy" for heroin, "Casper" for crack cocaine) as well as the more common descriptions of drugs, especially rock or crack cocaine.

Decker and Van Winkle (1996) solved this problem by parenthetically inserting conceptual tags in the text whenever talk of drug sales was found. This process allowed them to examine all of the statements made by gang members about a single concept (drug sales). As you can imagine, however, this still left the researchers with many pages of transcript material to analyze.

▣ What Ethical Issues Arise in Doing Qualitative Data Analysis?

The qualitative data analyst is never far from ethical issues and dilemmas. Throughout the analytic process, the analyst must consider how the findings will be used and how participants in the setting will react. Miles and Huberman (1994:204–205) suggest several specific questions that should be kept in mind:

Research integrity and quality. Is my study being conducted carefully, thoughtfully, and correctly in terms of some reasonable set of standards? Real analyses have real consequences, so you owe it to yourself and those you study to adhere strictly to the analysis methods that you believe will produce authentic, valid conclusions.

Ownership of data and conclusions. Who owns my field notes and analyses: I, my organization, my funders? And once my reports are written, who controls their dissemination? Of course, these concerns arise in any social research project, but the intimate involvement of the qualitative researcher with participants in the setting studied makes conflicts of interest between different stakeholders much more difficult to resolve. Working through the issues as they arise is essential.

Use and misuse of results. Do I have an obligation to help my findings be used appropriately? What if they are used harmfully or wrongly? It is prudent to develop understandings early in the project with all major stakeholders that specify what actions will be taken to encourage the appropriate use of project results and to respond to what is considered misuse of these results.

▣ Conclusion

The success of qualitative analyses may be difficult to judge, but Norman Denzin (2002) suggests that the following "interpretive criteria" questions could be asked:

- Does it illuminate the phenomenon as lived experience? In other words, do the materials bring the setting alive in terms of the people in that setting?

- Is it based on thickly contextualized materials? We should expect thick descriptions that encompass the social setting studied.

- Is it historically and relationally grounded? There must be a sense of the passage of time between events and the presence of relationships between social actors.

- Is the research processual and interactional? The researcher must have described the research process and his or her interactions within the setting.

- Does it engulf what is known about the phenomenon? This includes situating the analysis in the context of prior research and acknowledging the researcher's own orientation upon first starting the investigation. (pp. 362–363)

If the answers are yes, a study has achieved much of the promise of qualitative research.

Key Terms

Highlights

- Qualitative data analysis is guided by an emic focus of representing persons in the setting on their own terms, rather than by an etic focus on the researcher's terms.

- Ethnographers attempt to understand the culture of a group.

- Narrative analysis attempts to understand a life or a series of events as they unfolded in a meaningful progression.

- Grounded theory connotes a general explanation that develops in interaction with the data and is continually tested and refined as data collection continues.

- Special computer software can be used for the analysis of qualitative, textual, and pictorial data. Users can record their notes, categorize observations, specify links between categories, and count occurrences.

STUDENT STUDY SITE

The Student Study Site, available at **www.sagepub.com/chambliss4e,** includes useful study materials including web exercises with accompanying links, eFlashcards, videos, audio resources, journal articles, and encyclopedia articles, many of which are represented by the media links throughout the text. The site also features Interactive Exercises—represented by the green icon here—to help you understand the concepts in this book.

Exercises

Discussing Research

1. List the primary components of qualitative data analysis strategies. Compare and contrast each of these components with those relevant to quantitative data analysis. What are the similarities and differences? What differences do these make?

2. Does qualitative data analysis result in trustworthy results—in findings that achieve the goal of "authenticity"? Why would anyone question its use? What would you reply to the doubters?

3. Narrative analysis provides the "large picture" of how a life or event has unfolded, while conversation analysis focuses on the details of verbal interchange. When is each method most appropriate? How could one method add to the other?

4. Ethnography and grounded theory both refer to aspects of data analysis that are an inherent part of the qualitative approach. What do these approaches have in common? How do they differ? Can you identify elements of these two approaches in this chapter's examples of ethnomethodology, conversation analysis, and narrative analysis?

Finding Research

1. *The Qualitative Report* is an online journal about qualitative research. Inspect the table of contents for a recent issue (www .nova.edu/ssss/QR/index.html). Read one of the articles and write a brief article review.

2. Be a qualitative explorer! Go to this list of qualitative research websites and see what you can find that enriches your understanding of qualitative research (www.qualitativeresearch.uga. edu/QualPage/). Be careful to avoid textual data overload.

Critiquing Research

1. Read the complete text of one of the qualitative studies presented in this chapter and evaluate its analysis and conclusions for authenticity, using the criteria in this chapter.

Doing Research

1. Attend a sports game as an ethnographer. Write up your analysis and circulate it for criticism.

2. Write a narrative in class about your first date, car, college course, or something else you and your classmates agree on. Then collect all the narratives and analyze them in a "committee of the whole."

Follow the general procedures discussed in the example of narrative analysis in this chapter.

3. Try out the HyperRESEARCH tutorials that you can link to on the book Study Site. How might qualitative analysis software facilitate the analysis process? Might it hinder the analysis process in some ways? Explain your answers.

Ethics Questions

1. Pictures are worth a thousand words, so to speak, but is that 1,000 words too many? Should qualitative researchers (like yourself) feel free to take pictures of social interaction or other behaviors anytime, anywhere? What limits should an institutional review board place on researchers' ability to take pictures of others? What if the "after" picture of the Apache children in this chapter (Exhibit 10.11) also included Captain Pratt in a military uniform?

2. Participants in social settings often "forget" that an ethnographer is in their midst, planning to record what they say and do, even when the ethnographer has announced his role. New participants may not have heard the announcement, and everyone may simply get used to the ethnographer as if he was just "one of us." What efforts should an ethnographer take to keep people informed about his or her work in the setting under study? Consider settings such as a sports team, a political group, and a book group.

CHAPTER 11

Evaluation Research

D rug Abuse Resistance Education (DARE), as you probably know, is offered in elementary schools across America. For parents worried about drug abuse among youth and for many concerned citizens, the program has immediate appeal. It brings a special police officer into the schools once a week to talk to students about the hazards of drug abuse and to establish a direct link between local law enforcement and young people. You only have to check out bumper stickers or attend a few PTA meetings to learn that it's a popular program. It is one way many local governments have implemented antidrug policies.

And it is appealing. DARE seems to improve relations between the schools and law enforcement and to create a positive image of the police in the eyes of students.

It's a very positive program for kids . . . a way for law enforcement to interact with children in a nonthreatening fashion . . . DARE sponsored a basketball game. The middle school jazz band played. . . . We had families there. . . . DARE officers lead activities at the [middle school]. . . . Kids do woodworking and produce a play. (Taylor 1999:1, 11)

For some, the positive police-community relationships created by the program are enough to justify its continuation (Birkeland, Murphy-Graham, & Weiss 2005:248), but most communities are concerned with its value in reducing drug abuse among children. Does DARE lessen the use of illicit drugs among DARE students? Does it do so while they are enrolled in the program or, more important, after they enter middle or high school? Unfortunately, evaluations of DARE using social science methods led to the conclusion that students who participated in DARE were no less likely to use illicit drugs than comparable students who did not participate in DARE (Ringwalt et al. 1994; West & O'Neal 2004).

If, like us, you have a child who enjoyed DARE, or were yourself a DARE student, this may seem like a depressing way to begin a chapter on evaluation research. Nonetheless, it drives home an important point: To know whether social programs work, or how they work, we have to evaluate them systematically and fairly, whether we personally like the programs or not. And there's actually an optimistic conclusion to this introductory story: Evaluation research can make a difference. After the accumulation of evidence that DARE programs were ineffective (West & O'Neal 2004), a "new" DARE program was designed that engaged students more actively (Toppo 2002).

> Gone is the old-style approach to prevention in which an officer stands behind a podium and lectures students in straight rows. New DARE officers are trained as "coaches" to support kids who are using research-based refusal strategies in high-stakes peer-pressure environments. (DARE 2008)

Of course, the "new DARE" is now being evaluated, too. Sorry to say, one early quasi-experimental evaluation in 17 urban schools, funded by DARE America, found no effect of the program on students' substance use (Vincus, Ringwalt, Harris, & Shamblen 2010).

In this chapter, you will read about a variety of social program evaluations, alternative approaches to evaluation, and the different types of evaluation research and review ethical concerns. You should finish the chapter with a much better understanding of how the methods of applied social research can help improve society.

▣ What Is the History of Evaluation Research?

Evaluation research is not a method of data collection, like survey research or experiments; nor is it a unique component of research designs, like sampling or measurement. Instead, evaluation research is conducted for a distinctive purpose: to investigate social programs (such as substance abuse treatment programs, welfare programs, criminal justice programs, or employment and training programs). For each project, an evaluation researcher must select a research design and method of data collection that are useful for answering the particular research questions posed and appropriate for the particular program investigated.

Video Link 11.1
Watch a clip about evaluation research.

So, you can see why we placed this chapter after most of the others in the text. When you review or plan evaluation research, you have to think about the research process as a whole and how different parts of that process can best be combined.

The development of evaluation research as a major enterprise followed on the heels of the expansion of the federal government during the Great Depression and World War II. Large Depression-era government outlays for social programs stimulated interest in monitoring program output, and the military effort in World War II led to some of the necessary review and contracting procedures for sponsoring evaluation research. However, not until the Great Society programs of the 1960s did evaluation begin to be required when new social programs were funded (Dentler 2002; Rossi & Freeman 1989:34). The World Bank and International Monetary Fund (IMF) began to require evaluation of the programs they fund in other countries (Dentler:147). More than 100 contract research and development firms began in the United States between 1965 and 1975, and many federal agencies developed their own research units. The RAND Corporation expanded from its role as a U.S. Air Force planning unit into a major social research firm; SRI International spun off from Stanford University as a private firm; and Abt Associates in Cambridge, Massachusetts, which began in a garage in 1965, grew to employ more than 1,000 employees in five offices in the United States, Canada, and Europe.

With the decline of many Great Society programs in the early 1980s, many such evaluation research firms closed down. But recently, with more calls for government "accountability," the evaluation research enterprise has been growing again. The Community Mental Health Act Amendments of 1975 (Public Law 94–63) required quality assurance (QA) reviews, which often involve evaluation-like activities (Patton 2002:147–151). The Government Performance and Results Act of 1993 required some type of evaluation of all government programs (Office of Management and Budget n.d.). At century's end, the federal government was spending about $200 million annually on evaluating $400 billion in domestic programs, and the 30 major federal agencies had between them 200 distinct evaluation units (Boruch 1997). In 1999, the new Governmental Accounting Standards Board urged that more attention be given to "service efforts and accomplishments" in standard government fiscal reports (Campbell 2002).

The growth of evaluation research is also reflected in the social science community. The American Evaluation Association was founded in 1986 as a professional organization for evaluation researchers (merging two previous associations) and is the publisher of an evaluation research journal. In 1999, evaluation researchers founded the Campbell Collaboration to publicize and encourage systematic review of evaluation research studies. Their online archive contains 10,449 reports on randomized evaluation studies (Davies, Petrosino, & Chalmers 1999).

What Is Evaluation Research?

Exhibit 11.1 illustrates the process of evaluation research as a simple systems model. First, clients, customers, students, or some other persons or units—cases—enter the program as **inputs**. (Notice that this model regards programs as machines, with clients—people—seen as raw materials to be processed.) Students may begin a new school program, welfare recipients may enroll in a new job-training program, or

Researcher Interview Link 11.1
Watch a researcher describe a specific study that used evaluation research.

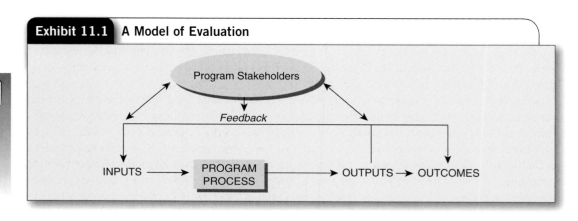

Exhibit 11.1 A Model of Evaluation

crime victims may be sent to a victim advocate. Resources and staff required by a program are also program inputs.

Next, some service or treatment is provided to the cases. This may be attendance in a class, assistance with a health problem, residence in new housing, or receipt of special cash benefits. This process of service delivery—the **program process**—may be simple or complicated, short or long, but it is designed to have some impact on the cases as inputs are consumed and outputs are produced.

Program **outputs** are the direct product of the program's service delivery process. They could include clients served, case managers trained, food parcels delivered, or arrests made. The program outputs may be desirable in themselves, but primarily they indicate that the program is operating.

Program **outcomes** indicate the impact of the program on the cases that have been processed. Outcomes can range from improved test scores or higher rates of job retention to fewer criminal offenses and lower rates of poverty. There are likely to be multiple outcomes of any social program, some intended and some unintended, some viewed as positive and others viewed as negative.

Through a **feedback** process, variation in outputs and outcomes can influence the inputs to the program. If not enough clients are being served, recruitment of new clients may increase. If too many negative side effects result from a trial medication, the trials may be limited or terminated. If a program does not lead to improved outcomes, clients may be sent elsewhere.

Evaluation research itself is really just a systematic approach to feedback; it strengthens the feedback loop through credible analyses of program operations and outcomes. Evaluation research also broadens this loop to include connections to parties outside of the program itself. A funding agency or political authority may mandate the research; outside experts may be brought in to conduct the research; and the evaluation research findings may be released to the public, or at least to funders, in a formal report.

The evaluation process as a whole, and the feedback in particular, can be understood only in relation to the interests and perspectives of program stakeholders. **Stakeholders** are those individuals and groups who have some basis of concern for the program. They might be clients, staff, managers, funders, or the public. The board of a program or agency, the parents or spouses of clients, the foundations that award program grants, the auditors who monitor program spending, the members of Congress—each is a potential program stakeholder, and each has an interest in the outcome of any program evaluation. Some may fund the evaluation, some may provide research data, and some may review—or even approve—the research report (Martin & Kettner 1996:3). Who the program stakeholders are, and what role they play in the program evaluation, can have tremendous consequences for the research.

Thus, there are real differences between traditional social science and evaluation research (Posavac & Carey 1997). Social science is motivated by theoretical concerns and is guided by the standards of research methods without consideration (ideally) for political factors. It examines specific organizations for what, in general, we can learn from them, not for improving that one organization. Practical ramifications, for particular programs, are not usually of any import. For evaluation research, on the other hand, the particular program and its impact are paramount. How the program works also matters—not to advance a theory but to improve the program. Finally, stakeholders of all sorts—not an abstract "scientific community"—have a legitimate role in setting the research agenda and may well intervene, even when they aren't supposed to. But overall, there is no sharp boundary between the two approaches: In their attempt to explain how and why the program has an impact and whether the program is needed, evaluation researchers often bring social theories into their projects—but for immediately practical aims.

Inputs: Resources, raw materials, clients, and staff that go into a program.

Program process: The complete treatment or service delivered by the program.

Outputs: The services delivered or new products produced by the program process.

Outcomes: The impact of the program process on the cases processed.

Feedback: Information about service delivery system outputs, outcomes, or operations that is available to any program inputs.

Stakeholders: Individuals and groups who have some basis of concern with the program.

Video Link 11.2
Watch a clip about the components of evaluation research.

What Are the Alternatives in Evaluation Designs?

Evaluation research tries to learn if, and how, real-world programs produce results. But that simple statement covers a number of important alternatives in research design, including the following:

- *Black box or program theory*—Do we care how the program gets results?
- *Researcher or stakeholder orientation*—Whose goals matter most?
- *Quantitative or qualitative methods*—Which methods provide the best answers?
- *Simple or complex outcomes*—How complicated should the findings be?

Black Box or Program Theory

Encyclopedia Link 11.1
Read more about theory-driven evaluation techniques.

Most evaluation research tries to determine whether a program has the intended effect. If the effect occurred, the program "worked"; if the effect didn't occur, then, some would say, the program should be abandoned or redesigned. In this simple approach, the process by which a program produces outcomes is often treated as a "black box" in which the inside of the program is unknown. The focus of such research is whether cases have changed as a result of their exposure to the program between the time they entered as inputs and when they exited as outputs (Chen 1990). The assumption is that program evaluation requires only the test of a simple input/output model, like that in Exhibit 11.1. There may be no attempt to "open the black box" of the program process.

But there are good reasons to open the black box and investigate how the process works (or doesn't work). Consider recent research on welfare-to-work programs. The Manpower Demonstration Research Corporation reviewed findings from research on these programs in Florida, Minnesota, and Canada (Lewin 2001a). In each location, adolescents with parents in a welfare-to-work program were compared to a control group of teenagers whose parents were also on welfare but were *not* enrolled in welfare-to-work. In all three locations, teenagers in the welfare-to-work program families did *worse* in school than those in the control group.

But why did requiring welfare mothers to get jobs hurt their children's schoolwork? Unfortunately, because the researchers had not investigated program process—had not opened the black box—we can't know for sure. Martha Zaslow, an author of the resulting research report, speculated (as cited in Lewin 2001a) that

> parents in the programs might have less time and energy to monitor their adolescents' behavior once they were employed. . . . Under the stress of working, they might adopt harsher parenting styles . . . The adolescents' assuming more responsibilities at home when parents got jobs was creating too great a burden. (p. A16)

Unfortunately, as Ms. Zaslow (as cited in Lewin 2001a) admitted, "We don't know exactly what's causing these effects, so it's really hard to say, at this point, what will be the long-term effects on these kids" (p. A16).

If an investigation of program process had been conducted, though, a **program theory** could have been developed. A program theory describes what has been learned about how the program has its effect. When a researcher has sufficient knowledge

Program theory: A descriptive or prescriptive model of how a program operates and produces effects.

before the investigation begins, outlining a program theory can help to guide the investigation of program process in the most productive directions. This is termed a **theory-driven evaluation.**

A program theory specifies how the program is expected to operate and identifies which program elements are operational (Chen 1990:32). In addition, a program theory specifies how a program is to produce its effects, thus improving the understanding of the relationship between the independent variable (the program) and the dependent variable (the outcome or outcomes). For example, Exhibit 11.2 illustrates the theory for an alcoholism treatment program. It shows that persons entering the program are expected to respond to the combination of motivational interviewing and peer support. A program theory also can decrease the risk of failure when the program is transported to other settings, because it will help to identify the conditions required for the program to have its intended effect.

> **Theory-driven evaluation:**
> A program evaluation guided by a theory that specifies the process by which the program has an effect.

Program theory can be either descriptive or prescriptive (Chen 1990). *Descriptive theory* specifies impacts that are generated and how this occurs. It suggests a causal mechanism, including intervening factors and the necessary context for the effects. Descriptive theories are generally empirically based. On the other hand, *prescriptive theory* specifies what *ought* to be done by the program and is not actually tested. Prescriptive theory specifies how to design or implement the treatment, what outcomes should be expected, and how performance should be judged. Comparison of the program's descriptive and prescriptive theories can help to identify implementation difficulties and incorrect understandings that can be fixed (Patton 2002:162–164).

Researcher or Stakeholder Orientation

Whose prescriptions direct the program? What outcomes it should achieve? Whom it should serve? Most social science assumes that the researcher decides. Research results are usually reported in professional journals or conferences, where scientific standards determine how it is judged. In program evaluation, however, the program sponsors or a government agency often set the research question; in consulting projects for businesses, the client—a manager, perhaps, or a division president—decides what question researchers will study. Research findings are reported to these authorities, who most often also specify the outcomes to be investigated. The primary evaluator of evaluation research, then, is the funding agency, not the professional social

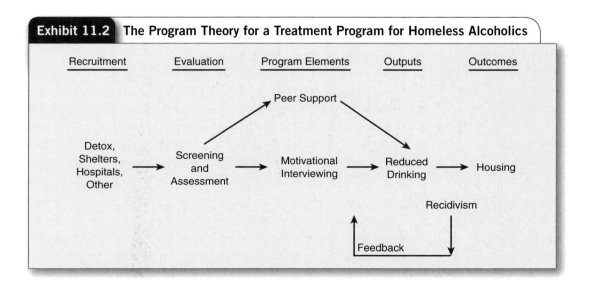

Exhibit 11.2 The Program Theory for a Treatment Program for Homeless Alcoholics

science community. Evaluation research is research for a client, and its results may directly affect the services, treatments, or even punishments (in the case of prison studies, for example) that program users receive. Who pays the piper, picks the tune.

Should the evaluation researcher insist on designing the project and specifying its goals? Or should she accept the suggestions and goals of the funding agency? What role should program staff and clients play? What responsibility does the researcher have to politicians and taxpayers when evaluating government-funded programs?

Various evaluation researchers have answered these questions through different—stakeholders, social science, and integrative—approaches (Chen 1990:66–68). **Stakeholder approaches** encourage researchers to be responsive to program stakeholders. Issues for study are to be based on the views of people involved with the program, and reports are to be made to program participants (Stake 1975). The researcher develops the program theory to clarify and develop the key stakeholders' theory of the program (Wholey 1987). In one stakeholder approach, termed *utilization-focused evaluation*, the evaluator forms a task force of program stakeholders, who help to shape the evaluation project so that they are most likely to use its results (Patton 2002:171–175). In evaluation research termed *action research* or *participatory research*, program participants are engaged with the researchers as coresearchers and help to design, conduct, and report the research. One research approach, termed *appreciative inquiry*, eliminates the professional researcher altogether in favor of a structured dialogue about needed changes among program participants themselves (Patton:177–185).

Egon Guba and Yvonna Lincoln (1989) argue for a stakeholder approach in their book, *Fourth Generation Evaluation:*

> The stakeholders and others who may be drawn into the evaluation are welcomed as equal partners in every aspect of design, implementation, interpretation, and resulting action of an evaluation—that is, they are accorded a full measure of political parity and control. . . . determining what questions are to be asked and what information is to be collected on the basis of stakeholder inputs. (p. 11)

Stakeholder approaches (to evaluation): An orientation to evaluation research that expects researchers to be responsive primarily to the people involved with the program.

Social science approaches (to evaluation): An orientation to evaluation research that expects researchers to emphasize the importance of researcher expertise and maintenance of autonomy from program stakeholders.

Integrative approaches (to evaluation): An orientation to evaluation research that expects researchers to respond to the concerns of people involved with the program stakeholders, as well as to the standards and goals of the social scientific community.

Social science approaches, in contrast, emphasize researcher expertise autonomy to develop the most trustworthy, unbiased program evaluation. They assume that "evaluators cannot passively accept the values and views of the other stakeholders" (Chen 1990:78). Instead, the researcher derives a program theory from information on how the program operates and current social science theory, not from the views of stakeholders. In one somewhat extreme form of this approach, *goal-free evaluation*, researchers do not even permit themselves to learn what goals the program stakeholders have for the program. Instead, the researcher assesses and then compares the needs of participants to a wide array of program outcomes (Scriven 1972). The goal-free evaluator wants to see the unanticipated outcomes and to remove any biases caused by knowing the program goals in advance.

Of course, there are disadvantages to both stakeholder and social science approaches to program evaluation. If stakeholders are ignored, researchers may find that participants are uncooperative, that their reports are unused, and that the next project remains unfunded. On the other hand, if social science procedures are neglected, standards of evidence will be compromised, conclusions about program effects will likely be invalid, and results are unlikely to be generalizable to other settings. These equally undesirable possibilities have led to several attempts to develop more integrated approaches to evaluation research.

Integrative approaches attempt to cover issues of concern to both stakeholders and evaluators (Chen & Rossi 1987:101–102). The emphasis given to either stakeholder or scientific concerns varies with the specific circumstances. Integrative approaches seek to balance responsiveness to stakeholders with objectivity and scientific validity. Evaluators negotiate regularly with key stakeholders during the planning of the research; preliminary findings are reported back to decision makers so they can make improvements; and when the final evaluation is conducted, the research team may operate more autonomously, minimizing intrusions from program stakeholders. Evaluators and clients thus work together.

Quantitative or Qualitative Methods

Quantitative and qualitative approaches to evaluation each have their strengths and appropriate uses. Quantitative research, with its clear percentages and numerical scores, allows quick comparisons over time and categories and, thus, is typically used in attempts to identify the effects of a social program. With numbers, you can systematically track change over time or compare outcomes between an experimental and a control group. Did the response times of emergency personnel tend to decrease? Did the students' test scores increase more in the experimental group than in the control group? Did housing retention improve for all subjects or just for those who were not substance abusers? Quantified results also can prevent distraction by the powerful anecdote, forcing you to see what happens in most cases, not just in the dramatic cases; they "force you to face reality," as a friend of ours puts it.

Journal Link 11.1
Read an article about a mixed-methods approach to evaluating classroom dynamics.

Qualitative methods, on the other hand, can add depth, detail, and nuance; they can clarify the meaning of survey responses and reveal more complex emotions and judgments people may have (Patton 2002). Perhaps the greatest contribution qualitative methods can make is in investigating program process—finding out what is "inside the black box." Quantitative measures, like staff contact hours or frequency of complaints, can track items such as service delivery, but finding out how clients experience the program is best accomplished by directly observing program activities and interviewing staff and clients intensively.

For example, Tim Diamond's (1992:17) observational study of work in a nursing home shows how the somewhat cool professionalism of new program aides was softened to include a greater sensitivity to interpersonal relations:

> The tensions generated by the introductory lecture and . . . ideas of career professionalism were reflected in our conversations as we waited for the second class to get under way. Yet within the next half hour they seemed to dissolve. Mrs. Bonderoid, our teacher, saw to that. . . . "What this [work] is going to take," she instructed, "is a lot of mother's wit." "Mother's wit," she said, not "mother wit," which connotes native intelligence irrespective of gender. She was talking about maternal feelings and skills.

Surveys could have asked the aides how satisfied they were with their training but would not have revealed the subtler side of "mother's wit."

Qualitative methods also can uncover how different individuals react to the treatment. For example, a quantitative evaluation of student reactions to an adult basic skills program for new immigrants relied heavily on the students' initial statements of their goals. However, qualitative interviews revealed that most new immigrants lacked sufficient experience in America to set meaningful goals; their initial goal statements simply reflected their eagerness to agree with their counselors' suggestions (Patton 2002:177–181).

Qualitative methods can, in general, help in understanding how social programs actually operate. In complex social programs, it is not always clear whether any particular features are responsible for the program's effect (or noneffect). Lisbeth B. Schorr, director of the Harvard Project on Effective Interventions, and Daniel Yankelovich, president of Public Agenda, put it this way: "Social programs are sprawling efforts with multiple components requiring constant mid-course corrections, the involvement of committed human beings, and flexible adaptation to local circumstances" (Schorr & Yankelovich 2000:A14). Schorr and Yankelovich pointed to the Ten Point Coalition, an alliance of black ministers that helped to reduce gang warfare in Boston through multiple initiatives, "ranging from neighborhood probation patrols to safe havens for recreation" (A14). Qualitative methods would help to describe a complex, multifaceted program like this. In general, the more complex the social program, the more value that qualitative methods can add to the evaluation process.

Simple or Complex Outcomes

Audio Link 11.1
Listen to examples
of different outcomes.

Few programs have only one outcome. Colleges provide not only academic education, for instance, but also—importantly—an amazingly efficient marketplace for potential spouses and lifetime friends. DARE programs may not reduce drug use, but they often seem to improve student–police relations. Some outcomes are direct and intended; others happen only over time, are uncertain, and may well not be desired. A decision to focus exclusively on a single outcome—probably the officially intended one—can easily cause a researcher to ignore even more important results.

Sometimes a single policy outcome is sought but is found not to be sufficient, either methodologically or substantively. When Sherman and Berk (1984) evaluated the impact of an immediate arrest policy in cases of domestic violence in Minneapolis, they focused on recidivism—repeating the offense—as the key outcome. Similarly, the reduction of recidivism was the single desired outcome of the prison boot camps that began opening in the 1990s. Boot camps were military-style programs for prison inmates that provided tough, highly regimented activities and harsh punishment for disciplinary infractions with the goal of scaring inmates "straight." But these single-purpose programs, both designed to reduce recidivism, turned out not to be quite so simple to evaluate. The Minneapolis researchers found that there were no adequate single sources for recidivism in domestic violence cases, so they had to hunt for evidence from court and police records, perform follow-up interviews with victims, and review family member reports. More easily measured variables, such as partners' ratings of the accused's subsequent behavior, received more attention. Boot camp research soon concluded that the experience did not reduce recidivism, but some participants felt that boot camps did have some beneficial effects:

> [A staff member] saw things unfold that he had never witnessed among inmates and their caretakers. . . . Profoundly affected the drill instructors and their charges . . . Graduation ceremonies routinely reduced inmates . . . sometimes even supervisors to tears. . . . Here, it was a totally different experience. (Latour, 2002:B7)

Some now argue that the failure of boot camps to reduce recidivism was due to the lack of postprison support rather than to failure of the camps to promote positive change in inmates. Looking at recidivism rates alone would ignore some important positive results.

So in spite of the difficulties, most evaluation researchers attempt to measure multiple outcomes (Mohr 1992). One such evaluation appears in Exhibit 11.3. Project New Hope was an ambitious experimental evaluation of the impact of guaranteeing jobs to poor people (DeParle 1999). It was designed to answer the

Exhibit 11.3   Outcomes in Project New Hope

Income and Employment (2nd program year)	New Hope	Control Group
Earnings	$6,602	$6,129
Wage subsidies	1,477	862
Welfare income	1,716	1,690
Food stamp income	1,418	1,242
Total income	11,213	9,915
% above poverty level	27%	19%
% continuously unemployed for 2 years	6%	13%
Hardships and Stress	**New Hope**	**Control Group**
% reporting:		
Unmet medical needs	17%	23%
Unmet dental needs	27%	34%
Periods without health insurance	49%	61%
Living in overcrowded conditions	14%	15%
Stressed much or all of the time	45%	50%
Satisfied or very satisfied with standard of living	65%	67%

following question: If low-income adults are given a job at a sufficient wage, above the poverty level, with child care and health care assured, how many would ultimately prosper?

In Project New Hope, 677 low-income adults in Milwaukee, Wisconsin, were offered a job involving work for 30 hours a week, as well as child care and health care benefits. A control group did not receive the guaranteed jobs. The outcome? Only 27% of the 677 stuck with the job long enough to lift themselves out of poverty, and their earnings as a whole were only slightly higher than those of the control group. Levels of depression were not decreased, nor was self-esteem increased by the job guarantee. But there were some positive effects: The number of people who never worked at all declined, and rates of health insurance and use of formal child care increased. Perhaps most important, the classroom performance and educational hopes of participants' male children increased, with the boys' test scores rising by the equivalent of 100 points on the SAT and their teachers ranking them as better behaved.

So did the New Hope program "work"? Clearly it didn't live up to initial expectations, but it certainly showed that social interventions can have some benefits. Would the boys' gains continue through adolescence? Longer-term outcomes would be needed. Why didn't girls (who were already performing better than the boys) benefit from their parents' enrollment in New Hope just as the boys did? A process analysis would add a great deal to the evaluation design. Collection of multiple outcomes, then, gives a better picture of program impact.

▣ What Can an Evaluation Study Focus On?

Research|Social Impact
Link 11.1
Read more about
evaluation studies.

Evaluation projects can focus on a variety of different questions related to social programs and their impact:

- What is the level of need for the program?
- Can the program be evaluated?
- How does the program operate?
- What is the program's impact?
- How efficient is the program?

The question asked will determine what research methods are used.

Needs Assessment

Needs assessment: A type of evaluation research that attempts to determine the needs of some population that might be met with a social program.

A **needs assessment** attempts, with systematic, credible evidence, to evaluate what needs exist in a population. Need may be assessed by social indicators, such as the poverty rate or the level of home ownership; interviews with local experts, such as school board members or team captains; surveys of populations potentially in need; or focus groups with community residents (Rossi & Freeman 1989).

Research in the News

PREDICTING CRIMINAL PROPENSITY

Nowhere is needs assessment more needed than in predicting the risk of recidivism among applicants for parole. In the 1920s, sociologist Ernest Burgess studied previously released inmates' criminal histories and classified them on the basis of 22 variables as *hobos, ne'er-do-wells, farm boys, drug addicts, gangsters,* and *recent immigrants.* Illinois used a classification like this for 30 years, while other states relied on "clinical judgment." Now most states use a quantitative risk assessment tool in which risk predictions are based on identified correlates of future criminality, such as age at first arrest, job and relationship history, history of drug abuse and gang activity, and behavior while incarcerated.

Source: Neyfakh, Leon. 2011. You will commit a crime in the future, true, false: Inside the science of predicting violence. *Boston Sunday Globe,* February 20:K1, K4.

It is not as easy as it sounds (Posavac & Carey 1997). Whose definitions of need should be used? How will we deal with ignorance of need? How can we understand the level of need without understanding the social context? (Short answer to that one: We can't!) What, after all, does "need" mean in the abstract?

The results of the Boston McKinney Project reveal the importance of taking a multidimensional approach to the investigation of need. The Boston McKinney Project evaluated the merits of providing formerly homeless mentally ill persons with staffed group housing as compared with individual housing (Schutt 2011). In a sense, you can think of the whole experiment as involving an attempt to answer the question "What type of housing do these persons 'need'?" Schutt and his colleagues first examined this question at the start of the project, by asking each project participant which type of housing he or she wanted (Schutt & Goldfinger 1996) and by independently asking two clinicians to estimate which of the two housing alternatives would be best for each participant (Goldfinger & Schutt 1996).

Exhibit 11.4 displays the findings. The clinicians recommended staffed group housing for 69% of the participants, whereas most of the participants (78%) sought individual housing. In fact, there was no correspondence between the housing recommendations of the clinicians and the housing preferences of the participants (who did not know what the clinicians had recommended for them). So which perspective reveals the level of need for staffed group housing as opposed to individual housing?

Of course, there's no objective answer. Policy makers' values, and their understanding of mental illness and homelessness, will influence which answer they prefer.

In general, it is a good idea to use multiple indicators of need. There is no absolute definition of need in this situation, nor is there in most projects. A good evaluation researcher will try to capture different perspectives on need and then help others make sense of the results.

Audio Link 11.2
Listen to an example of needs assessment.

Evaluability Assessment

Evaluation research is pointless if the program cannot be evaluated. Yes, some type of study is always possible, but to identify specifically the effects of a program may not be possible within the available time and resources. So researchers may conduct an **evaluability assessment** to learn this in advance, rather than expend time and effort on a fruitless project (Patton 2002:164).

Evaluability assessment: A type of evaluation research conducted to determine whether it is feasible to evaluate a program's effects within the available time and resources.

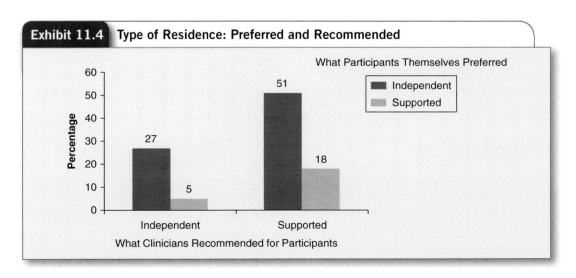

Exhibit 11.4 **Type of Residence: Preferred and Recommended**

Why might a social program not be evaluable?

Journal Link 11.2
Read about the initial findings of a program assessment which indicate and allow for further evaluation.

- Management only wants to have its superior performance confirmed and does not really care whether the program is having its intended effects. This is a very common problem.

- Staff are so alienated from the agency that they don't trust any attempt sponsored by management to check on their performance.

- Program personnel are just "helping people" or "putting in time" without any clear sense of what the program is trying to achieve.

- The program is not clearly distinct from other services delivered by the agency and so can't be evaluated by itself.

Because they are preliminary studies to "check things out," evaluability assessments often rely on qualitative methods. Program managers and key staff may be interviewed, or program sponsors may be asked about the importance they attach to different goals.

Sometimes an evaluability assessment can actually help to solve problems. Discussion with program managers and staff can result in changes in program operations. The evaluators may use the evaluability assessment to sensitize participants to the importance of clarifying their goals and objectives. The knowledge gained can be used to refine evaluation plans.

The President's Family Justice Center (FJC) Initiative was initiated in the administration of President George W. Bush to plan and implement comprehensive domestic violence services that would provide "one stop shopping" for victims in need of services. In 2004, the National Institute of Justice contracted with Abt Associates in Cambridge, Massachusetts to assess the evaluability of 15 pilot service programs that been awarded a total of $20 million and to develop an evaluation plan. In June 2005, Abt researchers Meg Townsend, Dana Hunt, Caity Baxter, and Peter Finn reported on their evaluability assessment.

Abt's assessment began with conversations to collect background information and perceptions of program goals and objectives from those who had designed the program. These conversations were followed by a review of the grant applications submitted by each of the 15 sites and phone conversations with site representatives. Site-specific data collection focused on the project's history at the site, its stage of implementation, staffing plans and target population, program activities and stability, goals identified by the site's director, apparent contradictions between goals and activities, and the state of data systems that could be used in the evaluation. Exhibit 11.5 shows the resulting logic model that illustrates the intended activities, outcomes and impacts for the Alameda County, California program. Although they had been able to begin the evaluability assessment process, Meg Townsend, Dana Hunt, and their colleagues concluded that in the summer of 2005, none of the 15 sites were far enough along with their programs to complete the assessment.

Process Evaluation

Journal Link 11.3
Read an evaluation of a treatment program for offenders.

What actually happens in a social program? In the New Jersey Income Maintenance Experiment, some welfare recipients received higher payments than others (Kershaw & Fair 1976): simple enough, and not too difficult to verify that the right people received the intended treatment. In the Minneapolis experiment on the police response to domestic violence (Sherman & Berk 1984), some individuals accused of assaulting their spouses were arrested, whereas others were just warned. This is a little bit more complicated, because the severity of the warning might have varied among police officers and, to minimize the risk of repeat harm, police officers were allowed to override the experimental assignment. To identify this deviation from the experimental design, the researchers would have had to keep track of the treatments delivered to each

Exhibit 11.5	Alameda Family Justice Center Logic Model

Inputs	Activities	Outcomes	Impacts	Goals
• On-site partners • Intake systems • Client management process • Space design • Site location	**FJC** • Case management • Assistance with restraining orders • Assistance with police reports • Legal assistance • Advocacy • Medical care • Forensic exams • Assessments and referral for treatment • Counseling • Safety planning • Emergency food/cash/transportation • Referral for shelter and other on-going care • Assistance with public assistance • 24-hour helpline • Parenting classes • Child care • Rape crises services • Faith-based services • Job training • Translation services	**Victims** • Increase likelihood to access services • Increase demand for services • Increase usage of services • Increase frequency of cross-referrals or use of multiple services	**Victims** • Reduce tendency to blame oneself for abuse • Reduce conditions that prevent women from leaving • Increase likelihood of reporting incident • Increase likelihood of request for temporary/permanent restraining orders • Increase likelihood of participating in prosecution	• Decrease incidents of DV ○ Decreased repeat victimizations ○ Decreased seriousness • Hold offenders accountable ○ Decrease repeat offenders • Break cycle of violence
	Community • Early intervention and prevention programming • FJC informational materials	**Community** • Increase knowledge of DV/SA/Elder Abuse • Increase awareness of services available	**Community** • Increase awareness of FJC • Decrease social tolerance for VAW	

(Continued)

Exhibit 11.5 (Continued)

Inputs	Activities	Outcomes	Impacts	Goals
	Systems • Collaboration between government and non-gov't providers • Improve access to batterer information	**Systems** • Improve DV policies and procedures • Increase understanding of each other's services • Increase coordination of services	**Systems** • Improve institutional response to DV • Decrease secondary trauma • Increase assurance of victim safety • Increase the number of successful criminal legal actions • Increase the number of successful civil legal actions	

Process evaluation: Evaluation research that investigates the process of service delivery.

accused spouse and collect information on what officers actually *did* when they warned an accused spouse. This would be **process evaluation**—research to investigate the process of service delivery.

Process evaluation is more important when more complex programs are evaluated. Many social programs comprise multiple elements and are delivered over an extended period of time, often by different providers in different areas. Due to this complexity, it is quite possible that the program as delivered is neither the same for all program recipients nor consistent with the formal program design.

The evaluation of DARE by Research Triangle Institute researchers Christopher Ringwalt and others (1994) included a process evaluation designed to address these issues:

- Assess the organizational structure and operation of representative DARE programs nationwide.

- Review and assess the factors that contribute to the effective implementation of DARE programs nationwide.

- Assess how DARE and other school-based drug prevention programs are tailored to meet the needs of specific populations.

The process evaluation (they called it an "implementation assessment") was an ambitious research project with site visits and informal interviews, discussions, and surveys of DARE program coordinators and advisers. These data indicated that DARE was operating as designed and was running relatively smoothly. Drug prevention coordinators in DARE school districts rated the program more highly than coordinators in districts with other alcohol and drug prevention programs rated theirs.

Process evaluation can also identify which specific part of the service delivery has the greatest impact. This, in turn, helps to explain why the program has an effect and which conditions are required for the effect.

(In Chapter 6, we described this as identifying the causal "mechanism.") In the DARE research, site visits revealed an insufficient number of officers and a lack of Spanish-language DARE books in a largely Hispanic school. At the same time, classroom observations indicated engaging presentations and active student participation (Ringwalt et al. 1994:69, 70).

Process analysis of this sort can also help to show how apparently clear findings may be incorrect. The apparently disappointing results of the Transitional Aid Research Project (TARP) provide an instructive lesson. TARP was a social experiment designed to determine whether financial aid during the transition from prison to the community would help released prisoners find employment and avoid returning to crime. Two thousand participants in Georgia and Texas were randomized to receive either a particular level of benefits over a particular period of time or no benefits (the control group). Initially, it seemed that the payments had no effect: The TARP treatment condition did not alter the rate of subsequent arrests for property or nonproperty crimes.

But this wasn't all there was to it. Peter Rossi tested a more elaborate causal model of TARP's effects, which is summarized in Exhibit 11.6. Participants who received TARP payments had more income to begin with and so had more to lose if they were arrested; therefore, they were less likely to commit crimes. However, TARP payments also created a disincentive to work and, therefore, increased the time available in which to commit crimes. Thus, the positive direct effect of TARP (more to lose) was cancelled out by its negative indirect effect (more free time).

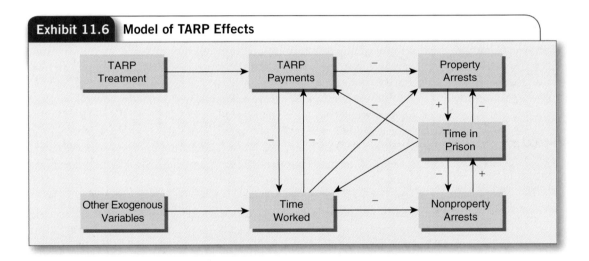

Exhibit 11.6 Model of TARP Effects

Formative evaluation occurs when the evaluation findings are used to help shape and refine the program (Rossi & Freeman 1989), for instance by being incorporated into the initial development of the service program. Evaluation may then lead to changes in recruitment procedures, program delivery, or measurement tools (Patton 2002:220).

> **Formative evaluation:** Process evaluation that is used to shape and refine program operations.

You can see the *formative* element in the following government report on the performance of the Health Care Finance Administration (HCFA):

> While HCFA's performance report and plan indicate that it is making some progress toward achieving its Medicare program integrity outcome, progress is difficult to measure because of continual goal changes that are sometimes hard to track or that are made with insufficient explanation. Of

the five fiscal year 2000 program integrity goals it discussed, HCFA reported that three were met, a fourth unmet goal was revised to reflect a new focus, and performance data for the fifth will not be available until mid-2001. HCFA plans to discontinue three of these goals. Although the federal share of Medicaid is projected to be $124 billion in fiscal year 2001, HCFA had no program integrity goal for Medicaid for fiscal year 2000. HCFA has since added a developmental goal concerning Medicaid payment accuracy. (U.S. Government Accounting Office 2001:7)

Process evaluation can employ a wide range of indicators. Program coverage can be monitored through program records, participant surveys, community surveys, and analysis of utilizers versus dropouts and ineligibles. Service delivery may be monitored through service records program staff complete, a management information system program administrators maintain, and program recipients' reports (Rossi & Freeman 1989).

Qualitative methods are often a key component of process evaluation studies, because they can be used to elucidate and understand internal program dynamics—even those that were not anticipated (Patton 2002:159; Posavac & Carey 1997). Qualitative researchers may develop detailed descriptions of how program participants engage with each other, how the program experience varies for different people, and how the program changes and evolves over time.

Research|Social Impact Link 11.2
Read more about how evaluations are used.

Impact Analysis

The core questions of evaluation research are these: Did the program work? Did it have the intended result? This kind of research is variously called **impact analysis**, **impact evaluation**, or **summative evaluation**. Formally speaking, impact analysis compares what happened after a program was implemented with what *would* have happened had there been no program at all.

> **Impact analysis (impact evaluation or summative evaluation):** Evaluation research that answers these questions: Did the program work? Did it have the intended result?

Think of the program—such as a new strategy for combating domestic violence or an income supplement—as an independent variable and the result it seeks as a dependent variable. The DARE program (independent variable), for instance, tries to reduce drug use (dependent variable). If the program is present, we should expect less drug use. In a more elaborate study, we might have multiple values of the independent variable, for instance, comparing conditions of "no program," "DARE program," and "other drug/alcohol education."

As in other areas of research, an experimental design is the preferred method for maximizing internal validity—that is, for making sure your causal claims about program impact are justified. Cases are assigned randomly to one or more experimental treatment groups and to a control group so that there is no systematic difference between the groups at the outset (see Chapter 6). The goal is to achieve a fair, unbiased test of the program itself so that differences between the types of people who are in the different groups do not influence judgment about the program's impact. It can be a difficult goal to achieve, because the usual practice in social programs is to let people decide for themselves whether they want to enter a program, as well as to establish eligibility criteria that ensure that people who enter the program are different from those who do not (Boruch 1997). In either case, a selection bias is introduced.

But sometimes researchers are able to conduct well-controlled experiments. Robert Drake et al. (1996) evaluated the impact of two different approaches to providing employment services for people diagnosed with severe mental disorders, using a randomized experimental design. One approach, group skills training (GST), emphasized pre-employment skills training and used separate agencies to provide vocational and mental health services. The other approach, individual placement and support (IPS), provided vocational and mental health services in a single program and placed people directly into jobs without pre-employment skills training. The researchers hypothesized that GST participants would be more likely to obtain jobs during the 18-month study period than would IPS participants.

Their experimental design is depicted in Exhibit 11.7. Cases were assigned randomly to the two groups, and then

1. both groups received a pretest;

2. one group received the experimental intervention (GST), and the other received the IPS approach; and

3. both groups received three posttests at 6, 12, and 18 months.

Contrary to the researchers' hypothesis, the IPS participants were twice as likely to obtain a competitive job as the GST participants. The IPS participants also worked more hours and earned more total wages. Although this was not the outcome Drake et al. had anticipated, it was valuable information for policy makers and program planners—and the study was rigorously experimental.

Program impact also can be evaluated with quasi-experimental designs (see Chapter 6), nonexperimental designs, or field research methods without a randomized experimental design. If program participants can be compared to nonparticipants who are reasonably comparable except for their program participation, causal conclusions about program impact can still be made. However, researchers must evaluate carefully the likelihood that factors other than program participation might have resulted in the appearance of a program effect. For example, when a study at New York's maximum-security prison for women found that "education [i.e., classes] is found to lower risk of new arrest," the conclusions were immediately suspect: The research design did not ensure that the women who enrolled in the prison classes were the same as those who were not, "leaving open the possibility that the results were due, at least in part, to self-selection, with the women most motivated to avoid reincarceration being the ones who took the college classes" (Lewin 2001b:A18). Such nonequivalent control groups are often our only option, but you should be alert to their weaknesses.

Impact analysis is an important undertaking that fully deserves the attention it has been given in government program funding requirements. However, you should realize that more rigorous evaluation designs

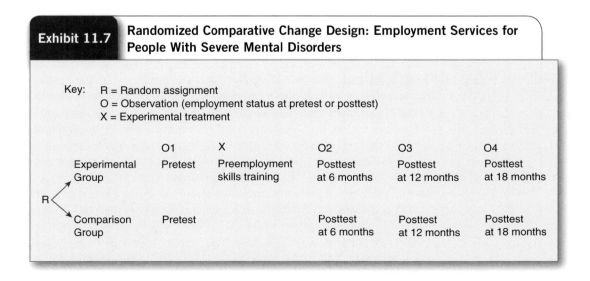

Exhibit 11.7 Randomized Comparative Change Design: Employment Services for People With Severe Mental Disorders

Key: R = Random assignment
       O = Observation (employment status at pretest or posttest)
       X = Experimental treatment

		O1	X	O2	O3	O4
	Experimental Group	Pretest	Preemployment skills training	Posttest at 6 months	Posttest at 12 months	Posttest at 18 months
R						
	Comparison Group	Pretest		Posttest at 6 months	Posttest at 12 months	Posttest at 18 months

are less likely to conclude that a program has the desired effect; as the standard of proof goes up, success is harder to demonstrate. The prevalence of "null findings" (or "we can't be sure it works") has led to a bit of gallows humor among evaluation researchers:

The Output/Outcome/Downstream Impact Blues

Donors often say,

And this is a fact,

Get out there and show us

Your impact

You must change peoples' lives

And help us take the credit

Or next time you want funding

You just might not get it.

So donors wake up

From your impossible dream.

You drop in your funding

A long way upstream.

The waters they flow,

They mingle, they blend

So how can you take credit

For what comes out in the end?

—Terry Smutylo, Director, Evaluation
International Development Research Centre
Ottawa, Canada (excerpt reprinted in Patton 2002:154; used by permission)

> **Cost-benefit analysis:** A type of evaluation research that compares program costs with the economic value of program benefits.
>
> **Cost-effectiveness analysis:** A type of evaluation research that compares program costs with actual program outcomes.
>
> **Efficiency analysis:** A type of evaluation research that compares program costs with program effects. It can be either a cost-benefit analysis or a cost-effectiveness analysis.

Efficiency Analysis

Finally, a program may be evaluated for how efficiently it provides its benefit; typically, financial measures are used. Are the program's financial benefits sufficient to offset the program's costs? The answer is provided by a **cost-benefit analysis**. How much does it cost to achieve a given effect? This answer is provided by a **cost-effectiveness analysis**. Program funders often require one or both of these types of **efficiency analysis**.

A cost-benefit analysis must (obviously) identify the specific costs and benefits to be studied, but my "benefit" may easily be your "cost." Program clients, for instance, will certainly have a different perspective on these issues than do taxpayers or program staff. Exhibit 11.8 lists factors that can be considered costs or benefits in a supported employment program from the standpoint of participants and taxpayers (Schalock & Butterworth 2000). Note that some anticipated impacts of the program (e.g., taxes and subsidies) are a cost to one group but a benefit to the other, and some impacts are not relevant to either.

Exhibit 11.8 Potential Costs and Benefits of a Social Program, by Beneficiary

Costs/Benefits	Perspective of Program Participants	Perspective of Rest of Society	Perspective of Entire Society*
Costs			
Operational costs of the program	0	–	–
Forgone leisure and home production	–	0	–
Benefits			
Earnings gains	+	0	+
Reduced costs of nonexperimental services	0	+	+
Transfers			
Reduced welfare benefits	–	+	0
Wage subsidies	+	–	0
Net benefits	±	±	±

After potential costs and benefits have been identified, they must be measured. This need is highlighted in recent government programs (Campbell 2002):

The Governmental Accounting Standards Board's (GASB) mission is to establish and improve standards of accounting and financial reporting for state and local governments in the United States. In June 1999, the GASB issued a major revision to current reporting requirements ("Statement 34"), which aims to provide information so citizens and other users can understand the financial position and cost of programs. (p. 1)

In addition to measuring services and their associated costs, a cost-benefit analysis must be able to make some type of estimation of how clients benefited from the program and what the economic value of this benefit was. A recent study of therapeutic communities provides a clear illustration. A *therapeutic community* is a method for treating substance abuse in which abusers participate in an intensive, structured living experience with other addicts who are attempting to stay sober. Because the treatment involves residential support as well as other types of services, it can be quite costly. Are those costs worth it?

Sacks, McKendrick, DeLeon, French, and McCollister (2002) conducted a cost-benefit analysis of a modified therapeutic community (TC) in which 342 homeless, mentally ill chemical abusers were randomly assigned to either a TC or a "treatment-as-usual" comparison group. Employment status, criminal activity, and utilization of health care services were each measured for the 3 months prior to entering treatment and the 3 months after treatment. Earnings from employment in each period were adjusted for costs incurred by criminal activity and utilization of health care services.

Was it worth it? The average cost of TC treatment for a client was $20,361. In comparison, the economic benefit (based on earnings) to the average TC client was $305,273, which declined to $273,698 after comparing post- to preprogram earnings. After adjusting for the cost of the program, the benefit was still $253,337. The resulting benefit-cost ratio was 13:1, although this ratio declined to only 5.2:1 after further adjustments (for cases with extreme values). Nonetheless, the TC program studied seems to have had a substantial benefit relative to its costs.

▣ Ethical Issues in Evaluation Research

Whenever you evaluate the needs of clients or analyze the impact of a program, you directly affect people's lives. Social workers want to believe their efforts matter; drug educators think they're preventing drug abuse. Homeless people have problems and may really appreciate the services an agency provides. Program administrators have bosses to please; foundations need big programs to fund; and domestic violence, for instance, is a real problem—and finding solutions to it matters. Participants and clients in social programs, then, are not just subjects eager to take part in your research; they care about your findings, deeply. This produces serious ethical as well as political challenges for the evaluation researcher (Boruch 1997:13; Dentler 2002:166).

There are many specific ethical challenges in evaluation research:

- How can confidentiality be preserved when the data are owned by a government agency or are subject to discovery in a legal proceeding?

- Who decides what burden an evaluation project can impose upon participants?

- Can a research decision legitimately be shaped by political considerations?

- Must findings be shared with all stakeholders or only with policy makers?

- Will a randomized experiment yield more defensible evidence than the alternatives?

- Will the results actually be used?

Is it fair to assign persons randomly to receive some social program or benefit? What fairer way to distribute scarce benefits than through a lottery? The State of Oregon has recently begun doing exactly this with some health care benefits (Yardley 2008). This is exactly the process that is involved in a randomized experimental design.

The Health Research Extension Act of 1985 (Public Law 99–158) mandated that the Department of Health and Human Services require all research organizations receiving federal funds to have an institutional review board (IRB) to assess all research for adherence to ethical practice guidelines. There are six federally mandated criteria (Boruch 1997):

- Are risks minimized?

- Are risks reasonable in relation to benefits?

- Is the selection of individuals equitable? (Randomization implies this.)

- Is informed consent given?

- Are the data monitored?

- Are privacy and confidentiality ensured? (pp. 29–33)

Evaluation researchers must consider these criteria before they even design a study. Subject confidentiality is particularly thorny because researchers, in general, are not usually exempted from providing evidence sought in legal proceedings. However, several federal statutes have been passed specifically to protect research data about certain vulnerable populations from legal disclosure requirements. For example, the Crime Control and Safe Streets Act (28CFR Part 11) includes the following stipulation (Boruch 1997):

> Copies of [research] information [about persons receiving services under the act or the subject of inquiries into criminal behavior] shall be immune from legal process and shall not, without the consent of the persons furnishing such information, be admitted as evidence or used for any purpose in any action, suit, or other judicial or administrative proceedings. (p. 60)

When ethical standards can't be met, modifications may be made in the study design. Several steps can be taken (Boruch 1997):

- Alter the group allocation ratios to minimize the number in the untreated control group.

- Use the minimum sample size required to be able to test the results adequately.

- Test just parts of new programs rather than entire programs.

- Compare treatments that vary in intensity (rather than presence or absence).

- Vary treatments between settings rather than among individuals within a setting. (pp. 67–68)

🔲 Conclusion

In social policy circles, hopes for evaluation research are high: Society would benefit from the programs that work well, that accomplish their goals, and that serve people who genuinely need them. At least that is the hope. Unfortunately, there are many obstacles to realizing this hope. Because social programs and the people who use them are complex, evaluation research designs can easily miss important outcomes or aspects of the program process. Because the many program stakeholders all have an interest in particular results from the evaluation, researchers can be subjected to an unusual level of cross-pressures and demands. Because the need to include program stakeholders in research decisions may undermine adherence to scientific standards, research designs can be weakened. Because program administrators may want to believe their programs really work well, researchers may be pressured to avoid null findings or, if they are not responsive, find their research reports ignored. Because the primary audience for evaluation research reports is program administrators, politicians, or members of the public, evaluation findings may need to be overly simplified, distorting the findings (Posavac & Carey 1997). Plenty of well-done evaluation research studies wind up in a recycling bin or hidden away in a file cabinet.

The rewards of evaluation research are often worth the risks, however. Evaluation research can provide social scientists with rare opportunities to study complex social processes, with real consequences, and to contribute to the public good. Although they may face unusual constraints on their research designs, most evaluation projects can also result in high-quality analyses and publications in reputable social science journals.

Encyclopedia Link 11.2
Read about the uses of evaluation research and its applications.

In many respects, evaluation research is an idea whose time has come. We may never achieve Donald Campbell's vision of an "experimenting society" (Campbell & Russo 1999) in which research is consistently used to evaluate new programs and to suggest constructive changes, but we are close enough to continue trying.

Key Terms

Highlights

- Evaluation research is social research that is conducted for a distinctive purpose: to investigate social programs.

- The development of evaluation research as a major enterprise followed on the heels of the expansion of the federal government during the Great Depression and World War II.

- The evaluation process can be modeled as a feedback system, with inputs entering the program, which generate outputs and then outcomes, which feed back to program stakeholders and affect program inputs.

- The evaluation process as a whole, and the feedback process in particular, can only be understood in relation to the interests and perspectives of program stakeholders.

- The process by which a program has an effect on outcomes is often treated as a "black box," but there is good reason to open the black box and investigate the process by which the program operates and produces, or fails to produce, an effect.

- A program theory may be developed before or after an investigation of the program process is completed. The theory can be either descriptive or prescriptive.

- Evaluation research is research for a client, and its results may directly affect the services, treatments, or punishments that program users receive.

- Evaluation researchers differ in the extent to which they attempt to orient their evaluations to program stakeholders.

- Qualitative methods are useful in describing the process of program delivery.

- Multiple outcomes are often necessary to understand program effects.

- There are five primary types of program evaluation: needs assessment, evaluability assessment, process evaluation (including formative evaluation), impact analysis (also termed summative evaluation), and efficiency (cost-benefit) analysis.

- Evaluation research raises complex ethical issues because it may involve withholding desired social benefits.

STUDENT STUDY SITE

The Student Study Site, available at **www.sagepub.com/chambliss4e,** includes useful study materials including web exercises with accompanying links, eFlashcards, videos, audio resources, journal articles, and encyclopedia articles, many of which are represented by the media links throughout the text. The site also features Interactive Exercises—represented by the green icon here—to help you understand the concepts in this book.

Exercises

Discussing Research

1. Would you prefer that evaluation researchers use a stakeholder or a social science approach? Compare and contrast these perspectives and list at least four arguments for the one you favor.

2. Propose a randomized experimental evaluation of a social program with which you are familiar. Include in your proposal a description of the program and its intended outcomes. Discuss the strengths and weaknesses of your proposed design.

3. Think of your primary health care provider as providing a "program" that should be evaluated. (If that makes you squeamish, you can focus on your college as the "program" instead.)

   a. How would you describe the contents of the "black box" of program operations?

   b. What program theory would specify how the program operates?

   c. What would be the advantages and disadvantages of using qualitative methods to evaluate this program?

   d. What would be the advantages and disadvantages of using quantitative methods?

   e. Which approach would you prefer and why?

Finding Research

1. Inspect the website maintained by the Governmental Accounting Standards Board (www.seagov.org). Read and report on performance measurement in government as described in one of the case studies.

2. Describe the resources available for evaluation researchers at one of the following three websites: www.wmich.edu/evalctr/, www.stanford.edu/~davidf/empowermentevaluation.html, or www.worldbank.org/oed/.

Critiquing Research

1. Read and summarize an evaluation research report published in the *Evaluation and Program Planning* journal. Be sure to identify the type of evaluation research that is described.

2. Select one of the evaluation research studies described in this chapter, read the original report (book or article) about it, and review its adherence to the ethical guidelines for evaluation research. Which guidelines do you feel are most important? Which are most difficult to adhere to?

Doing Research

1. Propose a randomized experimental evaluation of a social program with which you are familiar. Include in your proposal a description of the program and its intended outcomes. Discuss the strengths and weaknesses of your proposed design.

2. Identify the key stakeholders in a local social or educational program. Interview several stakeholders to determine their goals for the program and what tools they use to assess goal achievement. Compare and contrast the views of each stakeholder and try to account for any differences you find.

Ethics Questions

1. In the study of the housing alternatives by Schutt and colleagues (2011), an ethnographer learned that a house resident was talking seriously about cutting himself. If you were the ethnographer, would you have immediately informed house staff about this? Would you have told anyone? What if the resident asked you not to tell anyone? In what circumstances would you feel it is ethical to take action to prevent the likelihood of a subject's harming himself or herself or others?

2. Is it ethical to assign people to receive some social benefit on a random basis? Form two teams and debate the ethics of the TARP randomized evaluation of welfare payments described in this chapter.

Reviewing, Proposing, and Reporting Research

In a sense, we end this book where we began. As you begin writing up your findings, you can see the gaps in the research. While reviewing the literature—and finding where your own work fits in—you may discover more interesting possibilities or more exciting studies to be started. In the process of concluding each study, we almost naturally begin the next.

The primary goals of this chapter are to guide you in evaluating the research of other scholars, developing research proposals, and writing worthwhile reports of your own. We first discuss how to evaluate prior research—a necessary step before writing a research report or proposal. We then focus on writing research proposals and reports.

Comparing Research Designs

From different methods, we learn different things. Even when used to study the same social processes, the central features of experiments, surveys, qualitative methods, and evaluation research provide distinct perspectives. Comparing subjects randomly assigned to a treatment group and to a comparison group, asking standard questions of the members of a random sample, observing while participating in a natural social setting, or studying program impact involve markedly different decisions about measurement, causality, and generalizability. As you can see in Exhibit 12.1, not one of these methods can reasonably be graded as superior to the others in all respects, and each varies in its suitability to different research questions and goals. Choosing among them for a particular investigation requires consideration of the research problem, opportunities and resources, prior research, philosophical commitments, and research goals.

Experimental designs are strongest for testing nomothetic causal hypotheses (lawlike explanations that identify a common influence on a number of cases or events). These designs are most appropriate for studies of treatment effects (see Chapter 6). Research questions that are believed to involve basic social psychological processes are most appealing for laboratory studies, because the problem of generalizability is reduced. Random assignment reduces the possibility of preexisting differences between treatment and comparison groups to small, specifiable, chance levels, so many of the variables that might create a spurious association are controlled. Laboratory experiments permit unsurpassed control over conditions and are excellent for establishing internal validity (causality).

But experimental designs have weaknesses. For most laboratory experiments, people volunteer as subjects; since volunteers aren't like other people, generalizability is not good. Ethical and practical constraints limit your treatments (for instance, you can't randomly assign race or social class). Although some processes may be the same for all people, so that generalizing from volunteer subjects will work, it's difficult to know in advance which processes are really invariant. Field experiments, although apparently more generalizable studies, allow for less control than lab experiments; hence, treatments may not be delivered as intended, or other influences may intrude (see Chapter 9). Also, field experiments typically require unusual access (e.g., permission to revise a school curriculum or change police department policy) and can be very expensive.

Exhibit 12.1   Comparison of Research Methods[a]

Design	Measurement Validity	Generalizability	Causal Validity
Experiments	+	−	+
Surveys	+	+	−/+[b]
Participant Observation	−/+[c]	−	−

a. A plus (+) sign indicates where a method is strong; a minus (−) sign indicates where a method is weak.

b. Surveys are a weaker design for identifying causal effects than true experiments, but use of statistical controls can strengthen causal arguments.

c. Reliability is low compared to surveys, and systematic evaluation of measurement validity is often not possible.

Surveys, because of their probability sampling and standardized questions, are excellent for generalizable descriptive studies of large populations (see Chapter 7). They can include a large number of variables, unlike experiments, so that potential spuriousness can be statistically controlled; therefore, surveys can be used readily to test hypothesized causal relationships. And because many closed-ended questions are available that have been used in previous studies, it's easy to find reliable measures of commonly used variables.

But surveys, too, have weaknesses. Survey questionnaires can measure only what respondents are willing to say; they may not uncover behavior or attitudes that are socially unacceptable. Survey questions, being standardized, may miss the nuances of a respondent's feelings or the complexities of an attitude; they lump together what may be interestingly different responses. They rely on the truthfulness of respondents and on their accuracy in reporting (for instance, students are asked how many hours a week they study—Do they know? Is study time constant?).

Qualitative methods allow intensive measurement of new or developing concepts, subjective meanings, and causal mechanisms (see Chapter 9). In field research, a grounded theory approach helps you create and refine concepts and theories based on direct observation or in-depth interviewing. Interviewing reveals what people really mean by their ideas and allows you to explore their feelings at great length. How, exactly, social processes unfold over time can be explored using interviews and fieldwork. Qualitative methods can identify the multiple successive events that might have led to some outcome, thus identifying idiographic causal processes; they are excellent for studying new or poorly understood settings and populations that seek to remain hidden. When exploratory questions are posed or new groups studied, qualitative methods are preferred.

But such intensive study is time consuming, so fewer cases can be examined. Single or a few cases or unique settings are interesting but don't produce generalizable results. Also, most researchers can't spend 6 months away from home doing a project. Open-ended interviews take time—not just the 1 or 2 hours of the interview itself but time in scheduling, in missed appointments, in travel to reach your subjects, and so on.

When qualitative methods can find real differences in an independent variable—for example, several different management styles in a manufacturing company—you can test nomothetic causal hypotheses. But the impossibility of controlling numerous possible extraneous influences makes qualitative methods a weak approach to hypothesis testing.

Reviewing Research

A good literature review is the foundation for a research proposal, both in identifying gaps in current knowledge and in considering how to design a research project. It is also important to review the literature prior to writing an article about the research findings—the latest findings on your topic should be checked, and prior research on new issues should be consulted. This section helps you learn how to review the research that you locate. First, we focus on the process of reviewing single articles; then, we explain how to combine reviews of single articles into an overall literature review.

Exhibit 12.2 lists the questions you should ask when critiquing a social research study, and the following paragraphs provide an example. This particular critique does not answer all of the review questions, nor does it provide complete answers to all these questions, but it gives you the basic idea. In any case, remember that your goal is to evaluate research projects as integrated wholes. In addition to considering how valid the measures were and whether the causal conclusions were justified, you must consider how the *measurement*

> ### Exhibit 12.2 Questions to Ask About a Research Article
>
> In reading a research article, you want to know (a) What is the author's conclusion? and (b) Does the research presented adequately support that conclusion? The questions below will help you determine the answers.
>
> I. Overall assessment of the article
>
>   1. What is the basic question being posed?
>   2. Is the theoretical approach appropriate?
>   3. Is the literature review adequate?
>   4. Does the research design suit the question?
>   5. Is the study scientific in its fundamentals?
>   6. Are the ethical issues adequately addressed?
>   7. What are the key findings?
>
> II. Detailed assessment
>
>   1. What are the key concepts? Are they clearly defined?
>   2. What are the main hypotheses?
>   3. What are the main independent and dependent variables?
>   4. Are the measurements valid?
>   5. What are the units of analysis? Are they appropriate?
>   6. Are any causal relationships successfully established?
>   7. Is the effective sample (sampling plus response rate) representative?
>   8. Does context matter to the causal relationship?

approach might have affected the causal validity of the researcher's conclusions and how the *sampling strategy* might have altered the quality of measures. In other words, all the parts of a study affect each other. Our goal here is just to illustrate the process of critically thinking about a piece of research.

Case Study: "Night as Frontier"

A minor classic in sociological literature, Murray Melbin's 1978 article "Night as Frontier" compares 20th-century extension of human activity into nighttime hours with 19th-century geographic expansion into the American West. Melbin argues that just as there was a "frontier lifestyle" in the Old West of cowboys, a similar style of behavior, particularly toward strangers, prevails among late-night inhabitants of contemporary American cities. In developing this comparison of spatial frontiers with temporal frontiers, Melbin accomplished an insightful reconceptualization of how human beings live on a sparsely populated "frontier" of a different kind.

Suppose that you are a student of urban life and curious as to whether city dwellers, such as New Yorkers, are really as unfriendly and brusque as stereotypes portray them. Melbin's article describes a number of field experiments, conducted entirely in Boston, to discover whether people were more or less helpful to others at nighttime than during the day. Perhaps you could use his findings. But was his research properly conducted?

The Research Design

Melbin and his assistants conducted four different experiments, all designed to measure if time of day affected people's willingness to be "helpful or friendly" to strangers. He drew in part on a sizable literature in this area conducted by social psychologists, but his studies were simpler in design than most psychology experiments. In most cases, he had one independent variable—time of day—and one dependent variable—how likely people were to be helpful or friendly. Melbin's assistants, using a detailed sampling procedure (sampling

Encyclopedia Link 12.2
Read a review of
research designs.

both times of the day and subjects), approached random people on streets in Boston (also sampled). In one study, the researchers asked for directions; in another, they requested that subjects answer several interview questions. In a third study, they observed customers' interactions with cashiers at grocery stores. Finally, they left keys, tagged with "Please return" and an address, in various locations. In each case, the independent variable was time of day (for instance, when subjects were approached or the key was dropped on the street); the dependent variable was whether people were cooperative (directions, interviews), helpful (returning key), or friendly (smiling, conversational). A clear, simple coding scheme was used for all of these measures.

Analysis of the Design

Melbin's study was exploratory, designed to propose a new idea of how to understand nighttime in contemporary America. His experiments, therefore, were more in the manner of demonstrations—a first test of a new idea—than of continuing an established line of scientific research. Indeed, Melbin (1978) himself claimed to be advancing "the hypothesis that night is a frontier"; yet his experiments only test the idea that people at night are more "helpful and friendly" to strangers, which he argues is one of about a dozen characteristics of frontier communities.

But we can narrow our view to his specific question about helpfulness. His measures certainly have face validity, and in fact, in three of his four studies, people were indeed more friendly at night. And he didn't simply ask people if they would be helpful; he tested them in real situations in which they didn't know that it was an experiment. He also was open to surprises: In the "lost key" study, people were in fact *less* likely to return the key at night. Melbin realized that he had unintentionally slipped in another variable—whether the act of helpfulness was anonymous (the key study) or not (all the others). Only the community of face-to-face contact, he suggests, exists at night; help is not generally extended to those not part of the nighttime community. So the different trials also lend plausibility to his argument. He only studied city residents and only in Boston; it may be that the "nighttime community" exists only in urban settings, but an urban setting was a constant, not a variable, here.

There are at least two important problems in Melbin's design, despite its conscientious use of sampling, reliable coding procedures, and multiple measures. First, the studies don't really show that nighttime makes particular people more helpful and friendly; they show that people who are up at night—a self-selected group—are more helpful and friendly.

Perhaps the kind of people who prefer nightlife, not nighttime itself, is the true causal agent. And second, again, the studies were all conducted in a Northeastern city. Rural or suburban settings—a different context—could very well reveal different patterns.

An Overall Assessment

"Night as Frontier" certainly makes a persuasive argument with far more historical and theoretical detail than we've mentioned here. It tends to be research of the "exploratory" type, so its experiments are somewhat crude; neither the measures nor the studies themselves have been widely replicated. Ethically, the work is benign. Its main value may lie in the persuasiveness of the argument that nighttime is different than daytime and that the difference is much like the difference between densely settled areas and the old frontier West. For its conceptual insights, "Night as Frontier" deserves a respected place in the social science literature. In a detailed study of urban life and community, it may be helpful, but perhaps it is not fundamental.

Case Study: When Does Arrest Matter?

The goal of the literature review process is to integrate the results of your separate article reviews and develop an overall assessment of the implications of prior research. The integrated literature review should accomplish three goals (Hart 1998):

1. Summarize prior research.

2. Critique prior research.

3. Present pertinent conclusions. (186–187)

We'll discuss each of these goals in turn.

Summarize Prior Research

Your summary of prior research must focus on the particular research questions that you will address, but you may need also to provide some more general background. Carolyn Hoyle and Andrew Sanders (2000:14) begin their *British Journal of Criminology* research article about mandatory arrest policies in domestic violence cases with what they term a "provocative" question: What is the point of making it a crime for men to assault their female partners and ex-partners? They then review the different theories and supporting research that has justified different police policies: the "victim choice" position, the "pro-arrest" position, and the "victim empowerment" position. Finally, they review the research on the "controlling behaviors" of men that frames the specific research question on which they focus: how victims view the value of criminal justice interventions in their own cases (p. 15).

Ask yourself three questions about your summary of the literature (Pyrczak 2005):

1. *Have you been selective?* If there have been more than a few prior investigations of your research question, you will need to narrow your focus to the most relevant and highest-quality studies. Don't cite a large number of prior articles "just because they are there."

2. *Is the research up-to-date?* Be sure to include the latest research, not just the "classic" studies.

3. *Have you used direct quotes sparingly?* To focus your literature review, you need to express the key points from prior research in your own words. Use direct quotes only when they are essential for making an important point. (pp. 51–59)

Critique Prior Research

Evaluate the strengths and weaknesses of the prior research, answering the questions in Exhibit 12.2. You should select articles for review that reflect the work of credible authors in peer-reviewed journals who have been funded by reputable sources. Consider the following questions as you decide how much weight to give each article (Locke, Silverman, & Spirduso 1998):

1. *How was the report reviewed prior to its publication or release?* Articles published in academic journals go through a very rigorous review process, usually involving careful criticism and revision. Top "refereed" journals may accept only 10% of submitted articles, so they can be very selective. Dissertations go through a lengthy process of criticism and revision by a few members of the dissertation writer's home institution. A report released directly by a research organization is likely to have had only a limited review, although some research organizations maintain a rigorous internal review process. Papers presented at professional meetings may have had little prior review. Needless to say, more confidence can be placed in research results that have been subject to a more rigorous review.

2. *What is the author's reputation?* Reports by an author or team of authors who have published other work on the research question should be given somewhat greater credibility at the outset.

3. *Who funded and sponsored the research?* Major federal funding agencies and private foundations fund only research proposals that have been evaluated carefully and ranked highly by a panel of experts. They also

Audio Link 12.1
Listen to a clip about critiquing prior research.

often monitor closely the progress of the research. This does not guarantee that every such project report is good, but it goes a long way toward ensuring some worthwhile products. On the other hand, research that is funded by organizations that have a preference for a particular outcome should be given particularly close scrutiny. (pp. 37–44)

Present Pertinent Conclusions

Don't leave the reader guessing about the implications of the prior research for your own investigation. Present the conclusions you draw from the research you have reviewed. As you do so, follow several simple guidelines (Pyrczak 2005):

**Research|Social Impact
Link 12.1**
Read more about
presenting conclusions.

- Distinguish clearly your own opinion of prior research from conclusions of the authors of the articles you have reviewed.

- Make it clear when your own approach is based on the theoretical framework you are using rather than on the results of prior research.

- Acknowledge the potential limitations of any empirical research project. Don't emphasize problems in prior research that you can't avoid either. (pp. 53–56)

Explain how the unanswered questions raised by prior research or the limitations of methods used in prior research make it important for you to conduct your own investigation (Fink 2005:190–192).

A good example of how to conclude an integrated literature review is provided by an article based on the replication in Milwaukee of the Minneapolis Domestic Violence Experiment. For this article, Ray Paternoster, Robert Brame, Ronet Bachman, and Lawrence Sherman (1997) sought to determine whether police officers' use of fair procedures when arresting assault suspects would lessen the rate of subsequent domestic violence. Paternoster et al. concluded that there has been a major gap in the prior literature: "Even at the end of some seven experiments and millions of dollars, then, there is a great deal of ambiguity surrounding the question of how arrest impacts future spouse assault" (p. 164).

Specifically, they noted that each of the seven experiments focused on the effect of arrest itself but ignored the possibility that "particular kinds of police *procedure* might inhibit the recurrence of spouse assault" (Paternoster et al. 1997:165).

So Paternoster and his colleagues (1997) grounded their new analysis in additional literature on procedural justice and concluded that their new analysis would be "the first study to examine the effect of fairness judgments regarding a punitive criminal sanction (arrest) on serious criminal behavior (assaulting one's partner)" (p. 172).

Proposing New Research

Be grateful for people who require you to write a formal research proposal—and even more for those who give you constructive feedback. Whether your proposal is written for a professor, a thesis committee, an organization seeking practical advice, or a government agency that funds basic research, the proposal will force you to set out a problem statement and a research plan. Too many research projects begin without a clear problem statement or with only the barest of notions about which variables must be measured or what the analysis should look like. Such projects often wander along, lurching from side to side, and then collapse entirely or just peter out with a report that is ignored—and should be. Even in circumstances when a proposal is not required, you should prepare one and present it to others for feedback. Just writing your ideas down will help you to see how they can be improved, and feedback in almost any form will help you to refine your plans.

A well-designed proposal can go a long way toward shaping the final research report and will make it easier to progress at later research stages (Locke et al. 2000). Every research proposal should have at least six sections:

1. *An introductory statement of the research problem,* in which you clarify what it is that you are interested in studying

2. *A literature review,* in which you explain how your problem and plans build on what has already been reported in the literature on this topic

3. *A methodological plan,* detailing just how you will respond to the particular mix of opportunities and constraints you face

4. *A budget,* presenting a careful listing of the anticipated costs

5. *An ethics statement,* identifying human subjects issues in the research and how you will respond to them in an ethical fashion

6. *A statement of limitations,* reviewing weaknesses of the proposed research and presenting plans for minimizing their consequences

Journal Link 12.1
Read about how increased research restrictions gave way to implications for future research.

A research proposal also can be strengthened considerably by presenting a result of a pilot study of the research question. This might involve administering the proposed questionnaire to a small sample, conducting a preliminary version of the proposed experiment with a group of available subjects, or making observations over a limited period of time in a setting like that proposed for a qualitative study. Careful presentation of the methods used in the pilot study and the problems that were encountered will impress anyone who reviews the proposal.

If your research proposal will be reviewed competitively, it must present a compelling rationale for funding. The research problem that you propose to study is crucial; its importance cannot be overstated (see Chapter 2). If you propose to test a hypothesis, be sure that it is one for which there are plausible alternatives, so your study isn't just a boring report of the obvious (Dawes 1995:93).

Case Study: Community Health Workers and Cancer Clinical Trials

Particular academic departments, grant committees, and funding agencies will have specific proposal requirements. As an example, Exhibit 12.3 lists the primary required sections of the "Research Plan" for proposals to the National Institutes of Health (NIH), together with excerpts from a proposal by Russell Schutt and colleagues (2005) from two Harvard teaching hospitals submitted in this format to the National Cancer Institute (NCI) as part of a larger collaboration involving research and training at the University of Massachusetts Boston and the Dana Farber/Harvard Cancer Center. The research plan (which is excerpted) must be preceded by a proposed budget, biographical sketches of project personnel, and a discussion of the available resources for the project. Appendixes may include research instruments, prior publications by the authors, and findings from related work.

As you can see from the excerpts, the proposal was to study community health workers' (CHWs) knowledge of and orientations to cancer clinical trials and to then develop and test a training program for them

Audio Link 12.2
Listen to how research can be controversial.

| Exhibit 12.3 | A Grant Proposal to the National Cancer Institute |

Community Health Workers and Cancer Clinical Trials

Abstract

Disparities in cancer between subpopulations in the U.S. have been documented for several decades. One important area for intervention is the participation of underserved populations in cancer clinical trials. . . . Innovative community-based approaches are badly needed to affect these trends. This project will develop a clinical trials education training program for patient navigators and community health workers (CHWs). The primary goal of the training program is to help the CHWs effectively educate the communities they work with about the importance of clinical trials. An extensive program evaluation strategy has been included throughout the program development and implementation process. The evaluation will yield valuable information about CHWs' attitudes about clinical trials, how best to share this information with communities, about the effectiveness of community health workers to inform communities about clinical trials. . . . .

Research Plan

1. *Specific Aims*

    1. To develop a curriculum/program for training CHWs about clinical trials, so that they may educate the communities they work with about the importance of clinical trials

    2. To implement the training program with CHWs . . .

    3. To evaluate the impact of the training program . . .

2. *Background and Significance*

Risk, incidence, morbidity, and mortality for cancer in general and for some specific cancers are higher for blacks compared to whites, for poor persons compared to non-poor persons, and for rural residents compared to non-rural residents. Disparities have been documented across the cancer continuum ranging from risk factors and prevention to treatment and survival. The reasons for disparities in cancer treatment outcomes between different subpopulations are complex and many factors contribute. . . . One important area for intervention is the participation of underserved populations in cancer clinical trials.

Participation of minority populations in clinical trials is generally reported to be less frequent than participation of whites. . . .

Many barriers exist that prevent minority participation in clinical trials. . . . Most institutional committees charged with protecting human subjects do not adequately address all the concerns of these populations.

The federal government now requires that all persons involved in research with human subjects complete training on the principles of protection of human subjects. …many protections that have been instituted may actually serve as barriers to participation. For example, most IRBs now require extensive and highly detailed consent forms, which often use highly technical language and discuss procedures and concepts that are unfamiliar, overwhelming, and sometimes frightening.

Strategies to reverse the under-enrollment of minority and other underserved populations in clinical trials must address participant barriers, investigator barriers, and institutional barriers. We focus this proposal on an outreach strategy that will address some of the participant barriers ….

An untapped resource in addressing the clinical trials accrual problem among underserved populations is the increasing number of CHWs [Community Health Workers] employed in many communities. . . . In the proposed project, we will develop a curriculum about clinical trials and train CHWs involved in several cancer screening and outreach programs to use or adapt the curriculum to educate several key communities about cancer clinical trials.

3. *Progress Report/Preliminary Studies*

C.1. Collaborators: This program is a collaboration between Dana Farber Harvard Cancer Center, specifically the Brigham and Women's Hospital (BWH) and Massachusetts General Hospital (MGH), and the University of Massachusetts, Boston (UMB). The study team includes Dr. JudyAnn Bigby from Brigham and Women's Hospital and Harvard Medical School, Dr. Lidia Schapira from Massachusetts General Hospital and Harvard Medical School, and Dr. Russell Schutt from the University of Massachusetts, Boston.

The proposed project will build on a program that was implemented at the Massachusetts General Hospital as part of an effort to address language and referral barriers for underserved populations. . . . Dr. Schapira and colleagues designed and implemented training programs for interpreters to increase their knowledge and skills.

Dr. Russell Schutt, at UMass Boston, will oversee the evaluation components of the project. Dr. Schutt is Professor of Sociology and Director, Graduate Program in Applied Sociology at UMass Boston and he is also Lecturer on Sociology in the Department of Psychiatry (MMHC/BID) at the Harvard Medical School. Dr. Schutt has extensive experience in evaluation research and is the author of a leading research methods text in sociology (with versions adapted for social work, criminal justice, and undergraduate institutions). He has also designed ancillary training materials in research methods and has published more than 50 research articles and book chapters. He is co-investigator on the Women's Health Network Evaluation Project, an evaluation of the Mass. Department of Public Health case management program funded by the CDC's National Breast and Cervical Cancer Early Detection Program. He is also principal investigator . . ., "Reviewing the Past, Planning the Future" at the Harvard Medical School. This project is recruiting a large team of health policy experts to review research about the Women's Health Network project and to ensure the most effective program operations. Dr. Schutt plays a key role in this program, as evaluation activities are incorporated throughout the curriculum and training development and implementation process, and are iterative in nature. We view on-going evaluation as a critical component . . . .

4. *Research Design and Methods*

. . . During Year 1, the curriculum will be revised to meet the needs of a variety of CHWs, ... Representatives from these community programs will participate in the development of the curriculum. We will pilot test the training program, and then revise it as needed (year 2). . . . UMB will evaluate the development of the training, the training itself, . . . and conclude with an outcome analysis of the program's impact. These evaluation activities will help to design the program curriculum and to implement the most effective program components.

D.2. Curriculum Development and Training

We propose to develop a curriculum designed specifically to meet the learning needs of the CHWs, and to provide state-of-the-art knowledge of the process and language of clinical trials. . . .Our efforts to develop an effective training program will involve four steps: 1) needs assessment; 2) curriculum development; 3) pilot testing of the training program; and 4) revision of the curriculum and training program. Each of these steps is described below.

D.2.1. Needs Assessment: . . . The first phase of the project will include a needs assessment in order to identify the level of understanding and knowledge of community workers with respect to clinical trials. . . .First, we will conduct two focus groups with CHWs to probe the attitudes and beliefs about clinical trials, their experiences with community outreach, and their impressions of client orientations . . . . . . Second, . . . ten in-depth interviews (approximately one hour in length) will be conducted with selected health workers . . . . These interviews will be designed to provide more details about issues raised in the focus groups . . . . Third, a short structured survey will be designed to assess the backgrounds, attitudes and experiences of all CHWs involved in the project. This survey will include a measure of understanding of and attitudes toward clinical trials as well as information on the languages and cultural backgrounds of the CHWs. . . .

D.3.3. Program Evaluation: There will be several strategies utilized for evaluating the proposed program. First, an impact analysis will measure the change in CHW's understanding of and attitudes toward clinical trials. A structured survey . . .related to community education and clinical trials will be administered to participants prior to and following each training. . . . A measure of satisfaction with the training . . .

5. *Human Subjects*

E.1. Risks to Subjects: The risks of participation are minimal. The primary risk is the potential for loss of confidentiality.
E.2. Adequacy of Protection Against Risks: Confidentiality will be maintained by numerically coding data, . . . All information obtained from subjects will be accessible only to research staff.

(Continued)

Exhibit 12.3 (Continued)

E.3. Potential Benefits of the Proposed Research to Subjects: The proposed program evaluation will help . . . develop a community-based clinical trials education program . . . responsive to the needs . . . and reflects the language and values of the community.

E.4. Importance of the Knowledge to be Gained: . . . This project will help to address disparities in knowledge related to clinical trials, and . . . may impact on differential enrollment among minority cancer patients in clinical trials.

E.5. Women, Ethnic Minority, and Child Inclusion: All participants in the present investigation will be adults. We anticipate that the majority of participants will be women, ...

    E.5.1. Minority recruitment plan: We will work with all community health workers employed by specific programs . . . . The majority . . . are members of minority groups.

E.6. Risks Compared to Benefits: The benefits of the proposed study outweigh the potential risks. The knowledge gained will be substantial, and the risks are few and largely preventable. . .

E.7. Data Safety Monitoring: A data safety monitoring plan (DSMP) has been developed for this study. . . . All investigator-level staff members have completed the NIH human subject's certification as required. This is a minimal risk study, and thus we do not anticipate safety concerns.

**Research/Social Impact
Link 12.2**
Read more about
research proposals.

about clinical trials. The proposal included two types of evaluation research: a needs assessment to learn about CHWs and clinical trials and an outcome assessment to identify changes in CHWs' knowledge and orientations as a result of participation in the training program. The NCI review committee (composed of experts in these issues) approved the project and then after another administrative review it was awarded funds.

The reviewers recognized the proposal's strengths but also identified two issues that they believed had to be considered as the project was implemented. The issues were primarily methodological, related to validating the needs assessment tool and to using qualitative data.

> . . . The primary goal of the training program is to help the CHWs effectively educate the communities they work with about the importance of clinical trials. An extensive program evaluation strategy has been included throughout the program development and implementation process. The evaluation will yield valuable information about CHWs' attitudes about clinical trials, how best to share this information with communities, about the effectiveness of community health workers to inform communities about clinical trials. This collaboration between DF/HCC and UMB represents a unique opportunity to build on the strengths of each institution to address a pressing problem that influences the persistence of cancer-related disparities.
>
> Co-Leaders: Members of the investigative team have clearly delineated responsibilities based on their areas of expertise. . . .
>
> Institutional Environment: The institutional environment at HMS is excellent and several collaborations currently exist that will facilitate recruitment for this pilot project. . . .
>
> Merit/Importance: The purpose of this pilot is to take advantage of the popular community health worker (CHW) model to develop, implement, and evaluate a curriculum/program for training CHWs to educate the communities in which they work about the importance of clinical trials. The rationale

is that CHWs, with adequate training, could help community residents overcome certain barriers to clinical trials participation (e.g. lack of knowledge, mistrust, limited understanding, limited access to accurate/reliable information). This project builds on prior experiences training medical interpreters about clinical trials. The project will include 1) curriculum development (following a needs assessment via focus groups and in-depth interviews) that will include pilot testing and revisions, 2) implementation (training) and 3) program evaluation. The pilot is well described, with expected outcome and measurement strategies addressed. Examples of curricular content are provided. The evaluation plan will include both process and outcome measures. Plans to observe community education programs offered by the newly trained CHWs are also included. Potential challenges are acknowledged and incorporated into the training program (e.g., strategies to help CHWs maintain a focus on clinical trials education in their encounters and community education efforts).

Although the research plan is nicely laid out, there are a few remaining questions:

1. How will the survey designed to assess backgrounds, attitudes, and experience of CHWs be validated?

2. Will qualitative data from the CHWs be used to inform curricular development and, if so, in what ways?

. . . Future Potential: If successful, the curriculum could be implemented in other locations. The investigators also plan to evaluate the adaptability of the training to a train-the-trainer model. Given the popularity of the CHW model particularly in minority communities, this is a timely educational proposal.

Since NIH review committees reject most research proposals, require a revision before the others are recommended for funding, and do not actually fund many of even the meritorious proposals, NCI's decision about this proposal was very welcome news. If you get the impression that researchers cannot afford to leave any stone unturned in working through procedures in an NIH proposal, you are right. It is very difficult to convince a government agency that a research project is worth spending money on. And that is as it should be: Your tax dollars should be used only for research that has a high likelihood of yielding findings that are valid and useful. But even when you are proposing a smaller project to a more generous funding source—or even presenting a proposal to your professor—you should scrutinize the proposal carefully before submission and ask others to comment on it. Other people will often think of issues you neglected to consider, and you should allow yourself time to think about these issues and to reread and redraft the proposal. Besides, you will get no credit for having thrown together a proposal as best you could in the face of an impossible submission deadline.

When you develop a research proposal, it will help to work through each of the issues in Exhibit 12.4 (also see Herek 1995). It is too easy to omit important details and to avoid being self-critical while rushing to put a proposal together. However, it is painful to have a proposal rejected (or to receive a low grade). Better to make sure the proposal covers what it should and confronts the tough issues that reviewers (or your professor) will be sure to spot.

The points in Exhibit 12.4 can serve as a map to preceding chapters in this book and as a checklist of decisions that must be made throughout any research project. The points are organized in five sections, each concluding with a *checkpoint* at which you should consider whether to proceed with the research as planned, modify the plans, or stop the project altogether. The sequential ordering of these questions obscures a bit the way in which they should be answered: not as single questions, one at a time, but as a unit—first as five separate stages and then as a whole. Feel free to change your answers to earlier questions on the basis of your answers to later questions.

Exhibit 12.4 Decisions in Research Design

PROBLEM FORMULATION (Chapters 1–2)

1. Developing a research question
2. Assessing researchability of the problem
3. Consulting prior research
4. Relating to social theory
5. Choosing an approach: Deductive? Inductive? Descriptive?
6. Reviewing research guidelines

CHECKPOINT 1

Alternatives: • Continue as planned.
• Modify the plan.
• STOP. Abandon the plan.

RESEARCH VALIDITY (Chapters 3–5)

7. Establishing measurement validity
8. Establishing generalizability
9. Establishing causality
10. Data required: Longitudinal or cross-sectional?
11. Units of analysis: Individuals or groups?
12. What are major possible sources of causal invalidity?

CHECKPOINT 2

Alternatives: • Continue as planned.
• Modify the plan.
• STOP. Abandon the plan.

RESEARCH DESIGN (Chapters 6–8)

13. Choosing a research design, such as survey or participant observation
14. Specifying the research plan: Types of experiments, surveys, observations, etc.
15. Assessing ethical concerns

CHECKPOINT 3

Alternatives: • Continue as planned.
• Modify the plan.
• STOP. Abandon the plan.

DATA ANALYSIS (Chapter 9)

16. Choosing statistics, such as frequencies, cross-tabulation, etc.

CHECKPOINT 4

Alternatives: • Continue as planned.
• Modify the plan.
• STOP. Abandon the plan.

REVIEWING, PROPOSING, AND REPORTING RESEARCH (Chapter 10)

17. Organizing the text
18. Reviewing ethical and practical constraints

CHECKPOINT 5

Alternatives: • Continue as planned.
• Modify the plan.
• STOP. Abandon the plan.

A brief review of how the questions in Exhibit 12.4 might be answered with respect to the proposal to the National Cancer Institute by Schutt, Bigby, and Schapira (2005) should help you to review your own work. The research question concerned the need for and efficacy of a training program about cancer clinical trials, an evaluation research question (Question 1). This problem certainly was suitable for social research, and the funds we requested were judged to be adequate ($66,204 for the evaluation component) (Question 2). Prior research demonstrated a need for the investigation and the potential for our training program. Schutt's own prior research (Estabrook, Schutt, & Woodford 2008; Schutt, Cruz, & Woodford 2008; Schutt, Fawcett et al. 2010), helped to indicate the potential for the new proposed research (3). The proposal did not make a direct connection to social theory—a common deficit in evaluation research proposals—but did emphasize relevant prior research (4). The evaluation research plan had both inductive (needs assessment) and deductive (program impact) elements (5). The review of research guidelines continued up to the point of submission, and Schutt and his colleagues felt that their proposal took each into account (6). So it seemed reasonable to continue to develop the proposal (Checkpoint 1).

Measures would be developed through coding of qualitative data collected in focus groups and intensive interviews, analysis of survey data, and observations of training sessions. The specific measures in the quantitative survey instruments and in the observational protocol had been used in prior research and some evidence had been presented suggesting their validity (7). This pilot study was relatively weak in terms of generalizability, since Schutt and colleagues had to plan on studying what would be an availability sample of community health workers (8). Their needs assessment would involve only cross-sectional survey data, so they could only plan a strategy of multivariate statistical controls to test hypotheses about influences on knowledge and orientations. Their impact analysis was to include a before-and-after test to identify changes in individuals' knowledge and orientations, so their conclusion about an effect of the training program would have a somewhat stronger basis (9, 10, 11). Since they did not have a comparison for the impact analysis that was not exposed to the training they planned to develop, endogenous change and external events were potential sources of causal invalidity. There was also a special basis for concern about an interaction of selection and treatment, since those who agreed to participate in the training program could have been more open to change than those who didn't participate; without randomized assignment to the training program or a comparison group, the researchers could not be sure (12). In spite of some weaknesses, the potential value of the training program they were to develop, and the possibility of more rigorous tests of its value in the future, encouraged Schutt and his colleagues to continue with their plan (Checkpoint 2).

The use of mixed-methods design was very appropriate to the needs assessment portion of their research. A randomized experimental design would have been preferable for the impact analysis, but it was not possible to plan such a study within the limitations of their budget and time (13, 14). Neither Schutt and coresearchers nor the reviewers identified ethical concerns in the project, other than preserving the confidentiality of data collected. The noninvasive nature of their methods and their focus on issues concerning community health workers' job-related concerns meant that there was little potential for harm due to participation in their research. Neither the University of Massachusetts Boston's Institutional Review Board nor the Dana Farber/ Harvard Cancer Center's IRB found there to be ethical concerns about their plans (15). Implementing the research plan seemed justified (Checkpoint 3).

Schutt and his colleagues expected to use descriptive univariate and multivariate statistics for the analysis of their needs assessment data, as well as a grounded theory approach for the analysis of their qualitative data. They planned to use inferential statistics to test for differences in mean knowledge and orientations before and after the training program (16). They organized their proposal in the sections required by the National Institutes of Health. In order to report their results, they first wrote a comprehensive research report on the needs assessment (Schutt, Santiccioli, Maniates, Henlon, & Schapira 2008) and they subsequently published separate articles in peer-reviewed journals on the needs assessment (Schutt, Schapira et al. 2010) and on the impact analysis (Schapira & Schutt 2011) (17). They continued to review ethical and practical constraints throughout the project, but they encountered few unexpected obstacles and were able to overcome the challenges they did confront in recruitment for the training (18).

▣ Reporting Research

The goal of research is not just to discover something, but also to communicate that discovery to a larger audience: other social scientists, government officials, your teachers, the general public—perhaps several of these audiences. Whatever the study's particular outcome, if the research report enables the intended audience to comprehend the results and learn from them, the research can be judged a success. If the intended audience is not able to learn about the study's results, the research should be judged a failure—no matter how expensive the research, how sophisticated its design, or how much of yourself you invested in it.

Research in the News

LAWSUIT THREATENED OVER SCHOLARLY ARTICLE

Two researchers wrote an article criticizing a rating scale called the Psychopathy Checklist. This scale is widely used in criminal courts to determine whether a person is a psychopath, but the researchers argued that it was often being used inappropriately to measure the broader personality trait of psychopathy. But when the checklist author saw a prepublication copy of the paper, he felt that it distorted his statements and threatened to sue. Publication was delayed for three years, until the parties agreed on publication of the article followed by a rebuttal from the checklist author and a further response from the article authors.

Source: Carey, Benedict. 2010. Academic battle delays publication by 3 years. *The New York Times*, June 12:A14.

You began writing your research report when you worked on the research proposal, and you will find that the final report is much easier to write, and more adequate, if you write more material for it as you work out issues during the project. It is very disappointing to discover that something important was left out when it is too late to do anything about it. And, we don't need to point out that students (and professional researchers) often leave final papers (and reports) until the last possible minute (often for understandable reasons, including other coursework and job or family responsibilities). But be forewarned: *The last-minute approach does not work for research reports.*

Journal Link 12.2
Read an article whose conclusions are difficult to report because of sensitive issues and topics.

Writing and Organizing

A successful report must be well organized and clearly written. Getting to such a product is a difficult but not impossible goal. Consider the following principles formulated by experienced writers (Booth, Colomb, & Williams 1995):

- Respect the complexity of the task and don't expect to write a polished draft in a linear fashion. Your thinking will develop as you write, causing you to reorganize and rewrite.

- Leave enough time for dead ends, restarts, revisions, and so on and accept the fact that you will discard much of what you write.

- Write as fast as you comfortably can. Don't worry about spelling, grammar, and so on until you are polishing things up.

- Ask anyone you trust for reactions to what you have written.

- Write as you go along, so you have notes and report segments drafted even before you focus on writing the report. (pp. 150–151)

It is important to outline a report before writing it, but neither the organization of the report nor the first written draft should be considered fixed. As you write, you will get new ideas about how to organize the report. Try them out. As you review the first draft, you will see many ways to improve your writing. Focus particularly on how to shorten and clarify your statements. Make sure that each paragraph concerns only one topic. Remember the golden rule of good writing: Writing is revising!

You can ease the burden of writing in several ways:

Video Link 12.1
Watch how to present
social research results.

- Draw on the research proposal and on project notes. You aren't starting from scratch; you have all the material you've written during the course of the project.

- Refine your word processing skills on the computer so that you can use the most efficient techniques when reorganizing and editing.

- Seek criticism from friends, teachers, or other research consumers before you turn in the final product. They will alert you to problems in the research or the writing.

We often find it helpful to use *reverse outlining*. After you have written a first draft, read through the draft, noting down the key ideas as they come up. Do those notes reflect your original outline, or did you go astray? Are the paragraphs clean? How could your organization be improved?

Most important, leave yourself enough time so that you *can* revise, several times if possible, before turning in the final draft.

You can find more detailed reviews of writing techniques in Becker (1986), Booth et al. (1995), Mullins (1977), Strunk and White (1979), and Turabian (1967).

Your report should be clearly organized into sections, probably following a standard format that readers will immediately understand. Any research report should include an *introductory statement of the research problem*, a *literature review*, and a *methodology section*. These are the same three sections that should begin a research proposal. In addition, a research report must include a *findings section* with pertinent data displays. A *discussion section* may be used to interpret the findings and review the support for the study's hypotheses. A *conclusions section* should summarize the findings and draw implications for the theoretical framework used. Any weaknesses in the research design and ways to improve future research should be identified in this section. Compelling foci for additional research on the research question also should be noted. Most journals require a short abstract at the beginning that summarizes the research question and findings. A *bibliography* is also necessary. Depending on how the report is being published, *appendixes* containing the instruments used and specific information on the measures also may be included.

Exhibit 12.5 presents an outline of the sections in an academic journal article with some illustrative quotes. The article's introduction highlights the importance of the problem selected—the relation between marital disruption (divorce) and depression. The introduction also states clearly the gap in the research literature that the article is meant to fill—the untested possibility that depression might cause marital disruption rather than, or in addition to, marital disruption causing depression. The findings section (labeled "Results") begins by presenting the basic association between marital disruption and depression. Then it elaborates on this association by examining sex differences, the impact of prior marital quality, and various mediating and modifying effects. As indicated in the combined discussion and conclusions section, the analysis shows that marital disruption does indeed increase depression and specifies the time frame (3 years) during which this effect occurs.

Exhibit 12.5 | Sections in a Journal Article

Aseltine, Robert H. Jr. and Ronald C. Kessler. 1993. "Marital Disruption and Depression in a Community Sample." Journal of Health and Social Behavior 34(September):237–251.

INTRODUCTION
Despite 20 years of empirical research, the extent to which marital disruption causes poor mental health remains uncertain. The reason for this uncertainty is that previous research has consistently overlooked the potentially important problems of selection into and out of marriage on the basis of prior mental health. (p. 237)

SAMPLE AND MEASURES
Sample
Measures

RESULTS
The Basic Association Between Marital Disruption and Depression
Sex Differences
The Impact of Prior Marital Quality
The Mediating Effects of Secondary Changes
The Modifying Effects of Transitions to Secondary Roles

DISCUSSION [includes conclusions]
. . . According to the results, marital disruption does in fact cause a significant increase in depression compared to pre-divorce levels within a period of three years after the divorce. (p. 245)

These basic report sections present research results well, but many research reports include subsections tailored to the issues and stages in the specific study being reported. Lengthy applied reports on elaborate research projects may, in fact, be organized around the research project's different stages or foci.

The material that can be termed *the front matter* and the *back matter* of an applied report also is important. Applied reports usually begin with an executive summary: a summary list of the study's main findings, often with bullet points. Appendixes, the back matter, may present tables containing supporting data that were not discussed in the body of the report. Applied research reports also often append a copy of the research instrument(s).

For instance, Exhibit 12.6 outlines the sections in an applied research report. This particular report was mandated by the California State Legislature to review a state-funded program for the homeless mentally disabled. The goals of the report are described as both description and evaluation. The body of the report presents findings on the number and characteristics of homeless persons and on the operations of the state-funded program in each of 17 counties. The discussion section highlights service needs that are not being met. Nine appendixes then provide details on the study methodology and the counties studied.

An important principle for the researcher writing for a nonacademic audience is to make the findings and conclusions engaging and clear. You can see how Schutt did this in a report from a class research project designed with his graduate methods students (and in collaboration with several faculty knowledgeable about substance abuse) (Exhibit 12.7). These report excerpts indicate how he summarized key findings in an executive summary (Schutt et al. 1996:iv), emphasized the importance of the research in the introduction (Schutt et al. 1), used formatting and graphing to draw attention to particular findings in the body of the text (Schutt et al.:5), and tailored recommendations to his own university context (Schutt et al.:26).

A well-written research report requires (to be just a bit melodramatic) blood, sweat, and tears—and more time than you may at first anticipate. But writing one report will help you write the next report. And the issues you consider, if you approach your writing critically, will be sure to improve your subsequent research projects and sharpen your evaluations of other investigators' research projects.

Journal Link 12.3
Read an article about race relations that organizes data using charts and tables.

Exhibit 12.6 │ **Sections in an Applied Report**

Vernez, Georges, M. Audrey Burnam, Elizabeth A. McGlynn, Sally Trude, and Brian S. Mittman. 1988. Review of California's Program for the Homeless Mentally Disabled. *Santa Monica, CA: The RAND Corporation.*

SUMMARY
    In 1986, the California State Legislature mandated an independent review of the HMD programs that the counties had established with the state funds. The review was to determine the accountability of funds; describe the demographic and mental disorder characteristics of persons served; and assess the effectiveness of the program. This report describes the results of that review. (p. v)

INTRODUCTION
    Background
    California's Mental Health Services Act of 1985 . . . allocated $20 million annually to the state's 58 counties to support a wide range of services, from basic needs to rehabilitation. (pp. 1–2)
    Study Objectives
    Organization of the Report

HMD PROGRAM DESCRIPTION AND STUDY METHODOLOGY
    The HMD Program
    Study Design and Methods
    Study Limitations

COUNTING AND CHARACTERIZING THE HOMELESS
    Estimating the Number of Homeless People
    Characteristics of the Homeless Population

THE HMD PROGRAM IN 17 COUNTIES
    Service Priorities
    Delivery of Services
    Implementation Progress
    Selected Outcomes
    Effects on the Community and on County Service Agencies
    Service Gaps

DISCUSSION
    Underserved Groups of HMD
    Gaps in Continuity of Care
    A particularly large gap in the continuum of care is the lack of specialized housing alternatives for the mentally disabled. The nature of chronic mental illness limits the ability of these individuals to live completely independently. But their housing needs may change, and board-and-care facilities that are acceptable during some periods of their lives may become unacceptable at other times. (p. 57)
    Improved Service Delivery
    Issues for Further Research

Appendixes

A   SELECTION OF 17 SAMPLED COUNTIES
B. QUESTIONNAIRE FOR SURVEY OF THE HOMELESS
C. GUIDELINES FOR CASE STUDIES
D. INTERVIEW INSTRUMENTS FOR TELEPHONE SURVEY
E. HOMELESS STUDY SAMPLING DESIGN, ENUMERATION, AND SURVEY WEIGHTS
F. HOMELESS SURVEY FIELD PROCEDURES
G. SHORT SCREENER FOR MENTAL AND SUBSTANCE USE DISORDERS
H. CHARACTERISTICS OF THE COUNTIES AND THEIR HMD-FUNDED PROGRAMS
I.   CASE STUDIES FOR FOUR COUNTIES' HMD PROGRAMS

Exhibit 12.7 Student Substance Abuse, Report Excerpts

Executive Summary

- Rates of substance abuse were somewhat lower at UMass–Boston than among nationally selected samples of college students.
- Two-thirds of the respondents reported at least one close family member whose drinking or drug use had ever been of concern to them—one-third reported a high level of concern.
- Most students perceived substantial risk of harm due to illicit drug use, but just one-quarter thought alcohol use posed a great risk of harm.

Introduction

Binge drinking, other forms of alcohol abuse, and illicit drug use create numerous problems on college campuses. Deaths from binge drinking are too common and substance abuse is a factor in as many as two-thirds of on-campus sexual assaults (Finn, 1997; National Institute of Alcohol Abuse and Alcoholism, 1995). College presidents now rate alcohol abuse as the number one campus problem (Wechsler, Davenport, Dowdall, Moeykens, & Castillo, 1994) and many schools have been devising new substance abuse prevention policies and programs. However, in spite of increasing recognition of and knowledge about substance abuse problems at colleges as a whole, little attention has been focused on substance abuse at commuter schools.

Findings

The composite index identifies 27% of respondents as at risk of substance abuse (an index score of 2 or higher). One-quarter reported having smoked or used smokeless tobacco in the past two weeks.

27% of respondents were identified as at risk of substance abuse.

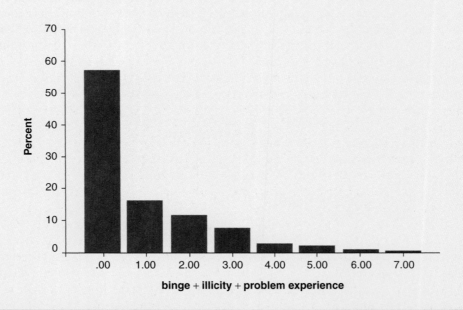

Exhibit 12.7 (Continued)

Recommendations

1. Enforce campus rules and regulations about substance use. When possible and where appropriate, communications from campus officials to students should heighten awareness of the UMass–Boston commitment to an alcohol- and drug-free environment.

2. Encourage those students involved in campus alcohol or drug-related problems or crises to connect with the PRIDE program.

3. Take advantage of widespread student interest in prevention by forming a university-wide council to monitor and stimulate interest in prevention activities.

Conclusion

Good critical skills are essential in evaluating research reports, whether your own or those produced by others. And it is really not just a question of sharpening your knives and going for the jugular. There are *always* weak points in any research, even published research. Being aware of the weaknesses, both in others' studies and in your own, is a major strength in itself. You need to be able to weigh the results of any particular research and to evaluate a study in terms of its contribution to understanding the social world—not in terms of whether it gives a definitive answer for all time, is perfectly controlled, or answers all questions.

This is not to say, however, that "anything goes." Much research lacks one or more of the three legs of validity—measurement validity, causal validity, or generalizability—and contributes more confusion than understanding about the social world. It's true that top scholarly journals maintain very high standards, partly because they have good critics in the review process and distinguished editors who make the final acceptance decisions. But some daily newspapers do a poor job of screening, and research reporting standards in many popular magazines, TV shows, and books are often abysmally poor. Keep your standards high when you read research reports but not so high or so critical that you dismiss studies that make tangible contributions to understanding the social world. And don't be so intimidated by high standards that you shrink from conducting research yourself.

Video Link 12.2
Watch how to present your data.

The growth of social science methods from infancy to adolescence, perhaps to young adulthood, ranks as a key intellectual accomplishment of the 20th century. Opinions about the causes and consequences of homelessness no longer need to depend on the scattered impressions of individuals, criminal justice policies can be shaped by systematic evidence of their effectiveness, and changes in the distribution of poverty and wealth in populations can be identified and charted. Employee productivity, neighborhood cohesion, and societal conflict can each be linked to individual psychological processes and to international economic strains. Systematic researchers looking at truly representative data can make connections and see patterns that no casual observer would ever discern.

Of course, social research methods are only helpful when the researchers are committed and honest. Research methods, like all knowledge, can be used poorly or well, for good purposes or bad, when appropriate or not. A claim that "We're basing this on research!" or "Our statistics prove it!" in itself provides no extra credibility. As you have learned throughout this book, we must first learn which methods were used, how they were applied, and whether final interpretations square with the evidence. But having done all that in good faith, we do emerge from confusion into clarity in our continuing effort to make sense of the social world.

Highlights

- Research reports should be evaluated systematically, using the review guide in Exhibit 12.2 and taking account of the interrelations among the design elements.

- Proposal writing should be a time for clarifying the research problem, reviewing the literature, and thinking ahead about the report that will be required. Tradeoffs between different design elements should be considered and the potential for mixing methods evaluated.

- Different types of reports typically pose different problems. Authors of student papers must be guided in part by the expectations of their professors. Thesis writers have to meet the requirements of different committee members but can benefit greatly from the areas of expertise represented on a typical thesis committee. Applied researchers are constrained by the expectations of the research sponsor; an advisory committee from the applied setting can help to avoid problems. Journal articles must pass a peer review by other social scientists and often are much improved in the process.

- Research reports should include an introductory statement of the research problem, a literature review, a methodology section, a findings section with pertinent data displays, and a conclusions section that identifies any weaknesses in the research design and points out implications for future research and theorizing. This basic report format should be modified according to the needs of a particular audience.

- All reports should be revised several times and critiqued by others before being presented in final form.

STUDENT STUDY SITE

The Student Study Site, available at **www.sagepub.com/chambliss4e,** includes useful study materials including web exercises with accompanying links, eFlashcards, videos, audio resources, journal articles, and encyclopedia articles, many of which are represented by the media links throughout the text. The site also features Interactive Exercises—represented by the green icon here—to help you understand the concepts in this book.

Exercises

Discussing Research

1. A good place to start developing your critical skills would be with one of the articles on the study site. Try reading one and fill in the answers to the article review questions in Exhibit 12.2. Do you agree with our answers to the other questions? Could you add some points to our critique or to the lessons on research design that we drew from these critiques?

2. How firm a foundation do social research methods provide for understanding the social world? Discuss the pro and con arguments, focusing on the variability of social research findings across different social contexts and the difficulty of understanding human subjectivity.

Finding Research

1. Go to the National Science Foundation's Sociology Program website (www.nsf.gov/funding/pgm_summ.jsp?pims_id=5369). What components does the National Science Foundation's Sociology Program look for in a proposed piece of research? Outline a research proposal to study a subject of your choice to be submitted to the National Science Foundation for funding.

2. The National Academy of Sciences wrote a lengthy report on ethics issues in scientific research. Visit the site and read the free executive summary you can obtain (www.nap.edu/catalog.php?record_id=10430). Summarize the information and guidelines in the report.

3. Search a social science journal to find five examples of social science research projects. Briefly describe each. How does each differ in its approach to reporting the research results? To whom do you think the author(s) of each is "reporting" (i.e., who is the audience)? How do you think the predicted audience has helped to shape the author's approach to reporting the results? Be sure to note the source in which you located your five examples.

Critiquing Research

1. A good place to start developing your critical skills would be with Murray Melbin's article that is reviewed in this chapter. Try reading it and fill in the answers to the article review questions that we did not cover (Exhibit 12.2). Do you agree with our answers to the other questions? Could you add some points to our critique or to the lessons about research designs that we drew from these critiques?

2. Read the journal article "Marital Disruption and Depression in a Community Sample" by Aseltine and Kessler in the September 1993 issue of *Journal of Health and Social Behavior.* How effective is the article in conveying the design and findings of the research? Could the article's organization be improved at all? Are there bases for disagreement about the interpretation of the findings?

3. Rate four journal articles for overall quality of the research and for effectiveness of the writing and data displays. Discuss how each could have been improved.

Doing Research

1. Call a local social or health service administrator or a criminal justice official and arrange for an interview. Ask the official about his or her experience with applied research reports and conclusions about the value of social research and the best techniques for reporting to practitioners.

2. Interview a student who has written an independent paper or thesis based on collecting original data. Ask the student to describe her or his experiences while writing the thesis. Review the decisions this student made in designing the research and ask about the stages of research design, data collection and analysis, and report writing that proved to be difficult.

3. Design a research proposal, following the outline and guidelines presented in this chapter. Focus on a research question that you could study on campus or in your local community.

Ethics Questions

1. Plagiarism is no joke. What are the regulations on plagiarism in class papers at your school? What do you think the ideal policy would be? Should this policy take into account cultural differences in teaching practices and learning styles? Do you think this ideal policy is likely to be implemented? Why or why not? Based on your experiences, do you believe that most student plagiarism is the result of misunderstanding about proper citation practices, or is it the result of dishonesty? Do you think that students who plagiarize while in school are less likely to be honest as social researchers?

2. Most journals now require full disclosure of funding sources, as well as paid consulting and other business relationships. Should researchers publishing in social science journals also be required to fully disclose all sources of funding, including receipt of payment for research done as a consultant? Should full disclosure of all previous funding sources be required in each published article? Write a short justification of the regulations you propose.

Glossary

Alternate-forms reliability: A procedure for testing the reliability of responses to survey questions in which subjects' answers are compared after the subjects have been asked slightly different versions of the questions or when randomly selected halves of the sample have been administered slightly different versions of the questions.

Anomalous: Unexpected patterns in data that do not seem to fit the theory being proposed.

Anonymity: Provided by research in which no identifying information is recorded that could be used to link respondents to their responses.

Association: A criterion for establishing a causal relationship between two variables: Variation in one variable is empirically related to variation in another variable.

Availability sampling: Sampling in which elements are selected on the basis of convenience.

Bar chart: A graphic for qualitative variables in which the variable's distribution is displayed with solid bars separated by spaces.

Base number (N): The total number of cases in a distribution.

Before-and-after design: A quasi-experimental design consisting of several before-and-after treatment comparisons of the same group of cases.

Belmont Report: Report in 1979 of the National Commission for the Protection of Human Subjects of Biomedical and Behavioral Research stipulating three basic ethical principles for the protection of human subjects: respect for persons, beneficence, and justice.

Beneficence: Minimizing possible harms and maximizing benefits.

Bias: Sampling bias occurs when some population characteristics are over- or underrepresented in the sample because of particular features of the method of selecting the sample.

Bimodal: A distribution in which two nonadjacent categories have about the same number of cases and these categories have more cases than any others.

Causal effect: The finding that change in one variable leads to change in another variable.

Causal (internal) validity: This type of validity exists when a conclusion that A leads to, or results in, B is correct.

Census: Research in which information is obtained through responses from or information about all available members of an entire population.

Central tendency: The most common value (for variables measured at the nominal level) or the value around which cases tend to center (for a qualitative variable).

Certificate of Confidentiality: Document issued by the National Institutes of Health to protect researchers from being legally required to disclose confidential information.

Ceteris paribus: Latin phrase meaning "other things being equal."

Chi-square: An inferential statistic used to test hypotheses about relationships between two or more variables in a cross-tabulation.

Closed-ended (fixed-choice) question: A survey question that provides preformatted response choices for the respondent to circle or check.

Cluster: A naturally occurring, mixed aggregate of elements of the population.

Cluster sampling: Sampling in which elements are selected in two or more stages, with the first stage being the random selection of naturally occurring clusters and the last stage being the random selection of elements within clusters.

Cognitive interview: A technique for evaluating questions in which researchers ask people test questions, and then probe with follow-up questions to learn how they understood the question and what their answers mean.

Cohort: Individuals or groups with a common starting point.

Cohort design: A longitudinal study in which data are collected at two or more points in time from individuals in a cohort.

Comparison groups: In an experiment, groups that have been exposed to different treatments, or values of the independent variable (e.g., a control group and an experimental group).

Compensatory rivalry (John Henry effect): A type of contamination in experimental and quasi-experimental designs that occurs when control group members are aware that they are being denied the treatment and modify their efforts by way of compensation.

Complete observation: A role in participant observation in which the researcher does not participate in group activities and is publicly defined as a researcher.

Complete (covert) participation: A role in field research in which the researcher does not reveal his or her identity as a researcher to those who are observed. Also called covert participation.

Computer-assisted personal interview (CAPI): A personal interview in which the laptop computer is used to display interview questions and to process responses that the interviewer types in, as well as to check that these responses fall within allowed ranges.

Computer-assisted qualitative data analysis: Analysis of textual, aural, or pictorial data using a special computer program that facilitates searching and coding text.

Concept: A mental image that summarizes a set of similar observations, feelings, or ideas.

Conceptualization: The process of specifying what we mean by a term. In deductive research, conceptualization helps to translate portions of an abstract theory into testable hypotheses involving specific variables. In inductive research, conceptualization is an important part of the process used to make sense of related observations.

Concurrent validity: The type of validity that exists when scores on a measure are closely related to scores on a criterion measured at the same time.

Confidentiality: Provided by research in which identifying information that could be used to link respondents to their responses is available only to designated research personnel for specific research needs.

Constant: A characteristic or value that does not change; quantitatively, a number that has a fixed value in a given situation.

Construct validity: The type of validity that is established by showing that a measure is related to other measures as specified in a theory.

Contamination: A source of causal invalidity that occurs when either the experimental and/or the comparison group is aware of the other group and is influenced in the posttest as a result.

Content analysis: A research method for systematically analyzing and making inferences from text.

Content validity: The type of validity that exists when the full range of a concept's meaning is covered by the measure.

Context: The larger set of interrelated circumstances in which a particular outcome should be understood.

Context effects: In survey research, refers to the influence that earlier questions may have on how subsequent questions are answered.

Contingent question: A question that is asked of only a subset of survey respondents.

Control group: A comparison group that receives no treatment.

Cost-benefit analysis: A type of evaluation research that compares program costs with the economic value of program benefits.

Cost-effectiveness analysis: A type of evaluation research that compares program costs with actual program outcomes.

Cover letter: The letter sent with a mailed questionnaire. It explains the survey's purpose and auspices and encourages the respondent to participate.

Criterion validity: The type of validity that is established by comparing the scores obtained on the measure being validated to those obtained with a more direct or already validated measure of the same phenomenon (the criterion).

Cross-population generalizability (external validity): Exists when findings about one group, population, or setting hold true for other groups, populations, or settings.

Cross-sectional research design: A study in which data are collected at only one point in time.

Cross-tabulation (crosstab): In the simplest case, a bivariate (two-variable) distribution showing the distribution of one variable for each category of another variable; can also be elaborated using three or more variables.

Data cleaning: The process of checking data for errors after the data have been entered in a computer file.

Debriefing: A researcher's informing subjects after an experiment about the experiment's purposes and methods and evaluating subjects' personal reactions to the experiment.

Deductive research: The type of research in which a specific expectation is deduced from a general premise and is then tested.

Demoralization: A type of contamination in experimental and quasi-experimental designs that occurs when control group members feel that they have been left out of some valuable treatment, performing worse than expected as a result.

Dependent variable: A variable that is hypothesized to vary depending on or under the influence of another variable.

Descriptive research: Research in which social phenomena are defined and described.

Descriptive statistics: Statistics used to describe the distribution of and relationship among variables.

Differential attrition (mortality): A problem that occurs in experiments when comparison groups become different because subjects in one group are more likely to drop out for various reasons compared to subjects in the other group(s).

Disproportionate stratified sampling: Sampling in which elements are selected from strata in proportions different from those that appear in the population.

Distribution of benefits: An ethical issue about how much researchers can influence the benefits subjects receive as part of the treatment being studied in a field experiment.

Double-barreled question: A single survey question that actually asks two questions but allows only one answer.

Double-blind procedure: An experimental method in which neither the subjects nor the staff delivering experimental treatments know which subjects are getting the treatment.

Double negative: A question or statement that contains two negatives, which can muddy the meaning of the question.

Ecological fallacy: An error in reasoning in which incorrect conclusions about individual-level processes are drawn from group-level data.

Efficiency analysis: A type of evaluation research that compares program costs with program effects. It can be either a cost-benefit analysis or a cost-effectiveness analysis.

Elaboration analysis: The process of introducing a third variable into an analysis to better understand (or elaborate) the bivariate (two-variable) relationship under consideration. Additional control variables also can be introduced.

Electronic survey: A survey that is sent and answered by computer, either through e-mail or on the web.

Elements: The individual members of the population whose characteristics are to be measured.

E-mail survey: A survey that is sent and answered through e-mail.

Emic focus: Representing a setting with the participants' terms.

Endogenous change: A source of causal invalidity that occurs when natural developments or changes in the subjects (independent of the experimental treatment itself) account for some or all of the observed change from the pretest to the posttest.

Illogical reasoning: The premature jumping to conclusions or arguing on the basis of invalid assumptions.

Ethnography: The study and systematic recording of human cultures.

Ethnomethodology: A qualitative research method focused on the way that participants in a social setting create and sustain a sense of reality.

Etic focus: Representing a setting with the researcher's terms.

Evaluability assessment: A type of evaluation research conducted to determine whether it is feasible to evaluate a program's effects within the available time and resources.

Evaluation research: Research that describes or identifies the impact of social policies and programs.

Exhaustive: Every case can be classified as having at least one attribute (or value) for the variable.

Expectancies of experiment staff (self-fulfilling prophecy): A source of treatment misidentification in experiments and quasi-experiments that occurs when change among experimental subjects is due to the positive expectancies of the staff who are delivering the treatment, rather than to the treatment itself; also called a self-fulfilling prophecy.

Experimental group: In an experiment, the group of subjects that receives the treatment or experimental manipulation.

Explanatory research: Seeks to identify causes and effects of social phenomena and to predict how one phenomenon will change or vary in response to variation in another phenomenon.

Exploratory research: Seeks to open new areas of inquiry, especially by understanding what meanings people give to situations.

Ex post facto control group design: A nonexperimental design in which comparison groups are selected after the treatment, program, or other variation in the independent variable has occurred.

Extraneous variable: A variable that influences both the independent and dependent variables so as to create a spurious association between them that disappears when the extraneous variable is controlled.

Face validity: The type of validity that exists when all inspection of items used to measure a concept suggests that they are appropriate "on their face."

Federal Policy for the Protection of Human Subjects: Federal regulations codifying basic principles for conducting research on human subjects; used as the basis for professional organizations' guidelines.

Feedback: Information about service delivery system outputs, outcomes, or operations that is available to any program inputs.

Fence-sitters: Survey respondents who see themselves as being neutral on an issue and choose a middle (neutral) response that is offered.

Field experiment: An experimental study conducted in a real-world setting.

Field notes: Notes that describe what has been observed, heard, or otherwise experienced in a participant observation study. These notes usually are written after the observational session.

Field research: Research in which natural social processes are studied as they happen and left relatively undisturbed.

Filter question: A survey question used to identify a subset of respondents who then are asked other questions.

Floaters: Survey respondents who provide an opinion on a topic in response to a closed-ended question that does not include a "Don't know" option but who will choose "Don't know" if it is available.

Focus groups: A qualitative method that involves unstructured group interviews in which the focus group leader actively encourages discussion among participants on the topics of interest.

Formative evaluation: Process evaluation that is used to shape and refine program operations.

Frequency polygon: A graphic for quantitative variables in which a continuous line connects data points representing the variable's distribution.

Gamma: A measure of association that is sometimes used in cross-tabular analysis.

Gatekeeper: A person in a field setting who can grant researchers access to the setting.

Generalizability: Exists when a conclusion holds true for the population, group, setting, or event that we say it does, given the conditions that we specify; it is the extent to which a study can inform us about persons, places, or events that were not directly studied.

Grounded theory: Systematic theory developed inductively, based on observations that are summarized into conceptual categories, reevaluated in the research setting, and gradually refined and linked to other conceptual categories.

Group-administered survey: A survey that is completed by individual respondents who are assembled in a group.

Group unit of analysis: A unit of analysis in which groups are the source of data and the focus of conclusions.

Hawthorne effect (effect of external events): A type of contamination in experimental and quasi-experimental designs that occurs when members of the treatment group change in terms of the dependent variable because their participation in the study makes them feel special.

Health Insurance Portability and Accountability Act (HIPAA): A U.S. federal law passed in 1996 that guarantees, among other things, specified privacy rights for medical patients, in particular those in research settings.

Histogram: A graphic for quantitative variables in which the variable's distribution is displayed with adjacent bars.

History effect (effect of external events): Events external to the study that influence posttest scores, resulting in causal invalidity.

Hypothesis: A tentative statement about empirical reality involving a relationship between two or more variables.

Impact analysis (impact evaluation or summative evaluation): Evaluation research that answers these questions: Did the program work? Did it have the intended result?

Independent variables: A variable that is hypothesized to cause, or lead to, variation in another variable.

Index: A composite measure based on summing, averaging, or otherwise combining the responses to multiple questions that are intended to measure the same concept.

Individual unit of analysis: A unit of analysis in which individuals are the source of data and the focus of conclusions.

Inductive reasoning: The type of reasoning that moves from the specific to the general.

Inductive research: The type of research in which general conclusions are drawn from specific data.

Inferential statistics: Statistics used to estimate how likely it is that a statistical result based on data from a random sample is representative of the population from which the sample is assumed to have been selected.

In-person interview: A survey in which an interviewer questions respondents face-to-face and record their answers.

Inputs: Resources, raw materials, clients, and staff that go into a program.

Institutional review board (IRB): A group of organizational and community representatives required by federal law to review the ethical issues in all proposed research that is federally funded, involves human subjects, or has any potential for harm to subjects.

Integrative approaches (to evaluation): An orientation to evaluation research that expects researchers to respond to the concerns of people involved with the program stakeholders, as well as to the standards and goals of the social scientific community.

Intensive (depth) interviewing: A qualitative method that involves open-ended, relatively unstructured questioning in which the interviewer seeks in-depth information on the interviewee's feelings, experiences, and perceptions.

Interactive voice response (IVR): A survey in which respondents receive automated calls and answer questions by pressing numbers on their touch-tone phones or speaking numbers that are interpreted by computerized voice recognition software.

Interitem reliability (internal consistency): An approach that calculates reliability based on the correlation among multiple items used to measure a single concept.

Interobserver reliability: When similar measurements are obtained by different observers rating the same persons, events, or places.

Interval level of measurement: A measurement of a variable in which the numbers indicating a variable's values represent fixed measurement units but have no absolute, or fixed, zero point.

Interpretive questions: Questions included in a questionnaire or interview schedule to help explain answers to other important questions.

Interquartile range: The range in a distribution between the end of the lst quartile and the beginning of the 3rd quartile.

Interview schedule: A survey instrument containing the questions asked by the interviewer in an in-person or phone survey.

Jottings: Brief notes written in the field about highlights of an observation period.

Justice: As used in human research ethics discussions, distributing benefits and risks of research fairly.

Key informant: An insider who is willing and able to provide a field researcher with superior access and information, including answers to questions that arise in the course of the research.

Level of measurement: The mathematical precision with which the values of a variable can be expressed. The nominal level of measurement, which is qualitative, has no mathematical interpretation; the quantitative levels of measurement—ordinal, interval, and ratio—are progressively more precise mathematically.

Longitudinal research design: A study in which data are collected that can be ordered in time; also defined as research in which data are collected at two or more points in time.

Mailed (self-administered) survey: A survey involving a mailed questionnaire to be completed by the respondent.

Matching: A procedure for equating the characteristics of individuals in different comparison groups in an experiment. Matching can be done on either an individual or an aggregate basis. For individual matching, individuals who are similar in terms of key characteristics are paired prior to assignment, and then the two members of each pair are assigned to the two groups. For aggregate matching, groups are chosen for comparison that are similar in terms of the distribution of key characteristics.

Matrix: A chart used to condense qualitative data into simple categories and provide a multidimensional summary that will facilitate subsequent, more intensive analysis.

Mean: The arithmetic, or weighted, average computed by adding up the value of all the cases and dividing by the total number of cases.

Measurement validity: Exists when an indicator measures what we think it measures.

Measure of association: A type of descriptive statistic that summarizes the strength of an association.

Mechanism: A discernible process that creates a causal connection between two variables.

Median: The position average, or the point, that divides a distribution in half (the 50th percentile).

Mode (probability average): The most frequent value in a distribution.

Multiple group before-and-after design: A type of quasi-experimental design in which several before-and-after comparisons are made involving the same independent and dependent variables but different groups.

Mutually exclusive: A variable's attributes (or values) are mutually exclusive when every case can be classified as having only one attribute (or value).

Narrative analysis: A form of qualitative analysis in which the analyst focuses on how respondents impose order on the flow of experience in their lives and so make sense of events and actions in which they have participated.

Needs assessment: A type of evaluation research that attempts to determine the needs of some population that might be met with a social program.

Netnography (cyberethnography or virtual ethnography): The use of ethnographic methods to study online communities.

Nominal level of measurement: Variables whose values have no mathematical interpretation; they vary in kind or quality but not amount.

Nonequivalent control group design: A quasi-experimental design in which there are experimental and comparison groups that are designated before the treatment occurs but are not created by random assignment.

Nonprobability sampling methods: Sampling methods in which the probability of selection of population elements is unknown.

Nonspuriousness: A criterion for establishing a causal relation between two variables; when a relationship between two variables is not due to variation in a third variable.

Normal distribution: A symmetric distribution shaped like a bell and centered around the population mean, with the number of cases tapering off in a predictable pattern on both sides of the mean.

Nuremberg war crime trials: Trials held in Nuremberg, Germany, in the years following World War II, in which the former leaders of Nazi Germany were charged with war crimes and crimes against humanity; frequently considered the first trials for people accused of genocide.

Obedience experiments (Milgram's): A series of famous experiments conducted during the 1960s by Stanley Milgram, a psychologist from Yale University, testing subjects' willingness to cause pain to another person if instructed to do so.

Omnibus survey: A survey that covers a range of topics of interest to different social scientists.

Open-ended question: A survey question to which the respondents reply in their own words, either by writing or by talking.

Operation: A procedure for identifying or indicating the value of cases on a variable.

Operationalization: The process of specifying the operations that will indicate the value of cases on a variable.

Ordinal level of measurement: A measurement of a variable in which the numbers indicating a variable's values specify only the order of the cases, permitting *greater than* and *less than* distinctions.

Outcomes: The impact of the program process on the cases processed.

Outlier: An exceptionally high or low value in a distribution.

Outputs: The services delivered or new products produced by the program process.

Overgeneralization: Occurs when we unjustifiably conclude that what is true for some cases is true for all cases.

Panel design: A longitudinal study in which data are collected from the same individuals—the panel—at two or more points in time.

Participant observation: A qualitative method for gathering data that involves developing a sustained relationship with people while they go about their normal activities.

Percentage: The relative frequency, computed by dividing the frequency of cases in a particular category by the total number of cases and multiplying by 100.

Periodicity: A sequence of elements (in a list to be sampled) that varies in some regular, periodic pattern.

Phone survey: A survey in which interviewers question respondents over the phone and record their answers.

Placebo effect: A source of treatment misidentification that can occur when subjects receive a treatment that they consider likely to be beneficial and improve as a result of the expectation rather than of the treatment itself.

Population: The entire set of individuals or other entities to which study findings are to be generalized.

Posttest: In experimental research, the measurement of an outcome (dependent) variable after an experimental intervention or after a presumed independent variable has changed for some other reason. The posttest is exactly the same "test" as the pretest, but it is administered at a different time.

Predictive validity: The type of validity that exists when a measure predicts scores on a criterion measured in the future.

Pretest: In experimental research, the measurement of an outcome (dependent) variable prior to an experimental intervention or change in a presumed independent variable for some other reason. The pretest is exactly the same "test" as the posttest, but it is administered at a different time.

Prison simulation study (Zimbardo's): Famous study from the early 1970s, organized by Stanford psychologist Philip Zimbardo, demonstrating the willingness of average college students quickly to become harsh disciplinarians when put in the role of (simulated) prison guards over other students; usually interpreted as demonstrating an easy human readiness to become cruel.

Probability of selection: The likelihood that an element will be selected from the population for inclusion in the sample. In a census of all the elements of a population, the probability that any particular element will be selected is 1.0. If half the elements in the population are sampled on the basis of chance (say, by tossing a coin), the probability of selection for each element is one half, or 0.5. As the size of the sample as a proportion of the population decreases, so does the probability of selection.

Probability sampling method: A sampling method that relies on a random, or chance, selection method so that the probability of selection of population elements is known.

Process analysis: A research design in which periodic measures are taken to determine whether a treatment is being delivered as planned, usually in a field experiment.

Process evaluation: Evaluation research that investigates the process of service delivery.

Program process: The complete treatment or service delivered by the program.

Program theory: A descriptive or prescriptive model of how a program operates and produces effects.

Progressive focusing: The process wherein when it appears that additional concepts need to be investigated or new relationships explored, the analysis adjusts the data collection.

Proportionate stratified sampling: Sampling method in which elements are selected from strata in exact proportion to their representation in the population.

Purposive sampling: A nonprobability sampling method in which elements are selected for a purpose, usually because of their unique position.

Qualitative data analysis: Techniques used to search and code textual, aural, and pictorial data and to explore relationships among the resulting categories.

Qualitative methods: Methods, such as participant observation, intensive interviewing, and focus groups, that are designed to capture social life as participants experience it rather than in categories the researcher predetermines. These methods typically involve exploratory research questions, inductive reasoning, an orientation to social context, and a focus on human subjectivity and the meanings participants attach to events and to their lives.

Quantitative data analysis: Statistical techniques used to describe and analyze variation in quantitative measures.

Quartiles: The points in a distribution corresponding to the first 25% of the cases, the first 50% of the cases, and the last 25% of the cases.

Quasi-experimental design: A research design in which there is a comparison group that is comparable to the experimental group in critical ways but subjects are not randomly assigned to the comparison and experimental groups.

Questionnaire: A survey instrument containing the questions in a self-administered survey.

Quota sampling: A nonprobability sampling method in which elements are selected to ensure that the sample represents certain characteristics in proportion to their prevalence in the population.

Random assignment (randomization): A procedure by which each experimental subject is placed in a group randomly.

Random digit dialing (RDD): The random dialing, by a machine, of numbers within designated phone prefixes, which creates a random sample for phone surveys.

Random number table: A table containing lists of numbers that are ordered solely on the basis of chance; it is used for drawing a random sample.

Random sampling: A method of sampling that relies on a random, or chance, selection method so that every element of the sampling frame has a known probability of being selected.

Range: The true upper limit in a distribution minus the true lower limit (or the highest rounded value minus the lowest rounded value, plus 1).

Ratio level of measurement: A measurement of a variable in which the numbers indicating the variable's values represent fixed measuring units *and* an absolute zero point.

Reactive effects: The changes in an individual or group behavior that are due to being observed or otherwise studied.

Reductionist fallacy (reductionism): An error in reasoning that occurs when incorrect conclusions about group-level processes are based on individual-level data.

Regression effect: A source of causal validity that occurs when subjects chosen because of their extreme scores on a dependent variable become less extreme on a posttest due to mathematical necessity, not the treatment.

Reliability: A measurement procedure yields consistent scores when the phenomenon being measured is not changing.

Repeated measures panel design: A quasi-experimental design consisting of several pretest and posttest observations of the same group.

Representative sample: A sample that "looks like" the population from which it was selected in all respects that are potentially relevant to the study. The distribution of characteristics among the elements of a representative sample is the same as the distribution of those characteristics among the total population. In an unrepresentative sample, some characteristics are overrepresented or underrepresented.

Research circle: A diagram of the elements of the research process, including theories, hypotheses, data collection, and data analysis.

Resistance to change: The reluctance to change our ideas in light of new information.

Respect for persons: In human subjects ethics discussions, treating persons as autonomous agents and protecting those with diminished autonomy.

Sample: A subset of a population that is used to study the population as a whole.

Sample generalizability: Exists when a conclusion based on a sample, or subset of a larger population holds true for that population.

Sampling frame: A list of all elements or other units containing the elements in a population.

Sampling interval: The number of cases between one sampled case and another in a systematic random sample.

Sampling units: Units listed at each stage of a multistage sampling design.

Saturation point: The point at which subject selection is ended in intensive interviewing because new interviews seem to yield little additional information.

Scale: A composite measure based on combining the responses to multiple questions pertaining to a common concept after these questions are differentially weighted, such that questions judged on some basis to be more important for the underlying concept contribute more to the composite score.

Science: A set of logical, systematic, documented methods for investigating nature and natural processes; the knowledge produced by these investigations.

Secondary data analysis: Analysis of data collected by someone other than the researcher or the researcher's assistants.

Selection bias: A source of internal (causal) invalidity that occurs when characteristics of experimental and comparison group subjects differ in any way that influences the outcome.

Selective (inaccurate) observation: Choosing to look only at things that are in line with our preferences or beliefs.

Serendipitous: Unexpected patterns in data, which stimulate new ideas or theoretical approaches.

Simple random sampling: A method of sampling in which every sample element is selected purely on the basis of chance through a random process.

Skewness: The extent to which cases are clustered more at one or the other end of the distribution of a quantitative variable, rather than in a symmetric pattern around its center. Skew can be positive (a right skew), with the number of cases tapering off in the positive direction, or negative (a left skew), with the number of cases tapering off in the negative direction.

Skip pattern: The unique combination of questions created in a survey by filter questions and contingent questions.

Snowball sampling: A method of sampling in which sample elements are selected as successive informants or interviewees identify them.

Social research question: A question about the social world that is answered through the collection and analysis of firsthand, verifiable, empirical data.

Social science: The use of scientific methods to investigate individuals, societies, and social processes; the knowledge produced by these investigations.

Social science approaches (to evaluation): An orientation to evaluation research that expects researchers to emphasize the importance of researcher expertise and maintenance of autonomy from program stakeholders.

Split-halves reliability: Reliability achieved when responses to the same questions by two randomly selected halves of a sample are about the same.

Spurious: Nature of a presumed relationship between two variables that is actually due to variation in a third variable.

Stakeholder approaches (to evaluation): An orientation to evaluation research that expects researchers to be responsive primarily to the people involved with the program.

Stakeholders: Individuals and groups who have some basis of concern with the program.

Standard deviation: The square root of the average squared deviation of each case from the mean.

Statistic: A numerical description of some feature of a variable or variables in a sample from a larger population.

Statistical significance: The mathematical likelihood that an association is not due to chance, judged by a criterion the analyst sets.

Stratified random sampling: A method of sampling in which sample elements are selected separately from population strata that the researcher identifies in advance.

Survey research: Research in which information is collected from a sample of individuals through their responses to a set of standardized questions.

Systematic random sampling: A method of sampling in which sample elements are selected from a list or from sequential files, with every nth element being selected after the first element is selected randomly.

Tacit knowledge: In field research, a credible sense of understanding of social processes that reflects the researcher's awareness of participants' actions, as well as their words, and of what they fail to state, feel deeply, and take for granted.

Target population: A set of elements larger than or different from the population sampled and to which the researcher would like to generalize study findings.

Tearoom Trade: Book by Laud Humphreys investigating the social background of men who engage in homosexual behavior in public facilities; controversially, he did not obtain informed consent from his subjects.

Test-retest reliability: A measurement showing that measures of a phenomenon at two points in time are highly correlated, if the phenomenon has not changed or has changed only as much as the phenomenon itself.

Theoretical sampling: A sampling method recommended for field researchers by Glaser and Strauss (1967). A theoretical sample is drawn in a sequential fashion, with settings or individuals selected for study as earlier observations or interviews indicate that these settings or individuals are influential.

Theory-driven evaluation: A program evaluation guided by a theory that specifies the process by which the program has an effect.

Time order: A criterion for establishing a causal relationship between two variables: The variation in the presumed cause (the independent variable) must occur before the variation in the presumed effect (the dependent variable).

Time series design: A quasi-experimental design consisting of many pretest and posttest observations of the same group.

Treatment misidentification: A problem that occurs in an experiment when not the treatment itself, but rather some unknown or unidentified intervening process is causing the outcome.

Trend (repeated cross-sectional) design: A longitudinal study in which data are collected at two or more points in time from different samples of the same population.

Triangulation: The use of multiple methods to study one research question.

True experiment: Experiment in which subjects are assigned randomly to an experimental group that receives a treatment or other manipulation of the independent variable and a comparison group that does not receive the treatment or receives some other manipulation. Outcomes are measured in a posttest.

Tuskegee syphilis study: Research study conducted by a branch of the U.S. government, lasting for roughly 50 years (ending in the 1970s), in which a sample of African American men diagnosed with syphilis were deliberately left untreated, without their knowledge, to learn about the lifetime course of the disease.

Unimodal: A distribution of a variable in which only one value is the most frequent.

Units of analysis: The entities being studied, whose behavior is to be understood.

Unobtrusive measure: A measurement based on physical traces or other data that are collected without the knowledge or participation of the individuals or groups that generated the data.

Validity: The state that exists when statements or conclusions about empirical reality are correct.

Variability: The extent to which cases are spread out through the distribution or clustered around just one value.

Variable: A characteristic or property that can vary (take on different values or attributes).

Variance: A statistic that measures the variability of a distribution as the average squared deviation of each case from the mean.

Web survey: A survey that is accessed and responded to on the World Wide Web.

Appendix A

Finding Information

Elizabeth Schneider, MLS

Russell K. Schutt, PhD

All research is conducted in order to "find information" in some sense, but the focus of this section is more specifically about finding information to inform a central research project. This has often been termed "searching the literature" but the popularity of the World Wide Web for finding information requires that we broaden our focus beyond the traditional search of the published literature. It may sound trite, but we do indeed live in an "information age," with an unprecedented amount of information of many types available to us with relatively little effort. Learning how to locate and use that information efficiently has become a prerequisite for social science.

▣ Searching the Literature

It is most important to search the literature before we begin a research study. A good literature review may reveal that the research problem already has been adequately investigated; it may highlight particular aspects of the research problem most in need of further investigation; or it may suggest that the planned research design is not appropriate for the problem chosen. It can highlight the strong and weak points of related theories. When we review previous research about our research question, we may learn about weaknesses in our measures, complexities in our research problem, and possible difficulties in data collection. The more of these problems that can be taken into account before, rather than after, data are collected, the better the final research product will be. Even when the rush to "find out" what people think or are doing creates pressure to just go out and ask or observe, it is important to take the time to search the literature and try to reap the benefit of prior investigations.

But the social science literature is not just a source for guidance at the start of an investigation. During a study, questions will arise that can be answered by careful reading of earlier research. After data collection has ceased, reviewing the literature can help to develop new insights into patterns in the data. Research articles published since a project began may suggest new hypotheses or questions to explore.

The best way of searching the literature will be determined in part by what library and bibliographic resources are available to you, but a brief review of some basic procedures and alternative strategies will help you get started on a productive search.

Preparing the Search

You should formulate a research question before you begin the search, although the question may change after you begin. Identify the question's parts and subparts and any related issues that you think might play an important role in the research. List the authors of relevant studies you are aware of, possible keywords that might specify the subject for your search, and perhaps the most important journals that you are concerned with checking. For example, if your research question is "What is the effect of informal social control on crime?" you might consider searching the literature electronically for studies that mentioned "informal social control" and "crime" or "crime rate" or "violence" and "arrest." If you are concerned with more specific aspects of this question, you should also include the relevant words in your list, such as "family" or "community policing" or even "Northeast."

Conducting the Search

Now you are ready to begin searching the literature. You should check for relevant books in your library and perhaps in the other college libraries in your area. This usually means conducting a search of an online catalog using a list of subject terms. But most scientific research is published in journal articles so that research results can quickly be disseminated to other scientists. The primary focus of your search must therefore be the journal literature. Fortunately, much of the journal literature can be identified online, without leaving your personal computer, and an increasing number of published journal articles can be downloaded directly to your own computer (depending on your particular access privileges). But just because there's a lot available online doesn't mean that you need to find it all. Keep in mind that your goal is to find reports of prior research investigations; this means that you should focus on scholarly journals that choose articles for publication after they have been reviewed by other social scientists—"refereed journals." Newspaper and magazine articles just won't do, although you may find some that raise important issues or even that summarize social science research investigations.

The social science literature should be consulted at both the beginning and the end of an investigation. Even while an investigation is in progress, consultations with the literature may help to resolve methodological problems or facilitate supplementary explorations. As with any part of the research process, the method you use will affect the quality of your results. You should try to ensure that your search method includes each of the following steps:

Specify your research question. Your research question should not be so broad that hundreds of articles are judged relevant, or so narrow that you miss important literature. "Is informal social control effective?" is probably too broad. "Does informal social control reduce rates of burglary)' in large cities?" is probably too narrow. "Is informal social control more effective in reducing crime rates than policing?" provides about the right level of specificity.

Identify appropriate bibliographic databases to search. Your school library may subscribe to *Sociological Abstracts* or *SocINDEX* and either of these similar databases of the sociological literature may meet your needs. You can limit your searches in these databases to articles written in English, articles that have been peer reviewed and so are likely to be of higher quality, and to articles in journals that your library owns. However, if you are studying a question about social factors in illness you should also search in *MEDLINE* or the slightly more comprehensive *PubMed,* the databases for searching the medical literature maintained by the National Library of Medicine. If your focus is on mental health, you'll also want to include a search in the psychological abstracts, with *PsycARTICLES* (or *PsycINFO,* if that is what your library offers). Searching in a database like *Academic OneFile* or *Google Scholar* will retrieve article abstracts across disciplines, but it will be important to review your results very carefully to ensure that the articles you focus on are appropriate for a sociological research paper. In order to find articles across the social sciences that have referred to a previous publication, like Sherman and Berk's study of the police response to domestic violence, the *Social Science Citation Index* (SSCI) will be helpful. SSCI has a unique "citation searching" feature that allows you to look up articles or books and see who else has cited them in their work. This is an excellent and efficient way to assemble a number of references that are highly relevant to your research and to find out which articles and books have had the biggest impact in a field. Unfortunately, some college libraries do not subscribe to SSCI, either in its print, CD-ROM, or online version, due to its expense, but if you have access to SSCI, you should consider using it whenever you want to make sure that you develop the strongest possible literature review for your topic.

Choose a search technology. For most purposes, an online bibliographic database that references the published journal literature will be all you need to find the relevant social science research literature. However, searches for more obscure topics or very recent literature may require that you also search websites or bibliographies of relevant books. You will also need to search websites when you need to learn about current debate about particular social issues or you are investigating current social programs.

Create a tentative list of search terms. List the parts and subparts of your research question and any related issues that you think are important: "informal social control," "policing," "influences on crime rates," and perhaps "community cohesion and crime." List the authors of relevant studies. Specify the most important journals that deal with your topic.

Narrow your search. The sheer number of references you find can be a problem. For example, searching for "social capital" in November 2011 resulted in 1,768 citations in *Sociological Abstracts* and 2,596 in *SocINDEX* to peer reviewed articles written in English scholarly journals. Depending on the database you are working with and the purposes of your search, you may want to limit your search to English language publications, to journal articles rather than conference papers or dissertations (both of which are more difficult to acquire), and to materials published in recent years. You should give most attention to articles published in the leading journals in the field. Your professor can help you identify them.

Refine your search. Learn as you go. If your search yields too many citations, try specifying the search terms more precisely. If you have not found much literature, try using more general terms. Whatever terms you search on first, don't consider your search complete until you have tried several different approaches and have seen how many articles you find. A search for "domestic violence" in *SocINDEX* on November 5, 2011, yielded 2,954 abstracts for peer reviewed journal articles in English; by adding "effects" OR "influences" as required search terms the number of hits dropped to 539. A good rule is to cast a net with your search terms that is wide enough to catch most of the relevant articles but not so wide that it identifies many useless citations. In any case, if you are searching a popular topic, you will need to spend a fair amount of time whittling down the list of citations.

Use Boolean search logic. It's often a good idea to narrow your search by requiring that abstracts contain combinations of words or phrases that include more of the specifics of your research question. Using the Boolean connector AND allows you to do this, while using the connector OR allows you to find abstracts containing different words that mean the same thing. Exhibit A.1 provides an example.

Exhibit A.1   Use of Boolean Connectors in a Literature Search

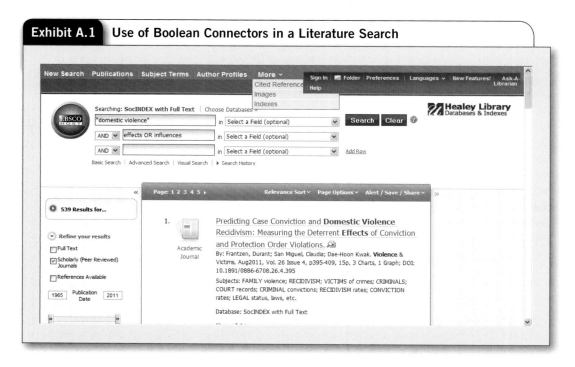

Use appropriate subject descriptors. Once you have found an article that you consider to be appropriate, take a look at the "Subject Terms" field in the citation (see Exhibit A.2). You can then redo your search after requiring that the articles be classified with some or all of these descriptor terms.

Exhibit A.2   Checking Standard Subject Matter Descriptors

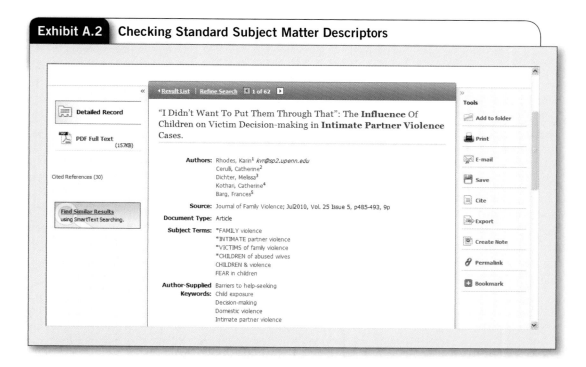

Check the results. Read the titles and abstracts you have found and identify the articles that appear to be most relevant. If possible, click on these article titles and generate a list of their references. See if you find more articles that are relevant to your research question but that you have missed so far. You will be surprised (we always are) at how many important articles your initial online search missed.

Read the articles. Now it is time to find the full text of the articles of interest. If you're lucky, many of the articles will be available to patrons of your library in online versions. If so, you'll be able to link to the full text just by clicking on a "full text" link. But many journals and/or specific issues of some journals will only be available in print, so you'll have to find them in your library (or order a copy through interlibrary loan). You may be tempted to write a "review" of the literature based on reading the abstracts or using only those articles available online, but you will be selling yourself short. Many crucial details about methods, findings, and theoretical implications will be found only in the body of the article and some important articles will not be available online. To understand, critique, and really learn from previous research studies, you must read the important articles, no matter how you have to retrieve them. But if you can't obtain the full text of an article, you'll just have to leave it out of your literature review and bibliography—reading the abstract just isn't enough.

Write the review. If you have done your job well, you will now have more than enough literature as background for your own research unless it is on a very obscure topic (see Exhibit A.3). (Of course, ultimately your search will be limited by the library holdings you have access to and by the time you have to order or find copies of journal articles, conference papers, and perhaps dissertations that you can't obtain online.) At this point, your main concern is to construct a coherent framework in which to develop your research question, drawing as many lessons as you can from previous research. You can use the literature to identify a useful theory and hypotheses to be reexamined, to find inadequately studied specific research questions, to explicate the disputes about your research question, to summarize the major findings of prior research, and to suggest appropriate methods of investigation.

Exhibit A.3   A Search in Sociological Abstracts on "Informal Social Control"

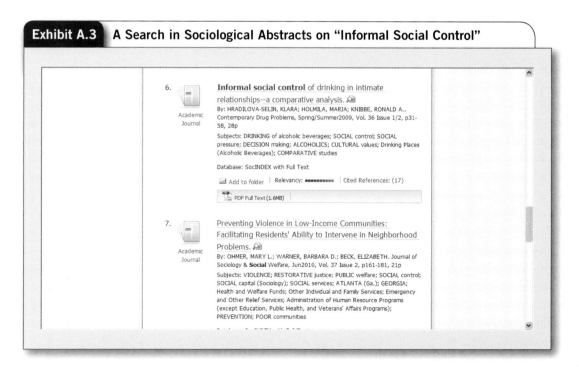

Be sure to take notes on each article you read, organizing your notes into standard sections: theory, methods, findings, conclusions. In any case, write the literature review so that it contributes to your study in some concrete way; don't feel compelled to discuss an article just because you have read it. Be judicious. You are conducting only one study of one issue; it will only obscure the value of your study if you try to relate it to every tangential point in related research.

Continue to search. Don't think of searching the literature as a one-time-only venture—something that you leave behind as you move on to your *real* research. You may encounter new questions or unanticipated problems as you conduct your research or as you burrow deeper into the literature. Searching the literature again to determine what others have found in response to these questions or what steps they have taken to resolve these problems can yield substantial improvements in your own research. There is so much literature on so many topics that it often is not possible to figure out in advance every subject you should search the literature for or what type of search will be most beneficial.

Another reason to make searching the literature an ongoing project is that the literature is always growing. During the course of one research study, whether it takes only one semester or several years, new findings will be published and relevant questions will be debated. Staying attuned to the literature and checking it at least when you are writing up your findings may save your study from being outdated. Of course, this does not make life any easier for researchers. For example, one of your authors was registered for a time with a service that every week sent citations of new journal articles on homelessness to his electronic mailbox. Most were not very important, and even looking over the abstracts for between 5 and 15 new articles each week is quite a chore—that's part of the price we pay for living in the information age!

Refer to a good book for even more specific guidance about literature searching. Arlene Fink's (2005) *Conducting Research Literature Reviews: From the Internet to Paper,* Sage Publications, is an excellent guide.

▣ Searching the Web

The World Wide Web provides access to vast amounts of information of many different sorts (O'Dochartaigh, 2002). You can search the holdings of other libraries and download the complete text of government reports, some conference papers, many books, and newspaper articles. You can find policies of local governments, descriptions of individual social scientists and particular research projects, and postings of advocacy groups. It's also hard to avoid finding a lot of information in which you have no interest, such as commercial advertisements, third-grade homework assignments, or college course syllabi. In November 2011 there were about 20 billion pages on the web (http://www.worldwidewebsize.com/).

After you are connected to the web with a browser like Microsoft Internet Explorer or Netscape Navigator, you can use three basic strategies for finding information: direct addressing—typing in the address, or URL, of a specific site; browsing—reviewing online lists of web sites; and searching—the most common approach. "Google" is currently the most popular search engine for searching the web. For some purposes, you will need to use only one strategy; for other purposes, you will want to use all three. End-of-chapter web exercises and the Sage Publications Study Site for this text both provide many URLs relevant to social science research.

Exhibit A.4 illustrates the first problem that you may encounter when searching the web: the sheer quantity of resources that are available. It is a much bigger problem than when searching bibliographic databases.

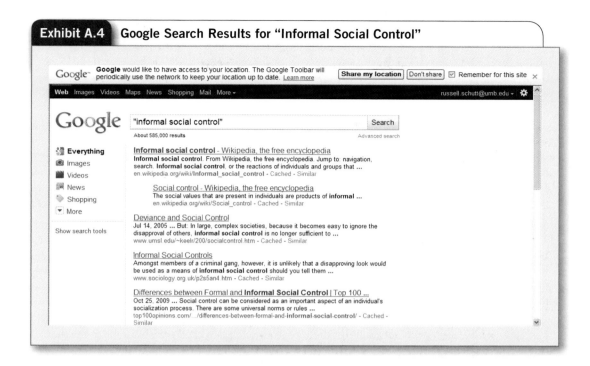

Exhibit A.4 **Google Search Results for "Informal Social Control"**

On the web, less is usually more. Limit your inspection of websites to the first few pages that turn up in your list (they're ranked by relevance). See what those first pages contain and then try to narrow your search by including some additional terms. Putting quotation marks around a phrase that you want to search will also help to limit your search—searching for "informal social control" on Google (on November 5, 2011) produced 585,000 sites, compared to the 6,060,000 sites retrieved when we omitted the quotes, so Google searched "informal" *and* "social" *and* "control."

Remember the following warnings when you conduct searches on the web:

- *Clarify your goals.* Before you begin the search, jot down the terms that you think you need to search for as well as a statement of what you want to accomplish with your search. This will help to ensure that you have a sense of what to look for and what to ignore.

- *Quality is not guaranteed.* Anyone can post almost anything, so the accuracy and adequacy of the information you find are always suspect. There's no journal editor or librarian to evaluate quality and relevance. You need to anticipate the different sources of information available on the web and to decide whether it is appropriate to use each of them for specific purposes. The sources you will find include:

- *Books*—Google is scanning the text of books that are out-of-print or no longer protected by copyright. In 2010, the total number of books scanned was over 15 million (out of perhaps 130 million books in the world) (http://en.wikipedia.org/wiki/Google_Books). When you search in Google Books, you will retrieve the pages in books that use the cited terms.

- *Newspaper articles*—These can range from local newspapers like the *Chicago Tribune* to national newspapers like the *New York Times.* Access to articles in these newspapers may be limited to subscribers.

- *Government policies*—You can find government policies and publications ranging from those done at the city or town level to those written by foreign governments.

- *Presented papers*—You may find the complete text of a formal presentation that was given at a meeting or conference.

- *Classroom lecture notes and outlines; listings from college catalogs*—These are pretty straightforward.

- *Commercial advertisements*—Advertising abounds on the web and it is especially prolific on search engine pages. Your search engine will even retrieve ads from the web and list them as results of your search! The boundaries between academic, nonprofit, and commercial information have become very porous, so you can't let your guard down.

- *Anticipate change.* Websites that are not maintained by stable organizations can come and go very quickly. Any search will result in attempts to link to some URLs that no longer exist.

- *One size does not fit all.* Different search engines use different procedures for indexing websites. Some attempt to be all-inclusive whereas others aim to be selective. As a result, you can get different results from different search engines (such as Google or Yahoo) even though you are searching for exactly the same terms.

- *Be concerned about generalizability.* You might be tempted to characterize police department policies by summarizing the documents you find at police department websites. But how many police departments are there? How many have posted their policies on the web? Are these policies representative of all police departments? In order to answer all these questions, you would have to conduct a research project just on the websites themselves.

- *Evaluate the sites.* There's a lot of stuff out there; so how do you know what's good? Some websites contain excellent advice and pointers on how to differentiate the good from the bad. You can find one example at http://olinuris.library.cornell.edu/ref/research/webeval.html.

- *Avoid web addiction.* Another danger of the extraordinary quantity of information available on the web is that one search will lead to another and to another and. . . . There are always more possibilities to explore and one more interesting source to check. Establish boundaries of time and effort to avoid the risk of losing all sense of proportion.

- *Cite your sources.* Using text or images from web sources without attribution is plagiarism. It is the same as copying someone else's work from a book or article and pretending that it is your own. Record the web address (URL), the name of the information provider, and the date on which you obtain material from the site. Include this information in a footnote to the material that you use in a paper.

Appendix B

Table of Random Numbers

Line/Col.	(1)	(2)	(3)	(4)	(5)	(6)	(7)	(8)	(9)	(10)	(11)	(12)	(13)	(14)
1	10480	15011	01536	02011	81647	91646	69179	14194	62590	36207	20969	99570	91291	90700
2	22368	46573	25595	85393	30995	89198	27982	53402	93965	34095	52666	19174	39615	99505
3	24130	48360	22527	97265	76393	64809	15179	24830	49340	32081	30680	19655	63348	58629
4	42167	93093	06243	61680	07856	16376	39440	53537	71341	57004	00849	74917	97758	16379
5	37570	39975	81837	16656	06121	91782	60468	81305	49684	60672	14110	06927	01263	54613
6	77921	06907	11008	42751	27756	53498	18602	70659	90655	15053	21916	81825	44394	42880
7	99562	72905	56420	69994	98872	31016	71194	18738	44013	48840	63213	21069	10634	12952
8	96301	91977	05463	07972	18876	20922	94595	56869	69014	60045	18425	84903	42508	32307
9	89579	14342	63661	10281	17453	18103	57740	84378	25331	12566	58678	44947	05585	56941
10	85475	36857	43342	53988	53060	59533	38867	62300	08158	17983	16439	11458	18593	64952
11	28918	69578	88231	33276	70997	79936	56865	05859	90106	31595	01547	85590	91610	78188
12	63553	40961	48235	03427	49626	69445	18663	72695	52180	20847	12234	90511	33703	90322
13	09429	93969	52636	92737	88974	33488	36320	17617	30015	08272	84115	27156	30613	74952
14	10365	61129	87529	85689	48237	52267	67689	93394	01511	26358	85104	20285	29975	89868
15	07119	97336	71048	08178	77233	13916	47564	81056	97735	85977	29372	74461	28551	90707
16	51085	12765	51821	51259	77452	16308	60756	92144	49442	53900	70960	63990	75601	40719
17	02368	21382	52404	60268	89368	19885	55322	44819	01188	65255	64835	44919	05944	55157
18	01011	54092	33362	94904	31273	04146	18594	29852	71585	85030	51132	01915	92747	64951
19	52162	53916	46369	58586	23216	14513	83149	98736	23495	64350	94738	17752	35156	35749
20	07056	97628	33787	09998	42698	06691	76988	13602	51851	46104	88916	19509	25625	58104
21	48663	91245	85828	14346	09172	30168	90229	04734	59193	22178	30421	61666	99904	32812

Line/Col.	(1)	(2)	(3)	(4)	(5)	(6)	(7)	(8)	(9)	(10)	(11)	(12)	(13)	(14)
22	54164	58492	22421	74103	47070	25306	76468	26384	58151	06646	21524	15227	96909	44592
23	32639	32363	05597	24200	13363	38005	94342	28728	35806	06912	17012	64161	18296	22851
24	29334	27001	87637	87308	58731	00256	45834	15398	46557	41135	10367	07684	36188	18510
25	02488	33062	28834	07351	19731	92420	60952	61280	50001	67658	32586	86679	50720	94953
26	81525	72295	04839	96423	24878	82651	66566	14778	76797	14780	13300	87074	79666	95725
27	29676	20591	68086	26432	46901	20849	89768	81536	86645	12659	92259	57102	80428	25280
28	00742	57392	39064	66432	84673	40027	32832	61362	98947	96067	64760	64584	96096	98253
29	05366	04213	25669	26422	44407	44048	37937	63904	45766	66134	75470	66520	34693	90449
30	91921	26418	64117	94305	26766	25940	39972	22209	71500	64568	91402	42416	07844	69618
31	00582	04711	87917	77341	42206	35126	74087	99547	81817	42607	43808	76655	62028	76630
32	00725	69884	62797	56170	86324	88072	76222	36086	84637	93161	76038	65855	77919	88006
33	69011	65797	95876	55293	18988	27354	26575	08625	40801	59920	29841	80150	12777	48501
34	25976	57948	29888	88604	67917	48708	18912	82271	65424	69774	33611	54262	85963	03547
35	09763	83473	73577	12908	30883	18317	28290	35797	05998	41688	34952	37888	38917	88050
36	91567	42595	27958	30134	04024	86385	29880	99730	55536	84855	29080	09250	79656	73211
37	17955	56349	90999	49127	20044	59931	06115	20542	18059	02008	73708	83317	36103	42791
38	46503	18584	18845	49618	02304	51038	20655	58727	28168	15475	56942	53389	20562	87338
39	92157	89634	94824	78171	84610	82834	09922	25417	44137	48413	25555	21246	35509	20468
40	14577	62765	35605	81263	39667	47358	56873	56307	61607	49518	89656	20103	77490	18062
41	98427	07523	33362	64270	01638	92477	66969	98420	04880	45585	46565	04102	46880	45709

(Continued)

(Continued)

Line/Col.	(1)	(2)	(3)	(4)	(5)	(6)	(7)	(8)	(9)	(10)	(11)	(12)	(13)	(14)
42	34914	63976	88720	82765	34476	17032	87589	40836	32427	70002	70663	88863	77775	69348
43	70060	28277	39475	46473	23219	53416	94970	25832	69975	94884	19661	72828	00102	66794
44	53976	54914	06990	67245	68350	82948	11398	42878	80287	88267	47363	46634	06541	97809
45	76072	29515	40980	07391	58745	25774	22987	80059	39911	96189	41151	14222	60697	59583
46	90725	52210	83974	29992	65831	38857	50490	83765	55657	14361	31720	57375	56228	41546
47	64364	67412	33339	31926	14883	24413	59744	92351	97473	89286	35931	04110	23726	51900
48	08962	00358	31662	25388	61642	34072	81249	35648	56891	69352	48373	45578	78547	81788
49	95012	68379	93526	70765	10593	04542	76463	54328	02349	17247	28865	14777	62730	92277
50	15664	10493	20492	38391	91132	21999	59516	81652	27195	48223	46751	22923	32261	85653
51	16408	81899	04153	53381	79401	21438	83035	92350	36693	31238	59649	91754	72772	02338
52	18629	81953	05520	91962	04739	13092	97662	24822	94730	06496	35090	04822	86772	98289
53	73115	35101	47498	87637	99016	71060	88824	71013	18735	20286	23153	72924	35165	43040
54	57491	16703	23167	49323	45021	33132	12544	41035	80780	45393	44812	12515	98931	91202
55	30405	83946	23792	14422	15059	45799	22716	19792	09983	74353	68668	30429	70735	25499
56	16631	35006	85900	98275	32388	52390	16815	69298	82732	38480	73817	32523	41961	44437
57	96773	20206	42559	78985	05300	22164	24369	54224	35083	19687	11052	91491	60383	19746
58	38935	64202	14349	82674	66523	44133	00697	35552	35970	19124	63318	29686	03387	59846
59	31624	76384	17403	53363	44167	64486	64758	75366	76554	31601	12614	33072	60332	92325
60	78919	19474	23632	27889	47914	02584	37680	20801	72152	39339	34806	08930	85001	87820
61	03931	33309	57047	74211	63445	17361	62825	39908	05607	91284	68833	25570	38818	46920
62	74426	33278	43972	10119	89917	15665	52872	73823	73144	88662	88970	74492	51805	99378

Line/Col.	(1)	(2)	(3)	(4)	(5)	(6)	(7)	(8)	(9)	(10)	(11)	(12)	(13)	(14)
63	09066	00903	20795	95452	92648	45454	09552	88815	16553	51125	79375	97596	16296	66092
64	42238	12426	87025	14267	20979	04508	64535	31355	86064	29472	47689	05974	52468	16834
65	16153	08002	26504	41744	81959	65642	74240	56302	00033	67107	77510	70625	28725	34191
66	21457	40742	29820	96783	29400	21840	15035	34537	33310	06116	95240	15957	16572	06004
67	21581	57802	02050	89728	17937	37621	47075	42080	97403	48626	68995	43805	33386	21597
68	55612	78095	83197	33732	05810	24813	86902	60397	16489	03264	88525	42786	05269	92532
69	44657	66999	99324	51281	84463	60563	79312	93454	68876	25471	93911	25650	12682	73572
70	91340	84979	46949	81973	37949	61023	43997	15263	80644	43942	89203	71795	99533	50501
71	91227	21199	31935	27022	84067	05462	35216	14486	29891	68607	41867	14951	91696	85065
72	50001	38140	66321	19924	72163	09538	12151	06878	91903	18749	34405	56087	82790	70925
73	65390	05224	72958	28609	81406	39147	25549	48542	42627	45233	57202	94617	23772	07896
74	27504	96131	83944	41575	10573	08619	64482	73923	36152	05184	94142	25299	84387	34925
75	37169	94851	39117	89632	00959	16487	65536	49071	39782	17095	02330	74301	00275	48280
76	11508	70225	51111	38351	19444	66499	71945	05422	13442	78675	84081	66938	93654	59894
77	37449	30362	06694	54690	04052	53115	62757	95348	78662	11163	81651	50245	34971	52924
78	46515	70331	85922	38329	57015	15765	97161	17869	45349	61796	66345	81073	49106	79860
79	30986	81223	42416	58353	21532	30502	32305	86482	05174	07901	54339	58861	74818	46942
80	63798	64995	46583	09765	44160	78128	83991	42865	92520	83531	80377	35909	81250	54238
81	82486	84846	99254	67632	43218	50076	21361	64816	51202	88124	41870	52689	51275	83556
82	21885	32906	92431	09060	64297	51674	64126	62570	26123	05155	59194	52799	28225	85762
83	60336	98782	07408	53458	13564	59089	26445	29789	85205	41001	12535	12133	14645	23541

(Continued)

(Continued)

Line/Col.	(1)	(2)	(3)	(4)	(5)	(6)	(7)	(8)	(9)	(10)	(11)	(12)	(13)	(14)
84	43937	46891	24010	25560	86355	33941	25786	54990	71899	15475	95434	98227	21824	19585
85	97656	63175	89303	16275	07100	92063	21942	18611	47348	20203	18534	03862	78095	50136
86	03299	01221	05418	38982	55758	92237	26759	86367	21216	98442	08303	56613	91511	75928
87	79626	06486	03574	17668	07785	76020	79924	25651	83325	88428	85076	72811	22717	50585
88	85636	68335	47539	03129	65651	11977	02510	26113	99447	68645	34327	15152	55230	93448
89	18039	14367	61337	06177	12143	46609	32989	74014	64708	00533	35398	58408	13261	47908
90	08362	15656	60627	36478	65648	16764	53412	09013	07832	41574	17639	82163	60859	75567
91	79556	29068	04142	16268	15387	12856	66227	38358	22478	73373	88732	09443	82558	05250
92	92608	82674	27072	32534	17075	27698	98204	63863	11951	34648	88022	56148	34925	57031
93	23982	25835	40055	67006	12293	02753	14827	22235	35071	99704	37543	11601	35503	85171
94	09915	96306	05908	97901	28395	14186	00821	80703	70426	75647	76310	88717	37890	40129
95	50937	33300	26695	62247	69927	76123	50842	43834	86654	70959	79725	93872	28117	19233
96	42488	78077	69882	61657	34136	79180	97526	43092	04098	73571	80799	76536	71255	64239
97	46764	86273	63003	93017	31204	36692	40202	35275	57306	55543	53203	18098	47625	88684
98	03237	45430	55417	63282	90816	17349	88298	90183	36600	78406	06216	95787	42579	90730
99	86591	81482	52667	61583	14972	90053	89534	76036	49199	43716	97548	04379	46370	28672
100	38534	01715	94964	87288	65680	43772	39560	12918	86537	62738	19636	51132	25739	56947

References

Adler, Patricia A. 1993. *Wheeling and dealing: An ethnography of an upper-level drug dealing and smuggling community,* 2nd ed. New York: Columbia University Press.

Adler, Patricia A. and Peter Adler. 2000. Intense loyalty in organizations: A case study of college athletics. Pp. 31–50 in *Qualitative studies of organizations,* ed. John Van Maanen. Thousand Oaks, CA: Sage.

Adorno, Theodor W., Nevitt Sanford, Else Frenkel-Brunswik, and Daniel Levinson. 1950. *The authoritarian personality.* New York: Harper.

Alfred, Randall. 1976. The church of Satan. Pp. 180–202 in *The new religious consciousness,* ed. Charles Glock and Robert Bellah. Berkeley: University of California Press.

Altheide, David L. and John M. Johnson. 1994. Criteria for assessing interpretive validity in qualitative research. Pp. 485–199 in *Handbook of qualitative research,* ed. Norman K. Denzin and Yvonna S. Lincoln. Thousand Oaks, CA: Sage.

Altman, Lawrence K. 1987. U.S. and France end rift on AIDS. *New York Times,* April 1. From http://www.nytimes.com (accessed May 28, 2007).

Altman, Lawrence K. 2008. Discoverers of AIDS and cancer viruses win Nobel. *New York Times,* October 7, 2008. From http://www.nytimes.com/2008/10/07/health/07nobel.htm (accessed October 23, 2008).

American Sociological Association (ASA). 1997. *Code of ethics.* Washington, DC: American Sociological Association.

Anderson, Elijah. 2000. *Code of the street: Decency, violence, and the moral life of the inner city,* reprint ed. New York: W. W. Norton.

Anderson, Elijah. 2003. Jelly's place: An ethnographic memoir (distinguished lecture). *Symbolic Interaction* 26(2): 217–237.

Anspach, Renee R. 1991. Everyday methods for assessing organizational effectiveness. *Social Problems* 38 (February): 1–19.

Aronson, Elliot and Judson Mills. 1959. The effect of severity of initiation on liking for a group. *Journal of Abnormal and Social Psychology* 59 (September): 177–181.

Arwood, Tracy and Sangeeta Panicker. 2007. *Assessing risk in social and behavioral sciences.* From Collaborative Institutional Training Initiative website: https://www.citiprogram.org/members/learners/ (accessed June 5, 2008).

Aseltine, Robert H., Jr. and Ronald C. Kessler. 1993. Marital disruption and depression in a community sample. *Journal of Health and Social Behavior* 34 (September): 237–251.

Associated Press. 2000. Researchers fear privacy breaches with online research. From http://www.pi.edu/News/TechNews/000919/onlineresearch.html (accessed September 28, 2008).

Ayres, Ian and Peter Siegelman. 1995. Race and gender discrimination in bargaining for a new car. *American Economic Review* 85(3): 304–321.

Ayres, Ian, Frederick E. Vars, and Nasser Zakariya. 2005. To insure prejudice: Racial disparities in taxicab tipping. *Yale Law Journal* 114: 1613.

Bachman, Ronet and Russell K. Schutt. 2007. *The practice of research in criminology and criminal justice,* 3rd ed. Thousand Oaks, CA: Sage.

Baumrind, Diana. 1964. Some thoughts on ethics of research: After reading Milgram's "Behavioral study of obedience." *American Psychologist* 19: 421–423.

Baumrind, Diana. 1985. Research using intentional deception: Ethical issues revisited. *American Psychologist* 40: 165–174.

Becker, Howard S. 1958. Problems of inference and proof in participant observation. *American Sociological* Review 23: 652–660.

Becker, Howard S. 1963. *Outsiders: Studies in the sociology of deviance.* New York: Free Press.

Becker, Howard S. 1986. *Writing for social scientists.* Chicago: University of Chicago Press. [This can be ordered directly from the American Sociological Association, 1722 N St. NW, Washington, DC 20036, (202) 833–3410.]

Bellah, Robert N., Richard Madsen, William M. Sullivan, Ann Swidler, and Steven M. Tipton. 1985. *Habits of the heart: Individualism and commitment in American life.* New York: Harper & Row.

Bendix, Reinhard. 1962. *Max Weber: An intellectual portrait.* Garden City, NY: Doubleday/Anchor.

Binder, Arnold and James W. Meeker. 1993. Implications of the failure to replicate the Minneapolis experimental findings. *American Sociological Review* 58: 886–888.

Birkeland, Sarah, Erin Murphy-Graham, and Carol Weiss. 2005. Good reasons for ignoring good evaluation: The case of the Drug Abuse Resistance Education (D.A.R.E.) program. *Evaluation and Program Planning* 28: 247–256.

Bogdewic, Stephan P. 1999. Participant observation. Pp. 47–70 in *Doing qualitative research,* 2nd ed., ed. Benjamin F. Crabtree and William L. Miller. Thousand Oaks, CA: Sage.

Booth, Wayne C., Gregory G. Colomb, and Joseph M. Williams. 1995. *The craft of research.* Chicago: University of Chicago Press.

Boruch, Robert F. 1997. *Randomized experiments for planning and evaluation: A practical guide.* Thousand Oaks, CA: Sage.

Bramel, Dana and Ronal Friend. 1981. Hawthorne, the myth of the docile worker, and class bias in psychology. *American Psychologist* 38 (September): 867–878.

Brewer, John and Albert Hunter. 1989. *Multimethod research: A synthesis of styles.* Newbury Park, CA: Sage.

Brown v. Board of Education (Brown I), 347 U.S. 483 (1954).

Butterfield, Fox. 1996a. After 10 years, juvenile crime begins to drop. *New York Times,* August 9, A1, A25.

Butterfield, Fox. 1996b. Gun violence may be subsiding, studies find. *New York Times,* October 14, A10.

Butterfield, Fox. 2000. As murder rates edge up, concern, but few answers. *New York Times,* June 18, A12.

Buzawa, Eve S. and Carl G. Buzawa, eds. 1996. *Do arrests and restraining orders work?* Thousand Oaks, CA: Sage.

Campbell, Donald T. and M. Jean Russo. 1999. *Social experimentation.* Thousand Oaks, CA: Sage.

Campbell, Donald T. and Julian C. Stanley. 1966. *Experimental and quasi-experimental designs for research.* Chicago: Rand McNally.

Campbell, Richard T. 1992. Longitudinal research. Pp. 1146–1158 in *Encyclopedia of sociology,* ed. Edgar F. Borgatta and Marie L. Borgatta. New York: Macmillan.

Cava, Anita, Reid Cushman, and Kenneth Goodman. 2007. *HIPAA and human subjects research.* From the University of Miami website: http://researchedu .med.miami.edu/x36.xml (accessed September 28, 2008).

Cave, Emma and Soren Holm. 2003. Milgram and Tuskegee: Paradigm research projects in bioethics. *Health Care Analysis* 11: 27–40.

Center for Survey Research, University of Massachusetts at Boston. 1987. Methodology: Designing good survey questions. *Newsletter,* April 3.

Chambliss, Daniel F. 1988. *Champions: The making of Olympic swimmers.* New York: Morrow.

Chambliss, Daniel F. 1989. The mundanity of excellence: An ethnographic report on stratification and Olympic swimmers. *Sociological Theory* 7(1): 70–86.

Chambliss, Daniel F. 1996. *Beyond caring: Hospitals, nurses, and the social organization of ethics.* Chicago: University of Chicago Press.

Chen, Huey-Tsyh. 1990. *Theory-driven evaluations.* Newbury Park, CA: Sage.

Chen, Huey-Tsyh and Peter H. Rossi. 1987. The theory-driven approach to validity. *Evaluation and Program Planning* 10: 95–103.

Coffey, Amanda and Paul Atkinson. 1996. *Making sense of qualitative data: Complementary research strategies.* Thousand Oaks, CA: Sage.

Cohen, Gary E. and Barbara A. Kerr. 1998. Computer-mediated counseling: An empirical study of a new mental health treatment. *Computers in Human Services* 15: 13–26.

Cohen, Susan G. and Gerald E. Ledford Jr. 1994. The effectiveness of self-managing teams: A quasi-experiment. *Human Relations* 47: 13–43.

Coleman, James S. and Thomas Hoffer. 1987. *Public and private high schools: The impact of communities.* New York: Basic Books.

Coleman, James S., Thomas Hoffer, and Sally Kilgore. 1982. *High school achievement: Public, Catholic, and private schools compared.* New York: Basic Books.

Collins, Randall. 1975. *Conflict sociology: Toward an explanatory science.* New York: Academic Press.

Converse, Jean M. 1984. Attitude measurement in psychology and sociology: The early years. Pp. 3–40 in *Surveying subjective phenomena,* vol. 2, ed. Charles F. Turner and Elizabeth Martin. New York: Russell Sage Foundation.

Cook, Thomas D. and Donald T. Campbell. 1979. *Quasi-experimentation: Design and analysis issues for field settings.* Chicago: Rand McNally.

Coontz, Stephanie. 1997. *The way we really are: Coming to terms with America's changing families.* New York: Basic Books.

Cooper, Kathleen B. and Michael D. Gallagher. 2004. *A nation online: Entering the broadband age.* Washington, DC: Economics and Statistics Administration and National Telecommunications and Information Administration, U.S. Department of Commerce.

Core Institute. 1994. *Core alcohol and drug survey: Long form.* Carbondale, IL: Fund for the Improvement of Postsecondary Education (FIPSE) Core Analysis Grantee Group, Core Institute, Student Health Programs, Southern Illinois University.

Costner, Herbert L. 1989. The validity of conclusions in evaluation research: A further development of Chen and Rossi's theory-driven approach. *Evaluation and Program Planning* 12: 345–353.

Couper, Mick P. 2000. Web surveys: A review of issues and approaches. *Public Opinion Quarterly* 64: 464–494.

Couper, Mick P. and Peter V. Miller. 2008. Web survey methods: Introduction. *Public Opinion Quarterly* 72: 831–835.

Cross, Harry, Genevieve Kenney, Jane Mell, and Wendy Zimmerman. 1990. *Employer hiring practices: Differential treatment of Hispanic and Anglo job seekers.* Washington DC: Urban Institute Press.

Curtin, Presser, and Singer. "Changes in Telephone in Telephone Survey Nonresponse over the past quarter century." *Public Opinion Quarterly,* 69:87–98

Dannefer, W. Dale and Russell K. Schutt. 1982. Race and juvenile justice processing in court and police agencies. *American Journal of Sociology* 87(March): 1113–1132.

D.A.R.E. 2008. *The "New" D.A.R.E. Program.* From www.dare.com/newdare.asp (accessed May 31, 2008).

Davies, Philip, Anthony Petrosino, and Iain Chalmers. 1999. *Report and papers from the exploratory meeting for the Campbell Collaboration.* London: School of Public Policy, University College.

Davis, James A. 1985. *The logic of causal order.* Sage University Paper Series on Quantitative Applications in the Social Sciences (series no. 07-055). Beverly Hills, CA: Sage.

Davis, James A. and Tom W. Smith. 1992. *The NORC General Social Survey: A user's guide.* Newbury Park, CA: Sage.

Dawes, Robyn. 1995. How do you formulate a testable exciting hypothesis? Pp. 93–96 in *How to write a successful research grant application: A guide for social and behavioral scientists,* ed. Willo Pequegnat and Ellen Stover. New York: Plenum Press.

Decker, Scott H. and Barrik Van Winkle. 1996. *Life in the gang: Family, friends, and violence.* New York: Cambridge University Press.

Dentler, Robert A. 2002. *Practicing sociology: Selected fields.* Westport, CT: Praeger.

Denzin, Norman K. 2002. The interpretive process. Pp. 349–368 in *The qualitative researcher's companion,* ed. A. Michael Huberman and Matthew B. Miles. Thousand Oaks, CA: Sage.

Denzin, Norman K. and Yvonna S. Lincoln. 1994. Introduction: Entering the field

of qualitative research. Pp. 1–17 in *Handbook of qualitative research,* ed. Norman K. Denzin and Yvonna S. Lincoln. Thousand Oaks, CA: Sage.

Denzin, Norman K. and Yvonna S. Lincoln. 2000. Introduction: The discipline and practice of qualitative research. Pp. 1–28 in *Handbook of qualitative research,* 2nd ed., ed. Norman Denzin and Yvonna S. Lincoln. Thousand Oaks, CA: Sage.

DeParle, Jason. 1999. Project to rescue needy stumbles against the persistence of poverty. *New York Times,* May 15, A 1, A10.

Diamond, Timothy. 1992. *Making gray gold: Narratives of nursing home care.* Chicago: University of Chicago Press.

Dillman, Don A. 1978. Mail and telephone surveys: *The total design method.* New York: Wiley.

Dillman, Don A. 1982/1991. Mail and other self-administered questionnaires. Pp. 637–638 in *Handbook of research design and social measurement,* 5th ed., ed. Delbert C. Miller. Newbury Park, CA: Sage.

Dillman, Don A. 2000. *Mail and Internet surveys: The tailored design method,* 2nd ed. New York: John Wiley & Sons.

Dillman, Don A. 2007. *Mail and Internet surveys: The tailored design method,* 2nd ed. Update with new Internet, visual, and mixed-mode guide. Hoboken, NJ: Wiley.

Donath, Judith S. 1999. Identity and deception in the virtual community. Pp. 29–59 in *Communities in cyberspace,* ed. Peter Kollock and Marc A. Smith. New York: Routledge.

Drake, Robert E., Gregory J. McHugo, Deborah R. Becker, William A. Anthony, and Robin E. Clark. 1996. The New Hampshire study of supported employment for people with severe mental illness. *Journal of Consulting and Clinical Psychology* 64: 391–399.

Duneier, Mitchell. 2006. Ethnography, the ecological fallacy, and the 1995 Chicago heat wave. *American Sociological Review* 71(4): 679–687.

Durkheim, Emile. 1951. *Suicide.* New York: Free Press.

Durkheim, Emile. 1984. *The division of labor in society.* Translated by W. D. Halls. New York: Free Press.

Ellis, Carolyn. 1986. *Fisher folk: Two communities on Chesapeake Bay.* Lexington: University Press of Kentucky.

Elliott, Andrew and Daniela Nesta. 2008. Romantic red: Red enhances men's attraction to women. *Journal of Personality and Social Psychology* 95(5): 1150–1164.

Emerson, Robert M., ed. 1983. *Contemporary field research.* Prospect Heights, IL: Waveland Press.

Emerson, Robert M., Rachel I. Fretz, and Linda L. Shaw. 1995. *Writing ethnographic fieldnotes.* Chicago: University of Chicago Press.

Erikson, Kai T. 1967. A comment on disguised observation in sociology. *Social Problems* 12: 366–373.

Erikson, Kai T. 1976. *Everything in its path: Destruction of community in the Buffalo Creek flood.* New York: Simon & Schuster.

Estabrook, Robin E., Russell K. Schutt, and Mary Lou Woodford. 2008. Translating research into practice: The participatory expert panel approach. *The Open Health Services and Policy Journal* 1: 19–26.

Favreault, Melissa. 2008. Discrimination and economic mobility. *The Urban Institute* April 3.

Fenno, Richard F., Jr. 1978. *Home style: House members in their districts.* Boston: Little, Brown.

Fink, Arlene. 2005. *Conducting research literature reviews: From the Internet to paper,* 2nd ed. Thousand Oaks, CA: Sage.

Fowler, Floyd J. 1988. *Survey research methods,* rev. ed. Newbury Park, CA: Sage.

Fowler, Floyd J. 1995. *Improving survey questions: Design and evaluation.* Thousand Oaks, CA: Sage.

Fowler, Floyd J. 1998. Personal communication, January 7. Center for Survey

Research, University of Massachusetts, Boston.

Fox, Nick and Chris Roberts. 1999. GPs in cyberspace: The sociology of a "virtual community." *The Sociological Review* 47:643–669.

Franklin, Mark N. 1996. Electoral participation. Pp. 216–235 in *Comparing Democracies: Elections and Voting in Global Perspective*, ed. Lawrence LeDuc, Richard G. Niemi, and Pippa Norris. Thousand Oaks, CA: Sage.

Freud, Sigmund. 1900/1999. *The interpretation of dreams*, 1st ed., ed. Ritchie Robertson, trans. Joyce Crick. Repr. Oxford, England: Oxford University Press.

Gaiser, Ted J. and Anthony E. Schreiner. 2009. *A guide to conducting online research.* Thousand Oaks, CA: Sage.

Gallup Organization. 2002. *Poll analyses, July 29, 2002: Bush job approval update.* From http://media.washingtonpost.com/wp-srv/politics/ssi/polls/postpoll_072307.html (accessed September 28, 2008).

Gallup Organization. 2008. *Election polls—Accuracy record in presidential elections.* From http://www.galIup.com/poll/9442/Election-Polls-Accuracy-Record-Presidential-EIections.aspx (accessed May 23, 2008).

Gallup. 2011. *Election Polls—Accuracy Record in Presidential Elections.* Retrieved March 17, 2011, from http://www.gallup.com/poll/9442/Election-Polls-Accuracy-Record-Presidential-Elections.aspx?version=print

Gibson, David R. 2005. Taking turns and talking ties: Networks and conversational interaction. *American Journal of Sociology* 110(6): 1561–1597.

Gilbert, Dennis (with Zogby International). 2002. *Hamilton College Muslim America poll.* Unpublished research report.

Gilchrist, Valerie J. and Robert L. Williams. 1999. Key informant interviews. Pp. 71–88 in *Doing qualitative research*, 2nd ed., ed. Benjamin F. Crabtree and William L. Miller. Thousand Oaks, CA: Sage.

Glaser, Barney G. and Anselm L. Strauss. 1967. *The discovery of grounded theory: Strategies for qualitative research.* London: Weidenfeld and Nicholson.

Goffman, Erving. 1959. *Presentation of self in everyday life.* New York: Doubleday Anchor.

Goffman, Erving. 1961. *Asylums: Essays on the social situation of mental patients and other inmates.* Garden City, NY: Doubleday.

Goffman, Erving, 1967. *Interaction Ritual.* New York: Doubleday Anchor Books.

Goldfinger, Stephen M. and Russell K. Schutt. 1996. Comparisons of clinicians' housing recommendations of homeless mentally ill persons. *Psychiatric Services* 47(4): 413–415.

Goleman, Daniel. 1993. Placebo effect is shown to be twice as powerful as expected. *New York Times*, August 17, C3.

Goode, Erich. 2002. Sexual involvement and social research in a fat civil rights organization. *Qualitative Sociology* 25(4): 501–504.

Gordon, Raymond. 1992. *Basic interviewing skills.* Itasca, IL: Peacock.

Governmental Accounting Standards Board (GASB). 2008. *Basic facts about service efforts and accomplishments reporting.* From http://www.gasb.org/SEA_fact_sheet_FINAL.pdf (accessed October 23, 2008).

Grossman, Lev. 2010. "Mark Zuckerberg." *Time*, December 15. From http://www.time.com/time/specials/packages/article/0,28804,2036683_2037183_2037185,00.html (accessed March 14, 2011).

Groves, Robert M. 1989. *Survey errors and survey costs.* New York: Wiley.

Groves, Robert M. and Mick P. Couper. 1998. *Nonresponse in household interview surveys.* New York: Wiley.

Groves, Robert M. and Robert L. Kahn. 1979/1991. *Surveys by telephone: A national comparison with personal interviews,* as adapted in Delbert C. Miller's *Handbook of research design and social measurement,* 5th ed. Newbury Park, CA: Sage.

Guba, Egon G. and Yvonna S. Lincoln. 1989. *Fourth generation evaluation.* Newbury Park, CA: Sage.

Gubrium, Jaber F. and James A. Holstein. 1997. *The new language of qualitative method.* New York: Oxford University Press.

Gubrium, Jaber F. and James A. Holstein. 2000. Analyzing interpretive practice. Pp. 487–508 in *The handbook of qualitative research,* 2nd ed., ed. Norman Denzin and Yvonna S. Lincoln. Thousand Oaks, CA: Sage.

Hadaway, C. Kirk, Penny Long Marler, and Mark Chaves. 1993. What the polls don't show: A closer look at U.S. church attendance. *American Sociological Review 58* (December): 741–752.

Hafner, Katie. 2005. In challenge to Google, Yahoo will scan books. *New York Times,* October 3, C1, C4.

Hagan, John. 1994. *Crime and disrepute.* Thousand Oaks, CA: Pine Forge Press.

Hage, Jerald and Barbara Foley Meeker. 1988. *Social causality.* Boston: Unwin Hyman.

Hampton, Keith N. and Neeti Gupta. 2008. Community and social interaction in the wireless city: Wi-Fi use in public and semi-public spaces. *New Media & Society* 10(6): 831–850.

Hampton, Keith, Oren Livio and Lauren Sessions Goulet. 2010. The social life of wireless urban spaces: Internet use, social networks, and the public realm. *Journal of Communication* 60(4): 701–722.

Haney, C., C. Banks, and Philip G. Zimbardo. 1973. Interpersonal dynamics in a simulated prison. *International Journal of Criminology and Penology* 1: 69–97.

Hargittai, E., 2007. Whose space? Among users and non-users of social network sites. *Journal of Computer-Mediated Communication* 13(1), article 14.

Hart, Chris. 1998. *Doing a literature review: Releasing the social science research imagination.* London: Sage.

Heckman, James J. and Peter Siegelman. 1993. The Urban Institute Audit Studies: Their methods and findings. Pp. 187–258 in *Clean and convincing evidence: Measurement of discrimination in America,* ed. M. Fix and R. J. Struyk. Washington DC: Urban Institute Press.

Herek, Gregory. 1995. Developing a theoretical framework and rationale for a research proposal. Pp. 85–91 in *How to write a successful research grant application: A guide for social and behavioral scientists,* ed. Willo Pequegnat and Ellen Stover. New York: Plenum Press.

Hite, Shere. 1987. *Women and love: A cultural revolution in progress.* New York: Alfred A. Knopf.

Hoyle, Carolyn and Andrew Sanders. 2000. Police response to domestic violence: From victim choice to victim empowerment. *British Journal of Criminology* 40: 14–26.

HRAF. 2005. *eHRAF collection of ethnography: Web.* New Haven, CT: Yale University. From www.yale.edu/hraf/collections_body_ethnoweb.htm (accessed July 3, 2005).

Huberman, A. Michael and Matthew B. Miles. 1994. Data management and analysis methods. Pp. 428–444 in *Handbook of qualitative research,* ed. Norman K. Denzin and Yvonna S. Lincoln. Thousand Oaks, CA: Sage.

Huff, Darrell. 1954. *How to lie with statistics.* New York: W. W. Norton.

Humphrey, Nicholas. 1992. A *history of the mind: Evolution and the birth of consciousness.* New York: Simon & Schuster.

Humphreys, Laud. 1970. *Tearoom trade: Impersonal sex in public places.* Chicago: Aldine.

Hunt, Morton. 1985. *Profiles of social research: The scientific study of human interactions.* New York: Russell Sage Foundation.

Internetworldstats.com. 2011. *Internet world stats: Usage and population statistics.* From http://www.internetworldstats.com/ (accessed March 19, 2011).

Inter-University Consortium for Political and Social Research. 1996. *Guide to resources and services 1995–1996.* Ann Arbor, MI: ICPSR.

Irvine, Leslie. 1998. Organizational ethics and field work realities: Negotiating ethical boundaries in codependents anonymous. Pp. 167–183 in *Doing ethnographic research: Fieldwork settings,* ed. Scott Grills. Thousand Oaks, CA: Sage.

James, Nalita and Hugh Busher. 2009. *Online interviewing.* Thousand Oaks, CA: Sage.

Jankowski, Martin Sanchez. 1991. *Islands in the street: Gangs and American urban society.* Berkeley: University of California Press.

Jones, James H. 1993. *Bad blood: The Tuskegee syphilis experiment.* New York: Free Press.

Kagay, Michael R. (with Janet Elder). 1992. Numbers are no problem for pollsters. Words are. *New York Times,* October 9, E5.

Kale-Lostuvali, Elif. 2007. Negotiating state provision: state-citizen encounters in the aftermath of the İzmit earthquake. *The Sociological Quarterly* 48: 745–767.

Kaufman, Sharon R. 1986. *The ageless self: Sources of meaning in late life.* Madison: University of Wisconsin Press.

Kenney, Charles. 1987. They've got your number. *Boston Globe Magazine,* August 30, 12, 46–56, 60.

Kershaw, David and Jerilyn Fair, eds. 1976. *Operations, surveys, and administration, vol. 2 of The New Jersey income-maintenance experiment.* New York: Academic Press.

King, Nigel and Christine Horrocks. 2010. *Interviews in qualitative research.* Thousand Oaks, CA: Sage.

King, Gary, Robert O. Keohane, and Sidney Verba. 1994. *Scientific inference in qualitative research.* Princeton, NJ: Princeton University Press.

Kinsey, Alfred C., Wardell B. Pomeroy, and Clyde E. Martin. 1948. *Sexual behavior in the human male.* Philadelphia: W. B. Saunders.

Kinsey, Alfred C., Wardell B. Pomeroy, Clyde E. Martin, and Paul H. Gebhard. 1953. *Sexual behavior in the human female.* Philadelphia: W. B. Saunders.

Klinenberg, Eric. 2006. Blaming the victims: Hearsay, labeling, and the hazards of quick-hit disaster ethnography. *American Sociological Review* 71(4): 690–698.

Koegel, Paul. 1987. *Ethnographic perspectives on homeless and homeless mentally ill women.* Washington, DC: Alcohol, Drug Abuse, and Mental Health Administration, Public Health Service, U.S. Department of Health and Human Services.

Korn, James H. (1997). *Illusions of reality: A history of deception in social psychology.* Albany: State University of New York Press.

Kozinets, Robert V. 2010. *Netnography: Doing ethnographic research online.* Thousand Oaks, CA: Sage.

Krauss, Clifford. 1996. New York crime rate plummets to levels not seen in 30 years. *New York Times,* December 20, A1, B4.

Kreuter, Frauke, Stanley Presser, and Roger Tourangeau. 2008. Social desirability bias in CATI, IVR, and web surveys: The effects of mode and question sensitivity. *Public Opinion Quarterly* 72: 847–865.

Krueger, Richard A. 1988. *Focus groups: A practical guide for applied research.* Newbury Park, CA: Sage.

Krueger, Richard A. and Mary Anne Casey. 2000. *Focus groups: A practical guide for applied research,* 3rd ed. Thousand Oaks, CA: Sage.

Kuzel, Anton J. 1999. Sampling in qualitative inquiry. Pp. 33–45 in *Doing qualitative research.* 2nd ed., ed. Benjamin F. Crabtree and William L. Miller. Thousand Oaks, CA: Sage.

Kvale, Steinar. 1996. *Interviews: An introduction to qualitative research interviewing.* Thousand Oaks, CA: Sage.

Labaw, Patricia J. 1980. *Advanced questionnaire design.* Cambridge, MA: ABT Books.

Larson, Calvin J. 1993. *Pure and applied sociological theory: Problems and issues.* New York: Harcourt Brace Jovanovich.

Latour, Francie. 2002. Marching orders: After 10 years, state closes prison boot camp. *Boston Sunday Globe,* June 16, Bl, B7.

Lavrakas, Paul J. 1987. *Telephone survey methods: Sampling, selection, and supervision.* Newbury Park, CA: Sage.

LeDuc, Lawrence, Richard G. Niemi, and Pippa Norris (Eds.). 1996. *Comparing democracies: Elections and voting in global perspective.* Thousand Oaks, CA: Sage.

Legal Information Institute. 2006. *U.S. Code Collection: Title 42, Chapter 7, Subchapter XI, Part C, §1320d.* From the Cornell University Law School website: http://www4.1aw.cornell.edu/uscode/html/uscode42/usc_sec_42_0000I320—d000-.html (accessed September 28, 2008).

Lelieveldt, Herman. 2003. March-April. *Increasing social capital through direct democracy? A case study of the "It's Our Neighbourhood's Turn" project.* Paper presented at the ECPR-Joint Sessions (Workshop 22: Bringing citizens back in: participatory democracy and political participation), Edinburgh, Scotland. From http://www.paltin.ro/biblioteca/Lelieveldt.pdf (accessed September 28, 2008).

Lelieveldt, Herman. 2004. Helping citizens help themselves: Neighborhood improvement programs and the impact of social networks, trust, and norms on neighborhood-oriented forms of participation. *Urban Affairs Review* 39: 531–551.

Lempert, Richard. 1989. Humility is a virtue: On the publicization of policy-relevant research. *Law & Society Review* 23(1): 145–161.

Lempert, Richard and Joseph Sanders. 1986. An invitation to law and social science: *Desert, disputes, and distribution.* New York: Longman.

Levy, Paul S. and Stanley Lemeshow. 1999. *Sampling of populations: Methods and applications,* 3rd ed. New York: Wiley.

Lewin, Tamar. 2001a. Surprising result in welfare-to-work studies. *New York Times,* July 31, A16.

Lewin, Tamar. 2001b. Income education is found to lower risk of new arrest. *New York Times,* November 16, A18.

Lewis, Kevin, Nicholas Christakis, Marco Gonzalez, Jason Kaufman, and Andreas Wimmer. 2008. Tastes, ties, and time: A new social network dataset using Facebook.com. *Social Networks* 30(4): 330–342.

Lieberson, Stanley. 1985. *Making it count: The improvement of social research and theory.* Berkeley: University of California Press.

Ling, Rich and Gitte Stald. 2010. Mobile communities: Are we talking about a village, a clan, or a small group? *American Behavioral Scientist* 53(8): 113–1147.

Lipset, Seymour Martin. 1990. *Continental divide: The values and institutions of the United States and Canada.* London: Routledge.

Litwin, Mark S. 1995. *How to measure survey reliability and validity.* Thousand Oaks, CA: Sage.

Locke, Lawrence F., Stephen J. Silverman, and Waneen Wyrick Spirduso. 1998. *Reading and understanding research.* Thousand Oaks, CA: Sage.

Locke, Lawrence F., Waneen Wyrick Spirduso, and Stephen J. Silverman. 2000. *Proposals That Work: A Guide for Planning Dissertations and Grant Proposals,* 4th ed. Thousand Oaks, CA: Sage.

Luker, Kristin. 1985. *Abortion and the politics of motherhood.* Berkeley: University of California Press.

Mangione, Thomas W. 1995. *Mail surveys: Improving the quality.* Thousand Oaks, CA: Sage.

Margolis, Eric. 2004. Looking at discipline, looking at labour: Photographic representations of Indian boarding schools. *Visual Studies* 19: 72–96.

Marini, Margaret Mooney and Burton Singer. 1988. Causality in the social sciences. Pp. 347–409 in *Sociological methodology,* vol. 18, ed. Clifford C. Clogg. Washington, DC: American Sociological Association.

Markoff, John. 2005. Transitions to democracy. Pp. 384–403 in *The handbook of political sociology: States, civil societies, and globalization,* ed. Thomas Janoski, Robert R. Alford, Alexander M. Hicks, and Mildred A. Schwartz. New York: Cambridge University Press.

Marshall, S. L. A. 1947/1978. Men against fire. Repr. Gloucester, MA: Peter Smith.

Martin, Lawrence L. and Peter M. Kettner. 1996. *Measuring the performance of human service programs.* Thousand Oaks, CA: Sage.

Maxwell, Joseph A. 1996. *Qualitative research design: An interactive approach.* Thousand Oaks, CA: Sage.

McPherson, Miller, Lynn Smith-Lovin, and Matthew E. Brashears. 2006. Social isolation in America: Changes in core discussion networks over two decades. *American Sociological Review* 71: 353–375.

Mead, Margaret. 1928/2001. *Coming of age in Samoa.* Repr. New York: HarperCollins Perennial.

Melbin, Murray. 1978. Night as frontier. *American Sociological Review* 43(1): 3–22.

Milbrath, Lester and M. L. Goel. 1977. *Political participation,* 2nd ed. Chicago: Rand McNally.

Miles, Matthew B. and A. Michael Huberman. 1994. *Qualitative data analysis,* 2nd ed. Thousand Oaks, CA: Sage.

Milgram, Stanley. 1963. Behavioral study of obedience. *Journal of Abnormal and Social Psychology* 67: 371–478.

Milgram, Stanley. 1964. Issues in the study of obedience: A reply to Baumrind. *American Psychologist* 19: 848–852.

Milgram, Stanley. 1965. Some conditions of obedience and disobedience to authority. *Human Relations* 18: 57–76.

Milgram, Stanley. 1974. *Obedience to authority: An experimental view.* New York: Harper & Row.

Miller, Susan. 1999. *Gender and community policing: Walking the talk.* Boston: Northeastern University Press.

Miller, Arthur G. 1986. *The obedience experiments: A case study of controversy in social science.* New York: Praeger.

Miller, Delbert C. 1991. *Handbook of research design and social measurement,* 5th ed. Newbury Park, CA: Sage.

Miller, Delbert C. and Nell J. Salkind. 2002. *Handbook of research design and social measurement,* 6th ed. Newbury Park, CA: Sage.

Miller, William L. and Benjamin F. Crabtree. 1999. The dance of interpretation. Pp. 127–143 in *Doing qualitative research,* ed. Benjamin F. Crabtree and William L. Miller. Thousand Oaks, CA: Sage.

Mitchell, Richard G. Jr. 1993. *Secrecy and fieldwork.* Newbury Park, CA: Sage.

Mohr, Lawrence B. 1992. *Impact analysis for program evaluation.* Newbury Park, CA: Sage.

Mooney, Christopher Z. and Mei Hsien Lee. 1995. Legislating morality in the American states: The case of abortion regulation reform. *American Journal of Political Science* 39: 599–627.

Moore, Spencer, Mark Daniel, Laura Linnan, Marci Campbell, Salli Benedict, and Andrea Meier. 2004. After Hurricane Floyd passed: Investigating the social determinants of disaster preparedness and recovery. *Family and Community Health* 27: 204–217.

Morrill, Calvin, Christine Yalda, Madeleine Adelman, Michael Musheno, and Cindy Bejarano. 2000. Telling tales in school: Youth culture and conflict narratives. *Law & Society Review* 34: 521–565.

Mullins, Carolyn J. 1977. *A guide to writing and publishing in the social and behavioral sciences.* New York: Wiley.

Myrdal, Gunnar. 1944/1964. *An American dilemma.* Repr. New York: McGraw-Hill.

National Geographic Society. 2000. *Survey 2000.* From http://business.clemson.edu/socio/s2k data211.html (accessed September 28, 2008).

National Opinion Research Center (NORC). 1992. *National data program for the social sciences: The NORC General Social Survey; Questions and answers* (mimeographed). Chicago: NORC.

National Opinion Research Center (NORC). 2006. *General social survey.* Chicago: National Opinion Research Center, University of Chicago.

National Technical Information Service, U.S. Department of Commerce. 1993. *Directory of U.S. government data files for mainframes and microcomputers.* Washington, DC: Federal Computer Products Center, National Technical Information Service, U.S. Department of Commerce.

Needleman, Carolyn. 1981. Discrepant assumptions in empirical research: The case of juvenile court screening. *Social Problems* 28 (February): 247–262.

Neumark, David. 1996. Sex discrimination in restaurant hiring: An audit study. *Quarterly Journal of Economics* 111: 915–41.

Newbury, Darren. 2005. Editorial: The challenge of visual studies. *Visual Studies* 20: 1–3.

Newman, Katherine S., Cybelle Fox, David J. Harding, Jal Mehta, and Wendy Roth. 2004. *Rampage: The social roots of school shootings.* New York: Basic Books.

Newport, Frank. 2000. *Popular vote in presidential race too close to call.* Princeton: The Gallup Organization. From the Gallup website: www.gallup.com/poll/releases/pr001107.asp (accessed December 13, 2000).

O'Dochartaigh, Niall. 2002. *The Internet research handbook: A practical guide for students and researchers in the social sciences.* Thousand Oaks, CA: Sage.

Office of Management and Budget. (n.d.). *Government Performance Results Act of 1993.* From www.whitehouse.gov/omb/mgmt-gpra/gplaw2m.html (accessed September 28, 2008).

Orcutt, James D. and J. Blake Turner. 1993. Shocking numbers and graphic accounts: Quantified images of drug problems in the print media. *Social Problems* 49: 190–206.

Orr, Larry L. 1999. *Social experiments: Evaluating public programs with experimental methods.* Thousand Oaks, CA: Sage.

Pager, Devah. 2003. The mark of a criminal record. *American Journal of Sociology* 108: 937–75.

Pager, Devah and Lincoln Quillian. 2005. Walking the talk? What employers say versus what they do. *American Sociological Review* 70: 355–380.

Panagopoulos, Costas. 2008. "Poll Accuracy in the 2008 Presidential Election." http://www.fordham.edu/images/academics/graduate_schools/gsas/elections_and_campaign_/poll%20accuracy%20in%20the%202008%20presidential%20election.pdf. Retrieved March 17, 2011.

Papineau, David. 1978. *For science in the social sciences.* London: Macmillan.

Parks, Malcolm R. and Kory Floyd. 1996. Making friends in cyberspace. *Journal of Communication* 46(1): 80–97.

Parks, Kathleen A., Ann M. Pardi, and Clara M. Bradizza. 2006. Collecting data on alcohol use and alcohol-related victimization: A comparison of telephone and web-based survey methods. *Journal of Studies on Alcohol* 67: 318–323.

Parlett, Malcolm and David Hamilton. 1976. Evaluation as illumination: A new approach to the study of innovative programmes. Pp. 140–157 in *Evaluation studies review annual*, vol. 1, ed. G. Class. Beverly Hills, CA: Sage.

Paternoster, Raymond, Robert Brame, Ronet Bachman, and Lawrence W. Sherman. 1997. Do fair procedures matter? The effect of procedural justice on spouse assault. *Law & Society Review* 31(1): 163–204.

Patton, Michael Quinn. 2002. *Qualitative research & evaluation methods*, 3rd ed. Thousand Oaks, CA: Sage.

Pew Internet & American Life Project. 2010. *Change in Internet access by age group, 2000-2009*. Washington DC: Pew Internet & American Life Project. From http://www.pewinternet.org/Infographics/2010/Internet-acess-by-age-group-over-time.aspx# (accessed May 15, 2011).

Phillips, David P. 1982. The impact of fictional television stories on U.S. adult fatalities: New evidence on the effect of the mass media on violence. *American Journal of Sociology* 87(May): 1340–1359.

Phoenix, Ann. 2003. Neoliberalism and masculinity: Racialization and the contradictions of schooling for 11- to 14-year-olds. *Youth & Society* 36: 227–246.

Pipher, Mary. 1994. *Reviving Ophelia: Saving the selves of adolescent girls*. New York: Ballantine Books.

Posavac, Emil J. and Raymond G. Carey. 1997. *Program evaluation: Methods and case studies*, 5th ed. Upper Saddle River, NJ: Prentice Hall.

Presley, Cheryl A., Philip W. Meilman, and Rob Lyerla. 1994. Development of the core alcohol and drug survey: Initial findings and future directions. *Journal of American College Health* 42: 248–255.

Price, Richard H., Michelle Van Ryn, and Amiram D. Vinokur. 1992. Impact of a preventive job search intervention on the likelihood of depression among the unemployed. *Journal of Health and Social Behavior* 33(June): 158–167.

Punch, Maurice. 1994. Politics and ethics in qualitative research. Pp. 83–97 in *Handbook of qualitative research*, ed. Norman K. Denzin and Yvonna S. Lincoln. Thousand Oaks, CA: Sage.

Putnam, Robert D. 2000. *Bowling alone: The collapse and revival of American community*. New York: Touchstone.

Pyrczak, Fred. 2005. *Evaluating research in academic journals: A practical guide to realistic evaluation*, 3rd ed. Glendale, CA: Pyrczak.

Radin, Charles A. 1997. Partnerships, awareness behind Boston's success. *Boston Globe*, February 19, A2, B7.

Ragin, Charles C. 1994. *Constructing social research*. Thousand Oaks, CA: Pine Forge Press.

Ramirez, Anthony. 2002. One more reason you're less likely to be murdered. *New York Times*, August 25, WK3.

Rankin, Bruce H. and James M. Quane. 2002. Social contexts and urban adolescent outcomes: The interrelated effects of neighborhoods, families, and peers on African-American youth. *Social Problems* 49: 79–100.

Reiss, Albert J., Jr. 1971. *The police and the public*. New Haven, CT: Yale University Press.

Reynolds, Paul Davidson. 1979. *Ethical dilemmas and social science research* San Francisco: Jossey-Bass.

Richards, Thomas J. and Lyn Richards. 1994. Using computers in qualitative research. Pp. 445–462 in *Handbook of qualitative research*, ed. Norman K. Denzin and Yvonna S. Lincoln. Thousand Oaks, CA: Sage.

Richardson, Laurel. 1995. Narrative and sociology. Pp. 198–221 in *Representation in ethnography*, ed. John Van Maanen. Thousand Oaks, CA: Sage.

Riessman, Catherine Kohler. 2002. Narrative analysis. Pp. 217–270 in *The qualitative researcher's companion*, ed. A. Michael Huberman and Matthew B. Miles. Thousand Oaks, CA: Sage.

Ringwalt, Christopher L., Jody M. Greene, Susan T. Ennett, Ronaldo Iachan, Richard R. Clayton, and Carl G. Leukefeld. 1994. *Past and future directions of the D.A.R.E. program: An evaluation review*. Research Triangle Park, NC: Research Triangle Institute.

Roman, Anthony. 2005. *Women's health network client survey: Field report*. Unpublished report. Boston: Center for Survey Research, University of Massachusetts.

Rookey, Bryan D., Steve Hanway, and Don A. Dillman. 2008. Does a probability-based household panel benefit from assignment to postal response as an alternative to Internet-only? *Public Opinion Quarterly* 72: 962–984.

Rosenberg, Morris. 1968. *The logic of survey analysis*. New York: Basic Books.

Rosenbloom, Stephanie. 2007. On Facebook, scholars link up with data. *New York Times*, December 17.

Rossi, Peter H. and Howard E. Freeman. 1989. *Evaluation: A systematic approach*, 4th ed. Newbury Park, CA: Sage.

Rossman, Gretchen B. and Sharon F. Rallis. 1998. *Learning in the field: An introduction to qualitative research*. Thousand Oaks, CA: Sage.

Rubin, Herbert J. and Irene S. Rubin. 1995. *Qualitative interviewing: The art of hearing data*. Thousand Oaks, CA: Sage.

Rueschemeyer, Dietrich, Evelyne Huber Stephens, and John D. Stephens. 1992. *Capitalist development and democracy*. Chicago: University of Chicago Press.

Sacks, Stanley, Karen McKendrick, George DeLeon, Michael T. French, and Kathryn E. McCollister. 2002. Benefit-cost analysis of a modified therapeutic community for mentally ill chemical abusers. *Evaluation & Program Planning* 25: 137–148.

Salisbury, Robert H. 1975. Research on political participation. *American Journal of Political Science* 19(May): 323–341.

Sampson, Robert J. and John H. Laub. 1994. Urban poverty and the family context of delinquency: A new look at structure and process in a classic study. *Child Development* 65: 523–540.

Sampson, Robert J. and Janet L. Lauritsen. 1994. Violent victimization and offending: individual-, situational-, and community-level risk factors. Pp. 1–114 in *Social Influences, vol. 3 of Understanding and preventing violence,* ed. Albert J. Reiss Jr. and Jeffrey A. Roth. Washington, DC: National Academy Press.

Sampson, Robert J. and Stephen W. Raudenbush. 1999. Systematic social observation of public spaces: A new look at disorder in urban neighborhoods. *American Journal of Sociology* 105: 603–651.

Scarce, Rik. 2005. *Eco-warriors: Understanding the radical environmental movement,* 2nd ed. Walnut Creek, CA: Left Coast Press.

Schaeffer, Nora Cate and Stanley Presser. 2003. The science of asking questions. *Annual Review of Sociology* 29: 65–88.

Schalock, Robert and John Butterworth. 2000. *A benefit-cost analysis model for social service agencies.* Boston: Institute for Community Inclusion.

Schapira, Lidia and Russell K. Schutt. 2011. Training community health workers about cancer clinical trials. *Journal of Immigrant and Minority Health.* From http://www.springerlink.com/content/g6p05g3113951q11/.

Schleyer, Titus K. L. and Jane L. Forrest. 2000. Methods for the design and administration of web-based surveys. *Journal of the American Medical Informatics Association* 7(4): 416–425.

Schober, Michael F. 1999. Making sense of survey questions. Pp. 77–94 in *Cognition and survey research,* ed. Monroe G. Sirken, Douglas J. Herrmann, Susan Schechter, Norbert Schwartz, Judith M. Tanur, and Roger Tourangeau. New York: Wiley.

Schorr, Lisbeth B. and Daniel Yankelovich. 2000. In search of a gold standard for social programs. *Boston Globe,* February 18, A19.

Schuman, Howard and Stanley Presser. 1981. *Questions and answers in attitude surveys: Experiments on question form, wording, and context.* New York: Academic Press.

Schutt, Russell K. 1988. *Working with the Homeless: the Backgrounds, Activities and Beliefs of Shelter Staff.* Boston: University of Massachusetts. Unpublished report.

Schutt, Russell K. (with Stephen M. Goldfinger). 2011. *Homelessness, housing and mental illness.* Cambridge, MA: Harvard University Press.

Schutt, Russell, JudyAnn Bigby, and Lidia Schapira. 2005. *Educating underserved communities about cancer clinical trials.* Proposal to the National Cancer Institute. Boston: University of Massachusetts Boston and Dana-Farber/Harvard Cancer Center.

Schutt, Russell K., Elizabeth Riley Cruz, and Mary Lou Woodford. 2008. Client satisfaction in a breast and cervical cancer early detection program: The influence of ethnicity and language, health, resources, and barriers. *Women & Health* 48: 283–302.

Schutt, Russell K., Xiaogang Deng, Gerald R. Garrett, Stephanie Hartwell, Sylvia Mignon, Joseph Bebo, Matthew O'Neill, Mary Aruda, Pat Duynstee, Pam DiNapoli, and Helen Reiskin. 1996. *Substance use and abuse among UMass Boston students.* Unpublished report, Department of Sociology, University of Massachusetts, Boston.

Schutt, Russell K. and Jacqueline Fawcett. 2005. *Case management in the women's health network.* Boston: University of Massachusetts. Unpublished report.

Schutt, Russell K., Jacqueline Fawcett, Gail B. Gall, Brooke Harrow, and Mary Lou Woodford. 2010. Case manager satisfaction in public health. *Professional Case Management* 15: 124–134.

Schutt, Russell K. and M. L. Fennell. 1992. Shelter staff satisfaction with services, the service network, and their jobs. *Current Research on Occupations and Professions* 7: 177–200.

Schutt, Russell K. and Stephen M. Goldfinger. 1996. Housing preferences and perceptions of health and functioning among mentally ill persons. *Psychiatric Services* 47(4): 381–386.

Schutt, Russell K., Jessica Santiccioli, Jennifer Maniates, Silas Henlon, Lidia Schapira. 2008. *Community health workers and cancer clinical trials: Experiences, knowledge, and concerns.* Research report for project funded by National Cancer Institute.

Schutt, Russell K., Lidia Schapira, Jennifer Maniates, Jessica Santiccioli, Silas Henlon, JudyAnn Bigby. 2010. Community health workers' support for cancer clinical trials: Description and explanation. *Journal of Community Health* 35: 417–422.

Scriven, Michael. 1972. Prose and cons about goal-free evaluation. *Evaluation Comment* 3: 1–7.

Selm, Martine Van and Nicholas W. Jankowski. 2006. Conducting online surveys. *Quality & Quantity* 40: 435–456.

Shepherd, Jane, David Hill, Joel Bristor, and Pat Montalvan. 1996. Converting an ongoing health study to CAPI: Findings from the national health and nutrition study. Pp. 159–164 in *Health Survey Research Methods Conference Proceedings,* ed. Richard B. Warnecke. Hyattsville, MD: U.S. Department of Health and Human Services.

Sherman, Lawrence W. 1992. *Policing domestic violence: Experiments and dilemmas.* New York: Free Press.

Sherman, Lawrence W. 1993. Implications of a failure to read the literature. *American Sociological Review* 58: 888–889.

Sherman, Lawrence W. and Richard A. Berk. 1984. The specific deterrent effects of arrest for domestic assault. *American Sociological Review* 49: 261–272.

Sieber, Joan E. 1992. *Planning ethically responsible research: A guide for students and internal review boards.* Thousand Oaks, CA: Sage.

Singer, Eleanor. 2006. Introduction: Nonresponse bias in household surveys. *Public Opinion Quarterly* 70(5): 637–645.

Sjoberg, Gideon, ed. 1967. *Ethics, politics, and social research.* Cambridge, MA: Schenkman.

Smith, Tom W. 1984. Nonattitudes: A review and evaluation. Pp. 215–255 in *Surveying subjective phenomena,* vol. 2, ed. Charles F. Turner and Elizabeth Martin. New York: Russell Sage Foundation.

Smith, Tom. 1987. That which we call welfare by any other name would smell sweeter: An analysis of the impact of question wording on response patterns. *Public Opinion Quarterly* 51(1): 75–83.

Smyth, Jolene D., Don A. Dillman, Leah Melani Christian, and Michael J. Stern. 2004. *How visual grouping influences answers to Internet surveys* (Technical Report #04–023). From Washington State University Social & Economic Sciences Research Center website: http://survey.sesrc.wsu.edu/dillman/papers.htm (accessed July 5, 2005). An extended version of paper presented at the 59th Annual Conference of the American Association for Public Opinion Research, Phoenix, AZ, May 13, 2004.

St. Pierre, Robert G. and Peter H. Rossi. 2006. Randomize groups, not individuals: A strategy for improving early childhood programs. *Evaluation Review* 30: 656–685.

Stake, Robert E., ed. 1975. *Evaluating the arts in education: A responsible approach.* Columbus, OH: Merrill.

Stake, Robert E. 1995. *The art of case study research.* Thousand Oaks, CA: Sage.

Stille, Alexander. 2000. A happiness index with a long reach: Beyond G.N.P. to subtler measures. *New York Times,* May 20, A17, A19.

Straus, Murray and Richard Gelles. 1988. *Intimate violence.* New York: Simon & Schuster.

Strickling, Lawrence E. 2010. *Digital nation: 21st century America's progress toward universal broadband Internet access. An NTIA research preview.* Washington, DC: National Telecommunications and Information Administration, U.S. Department of Commerce.

Strunk, William, Jr. and E. B. White. 1979. *The elements of style,* 3rd ed. New York: Macmillan.

Sudman, Seymour. 1976. *Applied sampling.* New York: Academic Press.

Survey.net. 2000. *Year 2000 presidential election survey.* From http://www.survey.net/sv-press.htm.

Survey on adultery: "I do" means "I don't." 1993. *The New York Times,* October 19, p. A20.

Tavernise, Sabrina. 2011. Youth, mobility, and poverty help drive cellphone-only status. *New York Times,* April 21, p. A13.

Taylor, Jerry. 1999. DARE gets updated in some area schools, others drop program. *Boston Sunday Globe,* May 16, 1, 11.

Thompson, Clive. 2007. Community urinalysis. *New York Times Magazine,* December 9. From http://www.nytimes.com/2007/12/09/magazine/09_11_urinalysis.html?ref=magazine (accessed September 28, 2008).

Thorne, Barrie. 1993. *Gender play: Girls and boys in school.* New Brunswick, NJ: Rutgers University Press.

Tjaden, Patricia and Nancy Thoennes. 2000. *Extent, nature, and consequences of intimate partner violence: Findings from the National Violence Against Women Survey* (NCJ 181867). Washington, DC: Office of Justice Programs, National Institute of Justice and the Centers for Disease Control and Prevention.

Toppo, Greg. 2002. Antidrug program backed by study. *Boston Globe,* October 29, A10.

Tourangeau, Roger, 2004. Survey research and societal change. *Annual Review of Psychology* 55: 775–801.

Tufte, Edward R. 1983. *The visual display of quantitative information.* Cheshire, CT: Graphics Press.

Turabian, Kate L. 1967. *A manual for writers of term papers, theses, and dissertations,* 3rd rev. ed. Chicago: University of Chicago Press.

Turner, Charles F. and Elizabeth Martin, eds. 1984. *Surveying subjective phenomena.* 2 vols. New York: Russell Sage Foundation.

Tuskegee Syphilis Study administrative records. 1929–1972. Atlanta, GA: Records of the Centers for Disease Control and Prevention, National Archives—Southeast Region.

U.S. Bureau of the Census. 1994. *Census catalog and guide, 1994.* Washington, DC: Department of Commerce, U.S. Bureau of the Census.

U.S. Bureau of the Census. 1999. *United States census 2000, updated summary: Census 2000 Operational Plan.* Washington, DC: U.S. Department of Commerce, Bureau of the Census, February.

U. S. Bureau of the Census. 2004–2005. *Statistical abstract of the United States.* Washington, DC: U.S. Department of Commerce, Bureau of the Census. From http://www.census.gov/prod/www/statistical-abstract-2001_2005.html (accessed September 28, 2008).

U.S. Bureau of the Census. 2010e. *Internet use in the United States: October 2009.* Washington, DC: U.S. Department of Commerce, Bureau of the Census. From http://www.census.gov/population/www/socdemo/computer/2009.html (accessed March 18, 2011).

U.S. Bureau of Economic Analysis, Communications Division. 2004. *Customer satisfaction survey report, FY 2004.* Washington, DC: U.S. Department of Commerce. From http://www.bea.gov/bea/about/cssr_2004_complete.pdf (accessed September 28, 2008).

U.S. Department of Health, Education, and Welfare. 1979. *The Belmont report: Ethical principles and guidelines for the protection of human subjects of research.* Washington, DC: Superintendent of Documents, U.S. Government Printing Office. (ERIC Document Reproduction Service No. ED 183582)

U.S. Bureau of Labor Statistics, Department of Labor. 1991. *Major programs of the bureau of labor statistics.* Washington, DC: U.S. Bureau of Labor Statistics, Department of Labor.

U.S. Bureau of Labor Statistics, Department of Labor. 1997a. *Employment and earnings.* Washington, DC: U.S. Bureau of Labor Statistics, Department of Labor.

U.S. Bureau of Labor Statistics, Department of Labor. 1997b. *Handbook of methods.* Washington, DC: U.S. Bureau of Labor Statistics, Department of Labor.

U.S. Government Accounting Office. 2001. *Health and human services: Status of achieving key outcomes and addressing major management challenges* (GAO-01-748). From http://www.gao.gov/new.items/d01748.pdf (accessed September 28, 2008).

Van Maanen, John. 1995. An end to innocence: The ethnography of ethnography. Pp. 1–35 in *Representation in ethnography,* ed. John Van Maanen. Thousand Oaks, CA: Sage.

Van Maanen, John. 2002. The fact of fiction in organizational ethnography. Pp. 101–117 in *The qualitative researcher's companion,* ed. A. Michael Huberman and Matthew B. Miles. Thousand Oaks, CA: Sage.

Verba, Sidney and Norman Nie. 1972. *Political participation: Political democracy and social equality.* New York: Harper & Row.

Vernez, Georges M., Audrey Burnam, Elizabeth A. McGlynn, Sally Trude, and Brian S. Mirttman. 1988. *Review of California's program for the homeless mentally disabled* (R-3631-CDMH). Santa Monica, CA: RAND.

Vincus, Amy A., Chris Ringwalt, Melissa S. Harris, Stephen R. Shamblen. 2010. A short-term, quasi-experimental evaluation of D.A.R.E.'s revised elementary school curriculum. *Journal of Drug Education* 40: 37–49.

Wageman, Ruth. 1995. Interdependence and group effectiveness. *Administrative Science Quarterly* 40:145–180.

Wallgren, Anders, Britt Wallgren, Rolf Persson, Ulf Jorner, and Jan-Aage Haaland. 1996. *Graphing statistics and data: Creating better charts.* Thousand Oaks, CA: Sage.

Webb, Eugene J., Donald T. Campbell, Richard D. Schwartz, and Lee Sechrest. 2000. *Unobtrusive measures,* rev. ed. Thousand Oaks, CA: Sage.

Weber, Max. 1930/1992. *The Protestant ethic and the spirit of capitalism,* trans. Talcott Parsons. Repr. London: Routledge.

Weber, Max. 1947/1997. *The theory of social and economic organization,* trans. A. M. Henderson and Talcott Parsons. Repr. New York: Free Press.

Weber, Robert Philip. 1985. *Basic content analysis.* Thousand Oaks, CA: Sage.

Weitzman, Eben and M. B. Miles. 1994. *Computer programs for qualitative data analysis.* Thousand Oaks, CA: Sage.

West, Steven L. and Ken K. O'Neal. 2004. Project D.A.R.E. outcome effectiveness revisited. *American Journal of Public Health* 94: 1027–1029.

Wholey, J. S. 1987. Evaluability assessment: Developing program theory. Pp. 77–92 in *Using program theory in evaluation: New directions for program evaluation* (no. 33), ed. Leonard Bickman. San Francisco: Jossey-Bass.

Whyte, William Foote. 1955. *Street corner society.* Chicago: University of Chicago Press.

Witte, James C., Lisa M. Amoroso, and Philip E. N. Howard. 2000. Research methodology: Method and representation in internet-based survey tools—mobility, community, and cultural identity in survey 2000. *Social Science Computer Review* 18: 179–195.

Wolcott, Harry F. 1995. *The art of fieldwork.* Walnut Creek, CA: AltaMira Press.

Yardley, William. 2008. Drawing lots for health care. *New York Times*, March 13. From http://www.nytimes.com/2008/03/13/us/13bend.html (accessed September 28, 2008).

Zimbardo, Philip G. 1973. *The Lucifer effect.* New York: Random House.

Credits

Chapter 1

Exhibit 1.4: Reprinted with permission of Simon & Schuster, Inc. from *Bowling Alone* by Robert D. Putnam. Copyright © 2000 Robert. D. Putnam.

Chapter 2

Exhibit 2.1: Data from Sherman and Berk, 1984:267.

Exhibit 2.5: The Gallup Organization. August 20, 2002. Poll Analyses, July 29, 2002. Bush Job Approval Update. http://www.gallup.com/poll/6478/Bush-Job-Approval-Update.aspx

Chapter 3

Exhibit 3.1: From the film *Obedience* © 1968 by Stanley Milgram, © Renewed 1993 by Alexandra Milgram, and distributed by Penn State Media Sales.

Exhibit 3.2: From the film *Obedience* © 1968 by Stanley Milgram, © Renewed 1993 by Alexandra Milgram, and distributed by Penn State Media Sales.

Exhibit 3.4: Tuskegee Syphilis Study Administrative Records. Records of the Centers for Disease Control and Prevention. National Archives—Southeast Region (Atlanta).

Exhibit 3.5: From *The Lucifer Effect* by Philip G. Zimbardo. Copyright 2007 by Philip G. Zimbardo, Inc. Used by permission of Random House, Inc., and Random House Group Ltd.

Chapter 4

Exhibit 4.2: Lenore Radloff, 1977. "The CES-D Scale: A Self-Report Depression Scale for Research in the General Population." *Applied Psycholgoical Measurement,* 1: 385–401. SAGE Publications, Inc.

Exhibit 4.4: Core Institute, Core Alcohol and Drug Survey, 1994. Carbondale, IL: Core Institute.

Exhibit 4.8: Based on Schutt (1988:7–10, 15, 16). Results reported in Schutt and Fennell (1992).

Chapter 5

Exhibit 5.3: Gallup (2011); Panagopoulos (2008).

Chapter 6

Exhibit 6.4: Ruth Wageman, 1995. "Interdependence and Group Effectiveness." *Administrative Science Quarterly,* 40:145–180. Published by Sage Publications on behalf of Johnson Graduate School of Management, Cornell University.

Exhibit 6.5: Adapted from Philips 1982:1347. Reprinted with permission from the University of Chicago Press.

Chapter 7

Exhibit 7.1: Filter Questions and Skip Patterns, Youth and Guns Survey, 2000.

Exhibit 7.2: Sample Interview Guide, Youth and Guns Survey, 2000.

Exhibit 7.5: Curtin, Presser, and Singer (2005:91).

Exhibit 7.6: Sample Interviewer Instructions, Youth and Guns Survey, 2000.

Exhibit 7.8: Dillman, 1978: 74–75. "Mail and Telephone Surveys: The Total Design Method." Reprinted with permission of John Wiley & Sons, Inc.

Chapter 8

Exhibit 8.1: General Social Survey, National Opinion Research Center 2010.

Exhibit 8.2: U.S. Bureau of Economic Analysis 2004:14.

Exhibit 8.3: General Social Survey, National Opinion Research Center 2010.

Exhibit 8.4: General Social Survey, National Opinion Research Center 2010.

Exhibit 8.5: General Social Survey, National Opinion Research Center 2010.

Exhibit 8.6: Adapted from Orcutt & Turner, 1993. Copyright 1993 by the Society for the Study of Social Problems. Reprinted by permission.

Exhibit 8.7: General Social Survey, National Opinion Research Center 2010.

Exhibit 8.8: General Social Survey, National Opinion Research Center 2010.

Exhibit 8.9: General Social Survey, National Opinion Research Center 2010.

Exhibit 8.13: General Social Survey, National Opinion Research Center 2010.

Exhibit 8.14: General Social Survey, National Opinion Research Center 2006.

Exhibit 8.15: General Social Survey, National Opinion Research Center 2006.

Exhibit 8.16: General Social Survey, National Opinion Research Center 2006.

Exhibit 8.18: General Social Survey, National Opinion Research Center 2010.

Chapter 9

Exhibit 9.5: Raudenbush and Sampson 1999:15.

Exhibit 9.6: St. Jean, Peter K.B. 2007. *Pockets of Crime: Broken Windows, Collective Efficacy, and the Criminal Point of View.* Chicago: University of Chicago Press.

Exhibit 9.7: Adapted from Richard A. Krueger and Mary Anne Casey, 2000. *Focus Groups: A Practical Guide for Applied Research,* 3rd ed. Copyright SAGE Publications. Used with permission.

Chapter 10

Exhibit 10.1: Miles & Huberman, 1994:53, Table 4.1. Copyright SAGE Publications. Used with permission.

Exhibit 10.2: Miles & Huberman, 1994:95, Table 5.2. Copyright SAGE Publications. Used with permission.

Exhibit 10.3: Patton, 2002:472.

Exhibit 10.4: Miles & Huberman, 1994:159, Figure 6.5. Copyright SAGE Publications. Used with permission.

Exhibit 10.5: Morrill et al., 2000:553, Table 1. Copyright 2000. Reprinted with permission of Blackwell Publishing Ltd.

Exhibit 10.6: Gibson, David R. 2005. Taking turns and talking ties: Networks and conversational interaction. *American Journal of Sociology* 110(6):1561–1597. Reprinted with permission from the University of Chicago Press.

Exhibit 10.7: Needleman 1981:248–256.

Exhibit 10.8: Reprinted by permission of the publisher from *Homelessness, Housing, and Mental Illness* by Russell K. Schutt, with Stephen M. Goldfinger, p. 135, Cambridge, Mass.: Harvard University Press, Copyright © 2011 by the President and Fellows of Harvard College.

Exhibit 10.9: Rueschemeyer et al. 1002:161. Reprinted with permission from the University of Chicago Press.

Exhibit 10.10: Reproduced by permission of International IDEA from *Turnout in the world—country by country performance (1945–1998). From Voter Turnout, A Global Survey* (http://www.idea.int/vt/survey/voter_turnout_pop2-2.cfm) (c) International Institute for Democracy and Electoral Assistance.

Chapter 11

Exhibit 11.1: Adapted from Martin and Kettner (1996).

Exhibit 11.3: Adapted from DeParle, Jason. 1999. "Project to Rescue Needy Stumbles Against the Persistence of Poverty." *The New York Times,* May 15, pp. A1, A10.; "New Hope for People with Low Incomes: Two-Year Results of a Program to Reduce Poverty and Reform Welfare" April 1999, Johannes M. Bos, et al. http://www.mdrc.org/publications/60/print.html

Exhibit 11.4: Based on Goldfinger and Schutt (1996).

Exhibit 11.5: Chen (1990:210). Reprinted with permission from SAGE Publications, Inc.

Exhibit 11.6: Drake et al. 1996:391–399. From the New Hampshire Study of Supported Employment for People With Severe Mental Illness in the *Journal of Consulting and Clinical Psychology* 64:391–399. Used with permission.

Exhibit 11.6: Chen (1990:210). Reprinted with permission from SAGE Publications, Inc.

Exhibit 11.7: Orr (1992:224, Table 6.5). Reprinted with permission from SAGE Publications, Inc.

Chapter 12

Exhibit 12.6: Vernez et al. (1988). Reprinted with permission.

Index

⑤SAGE research methods online

The essential tool for researchers . . .

. . . from the world's leading research methods publisher

Discover SRMO Lists— methods readings suggested by other SRMO users

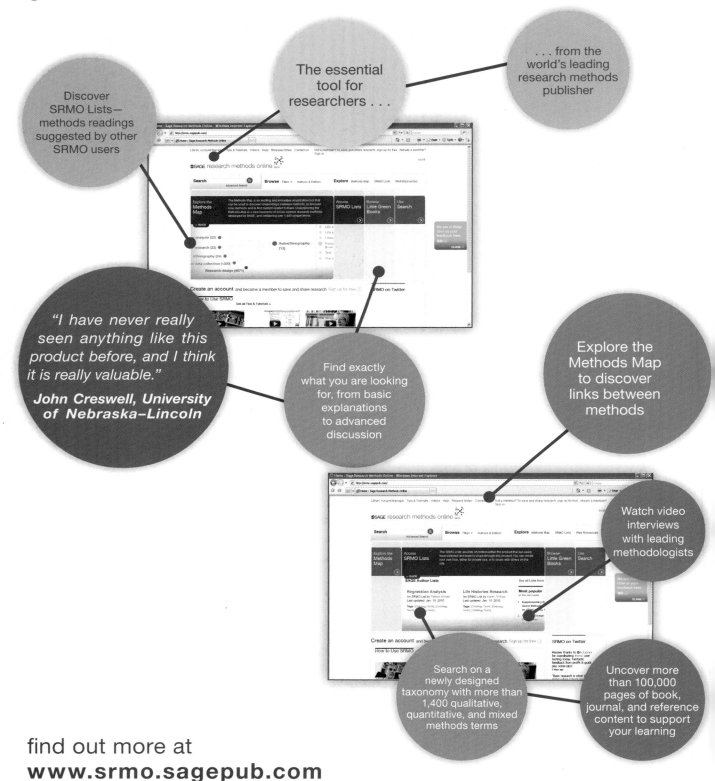

"I have never really seen anything like this product before, and I think it is really valuable."

John Creswell, University of Nebraska–Lincoln

Find exactly what you are looking for, from basic explanations to advanced discussion

Explore the Methods Map to discover links between methods

Watch video interviews with leading methodologists

Search on a newly designed taxonomy with more than 1,400 qualitative, quantitative, and mixed methods terms

Uncover more than 100,000 pages of book, journal, and reference content to support your learning

find out more at
www.srmo.sagepub.com